MICROWAVE IMAGING WITH LARGE ANTENNA ARRAYS

RADIO CAMERA PRINCIPLES AND TECHNIQUES

MICROWAVE IMAGING WITH LARGE ANTENNA ARRAYS

RADIO CAMERA PRINCIPLES AND TECHNIQUES

BERNARD D. STEINBERG
Professor of Electrical Engineering
and Director
Valley Forge Research Center
Moore School of Electrical Engineering
University of Pennsylvania

A Wiley-Interscience Publication
JOHN WILEY & SONS
New York Chichester Brisbane Toronto Singapore

Library of Congress Cataloging in Publication Data

Steinberg, Bernard D.
Microwave imaging with large antenna arrays.

"A Wiley-Interscience publication."
Includes bibliographical references and index.
1. Microwaves. 2. Imaging systems. I. Title.

TK7876.S736 1983 621.36′7 83-6705
ISBN 0-471-89173-8

Printed in the United States of America

10 9 8 7 6 5 4 3 2 1

This book is dedicated to my colleagues and students at the Valley Forge Research Center, without whom the work could not have been accomplished, and to the farsighted government sponsors and program managers whose loyalty during years of shrinking research budgets allowed us to develop the radio camera.

PREFACE

The work on which this book is based began more than 10 years ago. It was driven by the dream of breaking the barrier to aperture size of microwave antenna arrays. The goal was the development of means for designing and constructing phased arrays of arbitrary size and shape, including arbitrarily conformal arrays with arbitrary surface distortions and arbitrary electromagnetic couplings between antenna elements and the local environment. The primary research is now done and the techniques work. Self-cohering methods are used for beamforming. The general embodiment is called the radio camera. Aperture size can now be large enough to do high-resolution imaging with long wavelength microwave or radio frequency energy.

High-resolution microwave imaging is an important developing application area of electromagnetics and radar. Its main purpose is target classification, which is analogous to photo interpretation in optical photography. The book deals primarily with means of achieving arbitrarily high angular resolving power and secondarily with means for reducing the artifact content of the images. Although the terminology is identified with electromagnetics and radar, the techniques and procedures are equally applicable in sonar and ultrasound.

The objective of high-resolution microwave imaging is "picture taking." Although the relatively long wavelength precludes imagery with the detail of optical images, microwave images can provide enough to allow the user to count targets, observe their deployment, and even to distinguish between types of targets. This book does not reach that far; it concentrates on means of achieving the high angular resolution necessary for the desired picture-taking quality but largely ignores (with the exception of Chapter 3 and Section 4.5) sophisticated modifications in the signal processor that lead to optimized image quality. Image quality improvement is not covered because it is still in the research stage and belongs in a future volume; some discussion is given in Chapter 10. Also ignored is the entire field of digital picture processing or image enhancement; the subject of postprocessing image enhancement, sometimes called pixel-intensity processing, is not covered because there is an extensive literature in this field.

A note about the book and its use: aside from the first chapter, which is introductory to the subject and includes discussions on potential applications, the text is divided into chapters that are primarily theoretical (2, 3, 4, and 7) and chapters that deal with systems (5, 6, 8, and 9). The intended readers are engineers and physicists working on large aperture phased arrays and graduate students interested in large arrays, imaging, and spatial signal processing. At the University of Pennsylvania this work is taught in the second semester of a two-course sequence. In the first semester, the first text of this series, *Principles of Aperture and Array System Design*, (Wiley, 1976), is used. Student assignments in the second semester are two papers that deal roughly with the first half of this book (into Chapter 5) and the second half. Related topics in adaptive spatial signal processing, such as adaptive nulling antenna arrays, are encouraged.

I am grateful to the Moore School of Electrical Engineering of the University of Pennsylvania which invited me to join the faculty in 1971 so that this research effort could begin. The field site at Valley Forge, the Valley Forge Research Center, was indispensable for field testing the theories and techniques described in this book. I am indebted also to a host of colleagues on the faculty, to the engineers at the field site, and to my students, without whom this ambitious endeavor would have been a hopeless task. Noteworthy contributors were Professor C. N. Dorny, University of Pennsylvania, whose work on multiple phase-measurement concepts forms the basis for Section 4.4, Professor Y. Bar-Ness, who, when he made his contributions to the phase conjugation techniques reported in Section 8.6, was a visiting professor at the Moore School from Tel Aviv University, and Professor H. S. Taheri of Pahlavi University, Shirez, Iran, who contributed to the tolerance theory of the radio camera (Chapter 7) while on sabbatical at the Moore School. Three of my students who made valuable contributions were Dr. Eli Yadin (Jadlovker), whose master's thesis tested the theory of the effects of multipath and scattering on radio camera imagery (Chapter 7), and whose Ph.D. dissertation formed the basis for the airborne radio camera work in Chapter 9, Dr. Ajay K. Luthra, who contributed to the theory of the near-field diffraction pattern of an asymmetrical array (Chapter 2) and to the tolerance theory of the radio camera in a distorting medium (Chapter 7), and Mr. E. Hesham Attia, who developed much of the diversity theory of Chapter 3 and who demonstrated the inadequacy of the Muller-Buffington image feedback control theorem to radar imaging (Chapter 4).

Engineers at the field site, Donald Carlson, William Whistler, Bernard Meagher, and Dr. Earl N. Powers, were indispensable in the design of the experimental equipment and experiments which tested the theories described in the book, demonstrated their validities, and produced the microwave imagery. My deepest thanks go to Mrs. Shirley Levy, without whose meticulous care in the preparation of the manuscript the project would have been too confusing to undertake. To Mr. Sam H. Seeleman, the manager of the field site, who did the final editing, and to the many others who have been involved in the work of the Valley Forge Research Center my most sincere appreciation.

BERNARD D. STEINBERG

Philadelphia, Pennsylvania
June 1983

CONTENTS

MICROWAVE IMAGING WITH LARGE ANTENNA ARRAYS

RADIO CAMERA PRINCIPLES AND TECHNIQUES

CHAPTER ONE

Basics

1.1. THE RADIO CAMERA

Resolution Limit

A large aperture is essential to any microwave instrument designed for high angular resolution imaging. The minimum required size is approximately the ratio of the wavelength λ of the radiation to the resolution (in radians) desired in the image. Common optical instruments like the ordinary camera and the small telescope provide satisfactory resolving power because the optical wavelength is small (typically $\frac{1}{2}$ μm). The camera lens, when stopped down to the smallest aperture for use in bright sunlight, is about 10^4 optical wavelengths in diameter; hence the available resolution is 10^{-4} rad, which is three times finer than the human optical system. The optical telescope with a 2-in. objective lens has an aperture 10 times larger; its resolving power is 10^{-5} rad, which is fine enough for the spot size in the image to correspond to 3 km on the moon.

Required Aperture Size

The aperture size required for picture taking with microwaves is huge by comparison. Because the wavelengths at microwaves range from about 1 to 30 cm, the antennas can measure 100 m to 30 km. It is evident that special methods foreign to conventional antenna design practice must be considered. This book addresses these unusual problems and describes the means of overcoming them. A name has been coined for a microwave or radio wavelength imaging system with an aperture large enough to form high angular resolution images; that is, to "take pictures" of objects in much the same way that an optical camera takes pictures. The name given to this instrument is the *radio camera*.

Synthetic Aperture Radar

Unlike synthetic aperture radar (SAR), which is another technique for achieving uncommonly high angular resolution, the radio camera has a physically real antenna system with a large aperture. It may be located on the ground or in the air and be moving or stationary. The parts of the array may move independently of one another, but the resolving power is not constrained to any particular direction. Other constraints are introduced because of its large size.

Existing Large Antennas

The principles and techniques described in this book are designed for terrestrial imaging, although they are applicable to some degree in radio astronomy. The requirement for a huge antenna for picture taking readily explains why the radio camera has not yet been developed into a common instrument. The largest antennas have been built by the radio astronomers. The largest monolithic microwave dish, the radio astronomy antenna at Arecibo, Puerto Rico, is 305 m, but it is fixed in position and not readily scanned or pointed as a camera should be. Scanning in one dimension is accomplished by the earth's rotation and in the other by a limited movement of the antenna's feed structure. It is designed for astronomical imaging in which time is not important; its design is not what would normally be associated with a "picture-taking" camera.

The largest scanning microwave dishes are smaller still. Monolithic scanning structures experience changing gravitational stresses induced by scanning and changing thermal stresses induced by variations in local temperature and weather conditions. They are limited to modest sizes unless they are of homologous design, in which case gravity-induced deformations always produce parabolic surfaces, although of varying focal length. The diameter of the largest telescope at the National Radio Astronomy Observatory in West Virginia is 91.5 m. The fully steerable dish at the Max Planck Institute for Radioastronomie at Bonn is 100 m in diameter.

A Phased Array Is Required

Achievement of the antenna size required for terrestrial imaging demands a distributed antenna such as a phased array, the largest of which are the Cobra Dane and Pave Paws of the U. S. Air Force. They operate at L-band ($\lambda = 24$ cm) and UHF. The diameter of the Cobra Dane array is 95 ft, or 120 λ, which falls far short of the required aperture size for imaging.

High Cost of the Filled Array

Imaging is the measurement and replication of the intensity distribution of an active source or of the backscattering profile of a passive reflector or scene. It cannot be achieved with simple, interferometric apertures whose radiation patterns are characterized by closely spaced grating lobes. Filled apertures are preferable for imaging. Filling an aperture implies spacing the elements close enough to avoid grating lobes. This condition is satisfied under all conditions of scanning in an array in which the element spacing does not exceed one-half wavelength. However, for a resolving power of 10^{-5} to 10^{-4} rad a filled phased array with half-wavelength element spacing would require 4×10^8 to 4×10^{10} radiating and/or receiving circuits (consisting of antenna elements, phase shifters, etc.), which is clearly too large a number for practicality. Hence a compromise is required. It is calculated in Section 2.5 that the order of 10^3 to 10^4 elements will suffice for most imaging problems, provided that the elements are distributed randomly over the aperture to prevent the development of grating lobes. The techniques described in Chapter 3 give promise of a further reduction in element number. Because the elements are distributed randomly, the properties of the radiation pattern can be described only statistically. The statistical properties of the radiation patterns of these apertures have been derived [59], [81], [85] and are reviewed in Chapter 2.

Nondedicated Site

A drastic reduction in the number of elements is only one of several necessary engineering compromises that must be made by the radio camera designer. Although the smaller size (300-m) antenna required for 10^{-4} rad resolving power at X-band (3-cm wavelength) leads to a site size that could be acquired and dedicated solely to the use of the imaging system, the same is not true for a 30-km antenna. An area of 1000 km² is too large to be considered seriously for a dedicated array site. Thus the aperture of a radio camera must be a highly thinned antenna array in which the individual parts are separated from one another and distributed over an area as large as 10^3 km² (or larger if the wavelength is longer than 30 cm). The individual modules of this array would be placed on small, isolated parcels of ground, rooftops, or perhaps even moving vehicles, but rarely if ever would a field of this size be dedicated solely to the formation of the array, nor would it ever be needed. Instead, a communication subsystem (radio or cable) would interconnect the parts with a central processor that would control the system and deliver the images.

Arbitrary Surface

Another logical requirement in system design pertains to the shape or surface of the array. The aperture is so large and the sites of the individual array modules so varied that the design must circumvent the requirement that the array surface be planar or any other *a priori* shape. Practicality therefore demands that the system design procedure be sufficiently robust to permit the element coordinates (x, y, *and* z) to be arbitrarily chosen and possibly to vary with time. The latter property imposes still another design constraint, which leads to a most difficult and interesting condition; it denies the designer the *a priori* information needed to design the beamforming and scanning phase-shift program. Hence some low-order intelligence must be built into the system to allow beamforming and scanning to proceed in the absence of accurate knowledge of element position. In short, the system must *self-cohere*, *self-phase*, or *self-focus* or be *self-adaptive*. Means of accomplishing this task are the central thrusts of this book.

Self-Adaptivity Needed

The reason for this requirement of self-adaptivity is easily recognized. A telescope mirror or a microwave dish has an accurately shaped reflecting surface that focuses energy to a point; that is, the surface is shaped so that rays that leave a distant source simultaneously and arrive at the surface are reflected to reach a common point, called the focus or focal point, at the same time. Thus the shaped mirror equalizes the travel times of all rays from source to surface to focus. A lens does the same; its varying thickness with position differentially delays the light rays passing through it to provide a single point (the focus) at which light rays that left a distant source simultaneously will arrive simultaneously.

A phased array usually has a planar surface and does not depend on surface shape to time-align the signals passing through it. Instead, it has electrical circuits to delay the signals differentially from its different parts, so that simultaneity is achieved. Because the signals are relatively narrowband, phase shifters are used rather than the more complicated and more expensive time delays required to emulate the proper geometry; hence the term phased array is applied. To cause the array to focus on a source located at $\bar{r}$ it is necessary only to phase-shift the signal received at $\bar{r}'$ in the array by

$$\phi(\bar{r}', \bar{r}) = \frac{2\pi}{\lambda}|\bar{r}' - \bar{r}| + \phi_0 \tag{1.1}$$

where ϕ_0 is an arbitrary constant phase and all phases are measured modulo 2π. If, however, the designer does not know the array geometry with sufficient accuracy to calculate (1.1), the beam must be formed by a self-adaptive process.

Nonrigid Array

Other practical factors force the requirement for self-coherence. Even if an array is constrained to a single body or frame, it may be impossible to make the frame rigid enough when the array is large. An example is the airframe of an aircraft. In Chapter 9 the design of an airborne radio camera is discussed. The aperture size is as large as the aircraft. A commercial airliner like the Boeing 707 is 45 m long, which is huge for an airborne antenna array, although it is relatively small compared with the numbers quoted earlier. Nevertheless, the airframe is nonrigid at radar wavelengths. Hence an airframe-sized array in an airplane, particularly at the shorter wavelengths, would have to be self-adaptive. A giant array centered about an orbiting space vehicle would be nonrigid because payload limitations in booster technology would preclude lifting the enormous amount of structural material needed to make it rigid. Weather also has an effect on a large system. Differential weather (rain, snow, and sunshine) across a large array installed on the ground could vary the electromagnetic couplings between its elements and their local environments in an unpredictable manner. Weather could also cause the driving point impedances to vary across the array. Compensation for these variations must be closed-loop or self-adaptive because the designer has no way of knowing what they are. Another factor that imposes the need for self-cohering relates to the spatial and temporal variations in the integral of the dielectric constant along the ray paths to the different parts of a large array. Thus there are many reasons for the radio camera to introduce closed-loop or adaptive means for phase synchronizing its array.

Model of the Radio Camera

Figure 1.1 is a target to be imaged (*left*), scanned by a high resolution beam (*center*) formed by a radio camera (*right*). A pulsed transmitter illuminates the target region, the echoes of which are received by the antenna elements of a

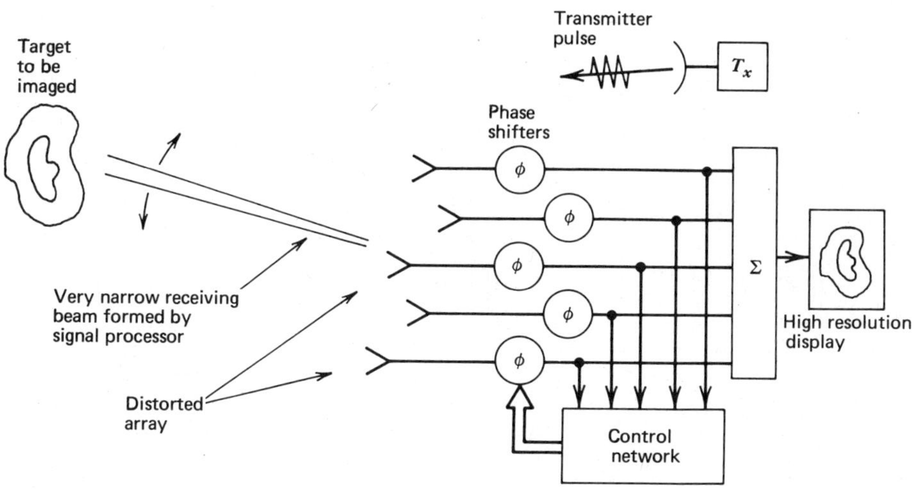

FIGURE 1.1. Giant distorted array self-coheres on target to form high-resolution image.

thinned and distorted receiving array. The receiving elements drive a bank of phase shifters; its phase settings are feedback controlled by measurements made within the system, and the phase-shifted received signals are summed to form a beam. The element signals are further phase shifted to scan the beam, as in a conventional phased array, and the scanned output is displayed.

1.2. USES AND ADVANTAGES OF THE RADIO CAMERA

Airport Application

Figure 1.2 illustrates an airport application. To the right is the search radar used by air traffic controllers, which transmits pulses to a target and receives the echoes. Also shown is an airport radar illuminating an airplane with a beam that is typically 3° in horizontal cross section. The antenna size is about 20 wavelengths. The echoes received by the radar receiver are displayed as intensities or blips on its plan position indicator (PPI) and disclose the target

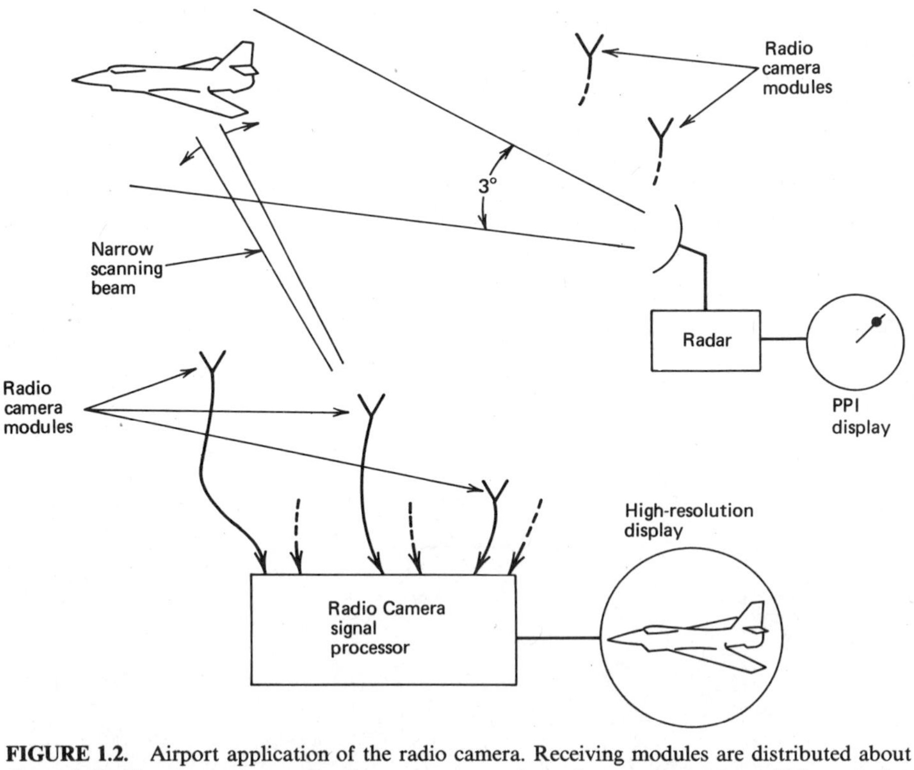

FIGURE 1.2. Airport application of the radio camera. Receiving modules are distributed about the airfield. Signal processor self-coheres to target echoes from radar to form high-resolution image.

coordinates. In addition, the reflections from the aircraft reach all other areas of the airport. Because most of the land within the confines of the airport is unused space, receiving modules having low gain, nonscanning antennas can be distributed about the airfield without interfering with operations. An airfield is commonly several kilometers in size; hence a beamwidth on the order of 10^{-5} rad is achievable, provided that a radio camera aperture covers the airport.

In Figure 1.2 receiving modules are shown distributed about the airport. The antenna elements are of low gain and are nonscanning. The signals received by them are combined in the signal processor of the instrument that self-coheres a beam focused on the target. This beam is narrow in cross section because the airport is huge compared to wavelength. After forming a tightly focused beam on the target the signal processor scans that beam across the target to obtain a high-magnification, high-resolution picture. Thus the controller has a conventional radar display that produces target coordinates *and* an auxiliary display with a high-resolution image of the target.

Airborne Application

Another example, shown in Figure 1.3, is the airborne application referred to in the preceding section. The radar antenna mounted in the nose of the aircraft is perhaps 10 times smaller than the conventional airport radar antenna. The wavelength of the airborne radar is also about 10 times smaller than that of airport radar. Hence the nose antenna is also about 20 λ. This antenna illuminates the terrain in conventional fashion. Because of its large beamwidth, only coarse ground features characterize its display. The size of the airframe is large enough, however, so that a radio camera with flush-mounted receiving elements, if organized successfully into a coherent, diffraction-limited aperture, could provide a resolving power comparable to the best human vision; that is, a few meters spot size on the ground from an altitude of 10 km. An air traveler by his own experience in seeing the ground from high altitude can recognize the value to a pilot or navigator of an all-weather, night and day, 360° continuous radar map of the terrain with resolution close to the human optical. Table 1.1 compares beamwidths of a radio camera the size of a Boeing 707 (45 m) at three radar wavelengths, and the human optical system.

SNR Improves

Not only is the resolving power of the microwave antenna greatly improved by distributing the aperture over the aircraft surface but the total area of the antenna can be increased. Because the radar detection range and signal-to-noise ratio of a target are monotonic with radar aperture, radar detection performance can be made to increase simultaneously. The tradeoffs are described in Chapter 9, in which the detection range of an *X*-band design for a fighter aircraft is increased by a factor of 4.

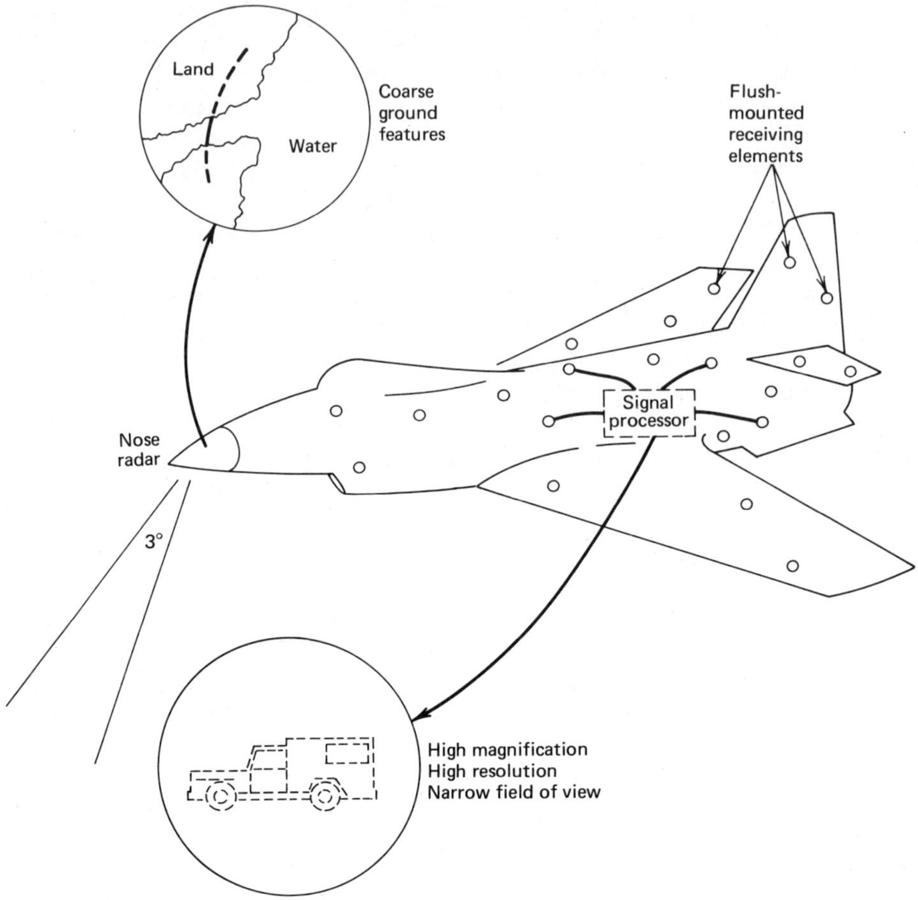

FIGURE 1.3. Airborne application of the radio camera. Antenna elements are distributed throughout the airframe to form a nonrigid phased array.

TABLE 1.1. Beamwidths of a 45-m Aperture and the Human Optical System

Physical Aperture	$\lambda(m)$	Beamwidth (mrad)
Boeing 707	3×10^{-1}	6.7
Boeing 707	3.2×10^{-2}	0.7
Boeing 707	8.6×10^{-3}	0.2
Human eye	5×10^{-7}	0.3

Basically Bistatic

Both the airport radio camera and airborne radio camera illustrate the essential bistatic nature of the radio camera. Unlike conventional radar, in which the same antenna is used for transmission and reception, the illuminator of the radio camera is generally separated from the receiving antenna. The illuminator is broad-beamed and provides the energy for imaging. The transmitter is the microwave equivalent of the sun in outdoor photography, and, as in photography, the camera can be completely separated from the illuminator. Figure 1.4 illustrates a case in which a wide-beam illuminator is completely separated from the receiving array. The latter is distributed over a region large enough to provide the capability for the desired narrow beam. Again, the signal processor coheres the array and scans the beam to form high-angular-resolution, high-magnification images.

Internal Motion Within the System

Nothing in the design procedure, which is described in later chapters, precludes the possibility of internal motion within the system. The set of receivers need not be on a rigid frame, and, in fact, the parts of the system may be in continuous motion with respect to one another. Figure 1.5 shows the bistatic system constructed from ships in a naval task force. One transmitter (or one at

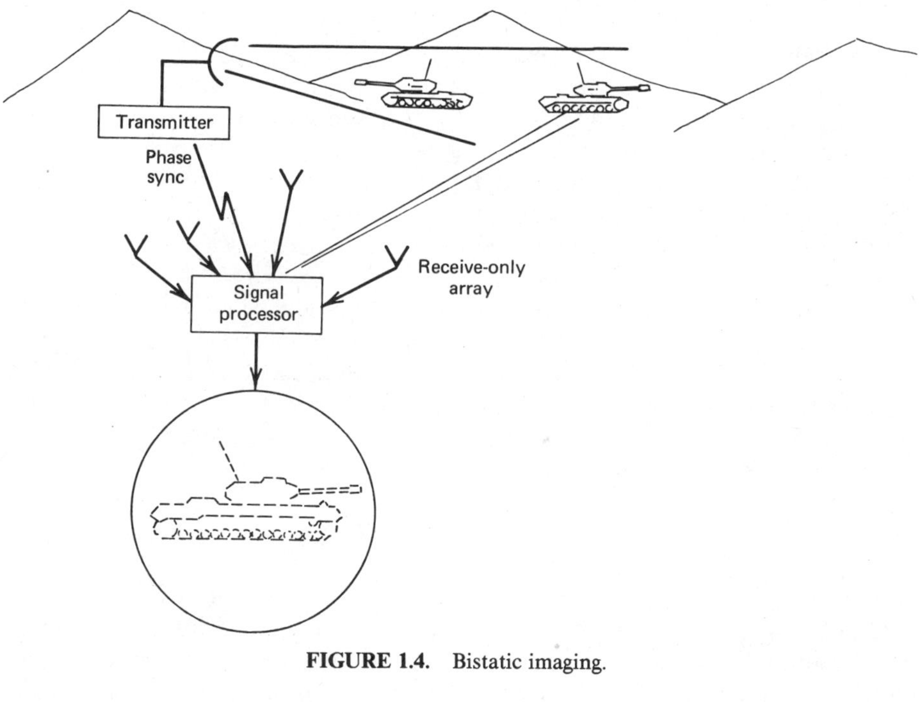

FIGURE 1.4. Bistatic imaging.

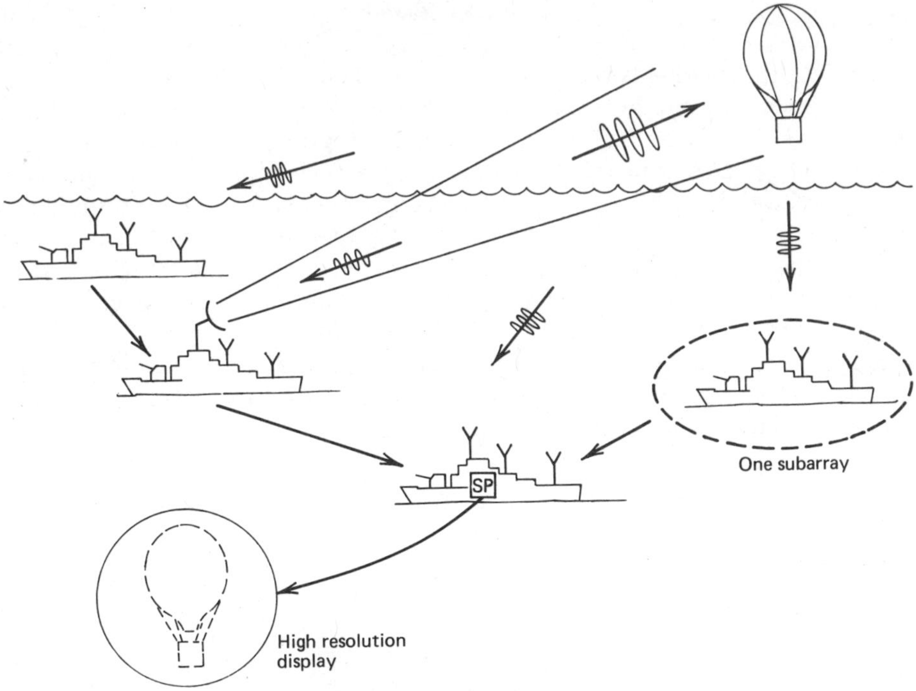

FIGURE 1.5. Radio camera that consists of distributed modules formed into subarrays which are in continuous motion in relation to each other.

a time) aboard one ship radiates pulses. Aboard each ship is a distributed receiving array that is itself one cluster or subarray in the task-force-sized instrument. The ships are interconnected by radio links. It is evident that the technique could be extended to use target-reflected energy from any transmission source, including illumination that does not originate within the task force. Provided only that its radiation also reaches it by a direct and steady path, the array may be phase-cohered by the illuminating wave to permit it to scan the target with its narrow beam.

Still another bistatic application is a giant, nonrigid array, many kilometers in size, located about a space station which may be in orbit at a geosynchronous or other altitude. The function of the equipment could be to image earth-surface echoes from a high-power transmitter in a lower orbit. A 10-km array operating in S-band ($\lambda = 10$ cm) in a 200-km altitude orbit would, if properly phase-synchronized, have a 2-m spot size on earth.

Applicable to Optics

The procedures involved in imaging, which are introduced in the next section, are not limited to microwaves, for they can be applied to any coherent radiation field. They are already being applied in coherent adaptive optics to create large telescopes that do not suffer the effects of atmospheric turbulence.

The multiple-mirror telescope at Kitts Peak, Arizona, is an example in which five 1-m telescopes are arranged in a pentagon with a sixth telescope in the center. The system is designed to coalign *and* cophase the six optical images by closed-loop circuits to form an image with the resolving power of a telescope as large as the entire instrument.

Underwater Sound

The techniques are applicable as well to acoustics, both in underwater sound and ultrasonics. An example of underwater sound is an array of large, sea-bottom-mounted and widely spaced acoustic subarrays. Although these subarrays would be rigidly fixed to the ocean bottom and their positions could be surveyed accurately, the variability in the refractive index of the ocean due to variations in depth, temperature, salinity, and "weather," as well as in multipath structure in the signal received across the array, renders conventional phased-array processing inefficient unless techniques similar to those described for the radio camera are used. An even more extreme example is the coherence into a diffraction-limited array of a collection of freely floating sonabuoys dropped from the air.

Ultrasonics

The imaging properties of the radio camera may prove to be applicable to imaging the interior of the human body with ultrasonic radiation in the 1-to-10-MHz region. By using self-cohered transducer arrays comparable in size to the distance being scanned within the body we can anticipate resolution in the order of wavelength in any arbitrary region of the body. Because the acoustic wavelength is a fraction of a millimeter in this frequency range, images with excellent resolving power should result.

1.3. RELATION TO SYNTHETIC APERTURE RADAR

Time-Shared System

Synthetic aperture radar is a well established technique for achieving fine angular resolution [19], [40], [14]. SAR, which dates to the early 1950s, is much older than the radio camera and the techniques are basically dissimilar. The synthetic aperture principle is illustrated in Figure 1.6. A radar in an aircraft that moves with constant velocity V transmits pulses spaced in time by τ or in position by $d = V\tau$. These locations in space define the array. The radar echoes from targets and terrain are received from almost the same locations, reception taking place after the aircraft has moved $2RV/c$ from each point of transmission. Because $R/c \ll \tau$ (R = range or distance to the target, c = speed of

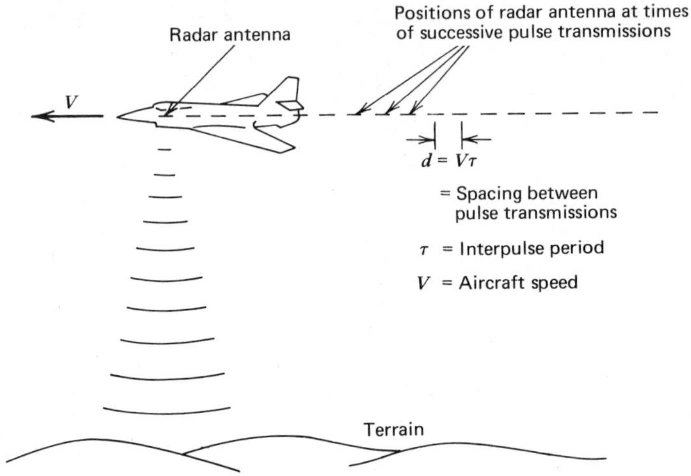

FIGURE 1.6. Synthetic aperture radar.

light), the displacement is small and for all practical purposes the transmitter and receiver are colocated.

Thus a single antenna and radar are time-shared over a large distance to synthesize a large aperture. The received echoes from each transmitted pulse are stored coherently in the signal processor for time T and summed. The processing may be optical or electronic. Because the antenna is moved in a straight line at constant speed and the radar pulses are uniformly spaced in time, a long aperture is formed along the flight path of the aircraft. The synthetic aperture length is $L = VT$ and the number of pulses summed is $N = T/\tau$; N is also the equivalent number of antenna elements in the synthetic array. The angular resolving power is finest broadside to the velocity vector of the aircraft, where it is $\lambda/2L$, and it degrades trigonometrically as the beam of the synthetic aperture scans toward the direction of flight. The factor of 2 in the denominator is due to the two-way phase delay in the pulse propagating to target and back; the effect is to double the wavenumber to $4\pi/\lambda$. Along the groundtrack the angular resolution is the endfire beamwidth of the long aperture. Because the signal processor design is based on this simple geometry, the aircraft flight path is constrained to straight and nonaccelerating motion. Lateral displacements due to wind gusts require measurement and compensation within the radar, and because the resolving power is best in the broadside direction synthetic aperture radar is used primarily for strip mapping or forming images at right angles to the direction of aircraft motion.

Different but Compatible

The radio camera, on the other hand, is basically a physically real phased array which may be ground-based or airborne. If airborne, no constraints are imposed on flight path or constancy of speed. The two techniques are not incompatible, however, and a radio camera can be formed from one or more synthetic aperture radars [24]. The element motion correction technique of the radio camera can be used to remove the nonaccelerating flight path constraint from the synthetic aperture radar. A time-multiplexed or synthetic aperture radio camera results. The acronym SARC is used for this system. A two-fold advantage combines the techniques. By applying the radio camera technique to the synthetic aperture radar the SAR can be freed of its usual flight path constraint, and by applying the synthetic aperture technique to the radio camera the cost of the radio camera can be reduced by reducing the number of its parts to a single transmitter and receiver. Thus the two techniques are natural companions. In Chapter 2 a SARC experiment is described.

1.4. RELATION TO RADIO ASTRONOMY TECHNIQUES

Huge Interferometers Are Becoming Common

Hugh microwave interferometers have been and are being built for radio astronomy. The Very Large Array (VLA) in New Mexico [42] consists of 27 dishes, each 25 m in diameter, deployed in a Y-configuration with arms 21, 21, and 19 km long. Still larger are the interferometers which have baselines continental and intercontinental in size. An exciting experiment in the last decade combined the Goldstone radar in California, the Haystack radar in Massachusetts, and equipment in the Crimea, USSR, into a three-element interferometer. The baseline between Goldstone and the Crimea is about 8×10^6 m and the aperture is nearly 10^8 wavelengths at 10-cm radiation. This resolving power far exceeds what is needed for picture taking. The planned Very Large Baseline Interferometer (VBLI) is an ambitious concept that spans North America and the Atlantic Ocean and extends into Europe [69].

Grating Lobes Can Be Overcome

Generally, however, an interferometer is useless for picture taking. Although it offers extraordinary resolving power, its angular response is highly ambiguous because of the grating lobes in its radiation pattern [85]. This objectional feature can be overcome in radio astronomy because in it the scene is unchanging in time and the interferometer baseline direction and length can be made to change. In this way a single interferometer can sample the radiation field from the scene in all combinations needed to form an image.

This technique is used extensively by radio astronomers under the general heading of "aperture synthesis" [41]. The "frozen" scene is the slowly changing extraterrestrial universe. Baseline rotation is accomplished naturally by the rotation of the earth, the baseline length by the use of multiple antennas.

Too Slow for Terrestrial Imaging

This practice, ingeniously applied in radio astronomy, is not particularly relevant to terrestrial imaging. The speed with which the image must be formed precludes its use. Instead, more elements must be added to make up for the loss of spacing variation provided by the earth's rotation. Three other important distinctions are made between the two activities. The first has to do with the nature of the scene, the second with the signal bandwidth, and the third with the energy source. All three make the interferometer more attractive to the radio astronomer than to the microwave imager. Imaging a complicated scene in an ordinary photograph is more difficult than the simpler ones typical of radio astronomy in which it is often of interest to resolve a few closely spaced objects or to measure the angular size of a relatively isolated object. Large-base interferometers perform these functions splendidly. The resolving power is approximately the reciprocal of the number of wavelengths along the projected baseline. Thus, as noted earlier, the resolving powers of these instruments are remarkable. The major drawback, however, is the ambiguity that results from the grating lobes of the radiation pattern. A two-element interferometer has lobes of nominally equal height spaced in angle by about the resolving power [85]. Additional elements can eliminate these grating lobes but the sidelobe level remains high (see Chapter 2). Large signal bandwidth helps reduce the level of the side radiation pattern (see Chapter 3). The bandwidths of radio sources are sufficient to clean up images automatically to a considerable extent. Of great significance is the fact that astronomical radio scenes are spatially noncoherent because radio sources in the field of view of a radio telescope are independent energy generators. The successful use of the aperture synthesis technique depends on spatial noncoherence. The contrary is true in microwave radar in which all targets reflect energy from a common transmitter, which results in a spatially coherent radiation field. Consequently, because of the considerable *a priori* knowledge of the "scenes" of interest to the radio astronomer, the large signal bandwidth available to him, and the spatial noncoherence between radio sources, he is able to use the remarkable resolving power of these instruments successfully without suffering seriously from angular ambiguities.

Terrestrial imaging of a complicated scene in present time with narrowband radiation originating from a common source and little or no *a priori* information is a different problem. It is this problem that radio camera technology addresses.

1.5. RELATION TO OPTICAL IMAGERY

High-resolution microwave images are different from optical images. The differences are due primarily to the longer wavelengths of microwaves in relation to optics and to the differences in image-forming techniques. Optical wavelengths are in the neighborhood of $\frac{1}{2}$ μm. The wavelengths used in infrared (IR) systems are about 20 times larger. Pictures taken with optical and IR imaging systems generally are in the micron range that extends from somewhat less than $\frac{1}{2}$ μm to longer than 10 μm.

The microwave region extends roughly from 1 to 30 cm, which is the primary wavelength region in which radio camera designs would be expected. Millimeter wave generators and receiving equipment may extend it to about 3 mm. Thus the shortest radio camera wavelength is likely to be three to four orders of magnitude larger than optical/IR wavelengths and the longer wavelength used for long-range search radar is close to six orders of magnitude larger than the shortest optical wavelength.

Optical image formers are usually lenses or mirrors made of single or multiple pieces of carefully shaped glass and are large compared with wavelength in the direction transverse to the optical axis. Because, however, the wavelength is small, the lens itself is small; therefore a lens is easily movable for the purpose of adjusting the direction of view or the focus.

The receivers of a radio camera, on the other hand, are distributed over a huge area and are separated by large distances. Because of the great size of the system, the shape of the surface that contains the antennas is arbitrary. The image-forming signal processor is electronic and must be powerful enough to accommodate the arbitrary and *often unknown a priori* aperture geometry. A random aperture almost invariably ensues directly as a result of the design constraints. The large spacing between elements, plus their random and sometimes time-varying locations, result in poor sidelobe properties, which seriously impacts on image quality. Furthermore, geometric asymmetries in the distribution of the parts of the system introduce certain biases in the two-dimensional diffraction pattern of the aperture.

The effects of these wavelength-related and image-forming-related differences on image properties is the subject of Chapter 2. None of these differences is advantageous to the radio camera. Image quality suffers from all of them. A major problem confronting the designer of a large instrument is the achievement of satisfactory image quality. Design parameters that relate to image contrast and dynamic range are derived later in Chapter 2, whereas in Chapter 3 several techniques for the improvement of image quality are discussed.

1.6. RELATION TO OPTICAL HOLOGRAPHY

Adaptive Microwave Holography

The radio camera technique may be viewed as adaptive microwave holography [88], which differs from optical holography in several important ways. In optical holography a recording is made of the interference between a radiation field produced by the source or scene to be imaged and a reference. The recording is the hologram. It records the intensity of the sum of the fields at

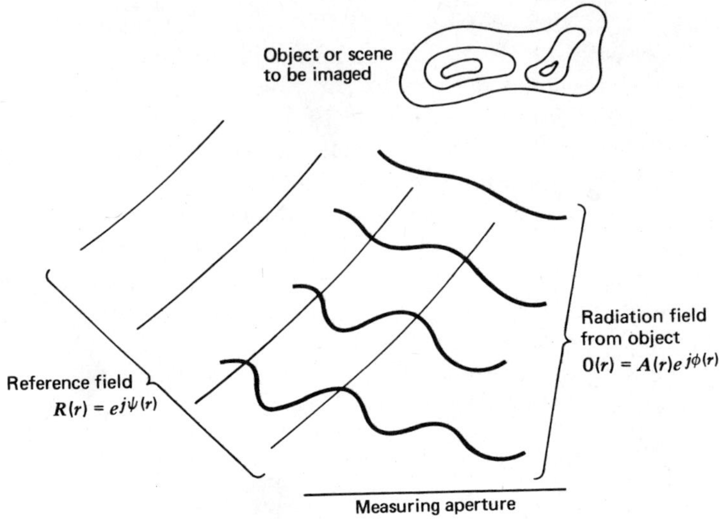

$$
\begin{aligned}
I &= |R + O|^2 \\
&= 1 + 2A \cos(\phi - \psi) + A^2 \\
&\simeq 1 + 2A \cos(\phi - \psi), \quad A \ll 1
\end{aligned}
$$

FIGURE 1.7. Conventional holography. Reference field interferes with object field; measuring aperture records unknown object field data (A, ϕ) from which the image may be reconstructed at a later time.

each point in the aperture in which the recording is made (Figure 1.7). Because the recording contains the amplitude and phase information in the wave field due to the source, an image of the source can be reconstructed from it.

Microwave holography performs the same functions and therefore follows the same overall rules. Because of the differences in wavelengths, bandwidths, and radiating and detecting components, however, the processes are performed in different ways. In microwaves, as in acoustics, a transducer converts the radiation field to an electrical current. The microwave radiation field is sampled by an antenna element such as a horn or dipole and funneled into a microwave circuit such as a waveguide. Currents are excited in the circuit by the field. Except for amplification, the frequency is too high for the signal to be processed directly; consequently the signal that represents the radiation field is heterodyned to a lower frequency in a nonlinear device called a mixer, which is usually a solid-state component. The current that represents the radiation field due to the source adds to the current in the same frequency band supplied by a local oscillator. Because of the nonlinear nature of the mixer, sum and difference frequency components, as well as the fundamental frequencies and their harmonics, appear. By frequency selection of the difference frequency a lower frequency signal is obtained. This lower or intermediate frequency (IF) signal retains the amplitude and phase information of the radio frequency (RF) signal, which, in turn, represents the interference field.

The optical interference field at each point in the aperture contains the intensity of the field produced by the source or scene and the phase difference between the two fields. The heterodyne or IF signal in the microwave receiver contains the amplitude of the RF field and the phase difference between the RF field and local oscillator. Thus the local oscillator can replace the reference field, provided only that the phase of the local oscillator is made equal to the phase of the reference field at each point in the aperture. In this way it is possible to eliminate the reference field entirely.

Reference Field May Be Eliminated

This is the first essential difference between optical and microwave holography; as illustrated in Figure 1.8, the reference field may be eliminated and replaced by a correctly phased local oscillator within the microwave receiver.

In that figure the ith antenna element located at r_i delivers an RF signal to a microwave mixer. The instantaneous signal is $A_i \exp[j(\omega_0 t + \phi_i)]$. The local oscillator wave with which it mixes has constant amplitude and instantaneous phase $\omega_{LO} t + \psi_i$. The IF wave is $A_i \exp[j(\omega_{IF} t + \phi_i - \psi_i)]$; A_i and ϕ_i are the amplitude and phase of the source radiation field at the ith element; ψ_i is the proper phase for the reference field which may be obtained from the field or from the local oscillator (see Figure 1.8). The IF wave is demodulated coherently in quadrature demodulators and the resolved components of the IF wave are sampled and stored. The stored data constitute the hologram.

Image reconstruction follows application of the appropriate integral transform to the holographic data. The two-dimensional Fourier transform is the correct one when the array is planar and the source to be imaged is in the far field of the array.

Simultaneity of Object and Reference Field Not Required

A second basic difference is that the reference wave need not coexist in time with the signal field. This is evident from the fact that the phase difference

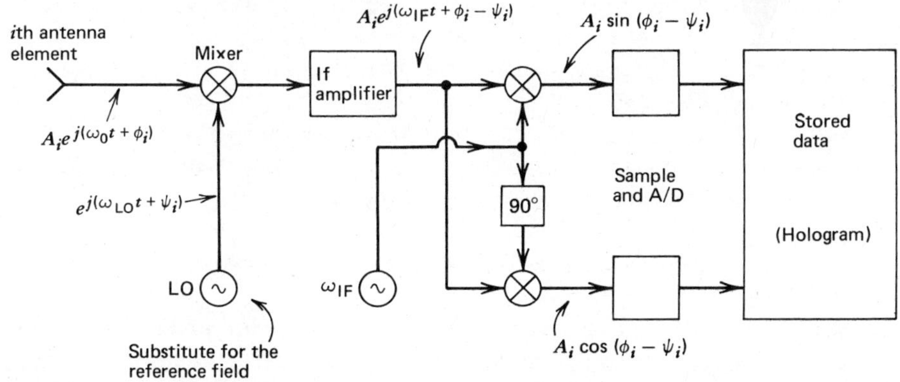

FIGURE 1.8. Microwave holographic receiver. Reference field may be eliminated and replaced instead by a correctly phased local oscillator within the microwave receiver.

$\phi_i - \psi_i$ constitutes the phase information for the hologram. The two constituents ϕ_i and ψ_i, however, need not reach the data store at the same time. If the local oscillator phase is zero, the phase information that reaches the store is ϕ_i. The reference phase ψ_i can be added to the store at any other time, after which the phase difference can be obtained. Thus there is no requirement for simultaneity of the signal and reference fields.

Simultaneity of Measurement of Object Field Across Array Not Required

A third basic difference is that the object field need not be measured simultaneously across the array; for example, let the ith antenna element of Figure 1.8 be the *only* physical element in the array and let it be moved to a new position before each successive radar transmission. These several positions constitute the array. At each new location the echo sequence is received, passed through the receiver, and stored. Provided that the terrain and other targets have not moved during the time that the antenna element occupies its successive positions, the stored data are identical to those that would have been obtained had the array been a conventional design. Thus a time-shared system can produce the microwave hologram. The technique just described is used in airborne synthetic aperture radar, described in Section 1.3.

Reference Wave Essential in a Large System

The system in Figure 1.8 is appropriate when the phase of the local oscillator can be calculated *a priori*. For this condition to obtain the coordinates of the antenna elements must be known at all times, the refractive index of the propagation medium must be constant across the array, the mutual coupling between the elements must be the same or at least constant and predictable, and the coupling of the antenna elements to their local electromagnetic environments must be the same. Only in relatively small antenna arrays will these conditions be met. In a microwave array large enough to provide image-quality resolving power the signal phase is likely to have an unpredictable, random component across the array, the reasons for which were discussed in Section 1.1. When these things happen it is impossible to specify *a priori* the desired reference phase ψ_i. Instead, the phase must be deduced by the system itself from the measurements of a radiation field induced by a source of known properties. This system is called adaptive or self-cohering. Adaptive beamforming is central to the operation of the radio camera.

External Source Generates Reference Field

Required is an adaptive beamforming or phase-synchronizing source. Its function is to generate the reference field. The ideal phase-synchronizing source is a point source radiating with high signal strength in free space. These three properties ensure that the reference field is planar across the entire array (or spherical if it is in the near field of the array) and that the phase measurements are not contaminated by noise or other interference. An active source like a radar beacon with a small, omnidirectional transmitting antenna is one example. Another is a corner reflector, which is an excellent passive reference source that provides a stable pulse echo from a radar transmitter. Many targets of opportunity (TOO), which exhibit the desired reflecting properties for a reference source, are found in radar and are discussed in Chapter 6. Examples are houses and water towers.

The Radar Application of the Radio Camera

Radar is designed for short pulse operation to permit the differentiation of targets according to range. This feature makes it possible to distinguish between echoes from targets on the basis of their arrival times after the transmitted pulse. Consequently the signal processor has a variety of radiation fields from which to select a reference. Simple algorithms can test the echo sequence across the array from successive range bins to determine which echo provides the most satisfactory reference field. The important fact is that simultaneity of the reference and object fields is not required, which permits the target scene and beamforming source to occur at different and arbitrary distances from the radar. It is necessary only that phase measurements be made from both range bins to allow their differences to be taken.

1.7. PROCEDURES INVOLVED IN IMAGING

Figure 1.9 contains the basic requirements of a radio camera in the context of the airport application of Figure 1.2. The earlier illustration is expanded to show the functions performed by the signal processor. The size of the radar antenna is 20 wavelengths, as discussed earlier. Its 1/20 rad or 3° beam is 1 km in cross section at a distance of 20 km, and even when an approaching target aircraft is at that distance the beamwidth is immense compared with the size of the aircraft. When the target is at longer ranges, the beam cross section is still larger. Even on the final approach to the airport the beam is wider than the target. Thus there is no possibility of obtaining fine imaging information from the airport radar because its beam cross section is always larger than the target itself. The echoes from the target can, however, be received by antennas distributed over the entire airport. A runway large enough to accommodate a jumbo jet is at least 3 km long. A major airport will be at least 30,000 wavelengths at *S*-band (10 cm) and many will be twice this size and larger. Hence an airport-sized radio camera would be in the 10^4 to 10^5 λ region and its beamwidth, even at a distance of 20 km, would be small (at most 2 m) compared with the target aircraft.

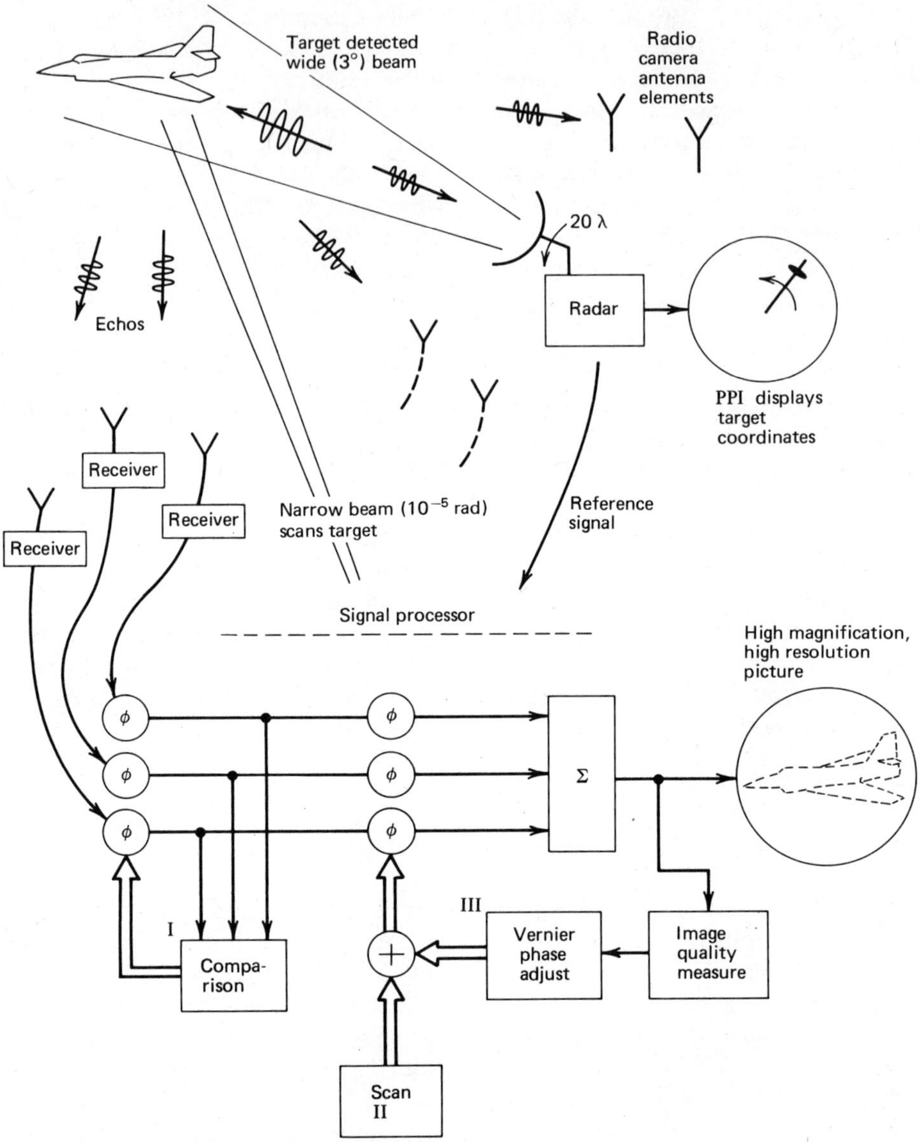

FIGURE 1.9. Radio camera for airport application showing radar transmitter, distributed array, and high-angular-resolution signal processor.

Beamforming

The upper portion of the illustration shows a radar illuminating a target with a strong pulse and receiving a weak echo which, after detection and processing in the receiver, is delivered to a radar display that portrays the echo at the target coordinates. The same target also scatters energy to the widely dispersed and randomly located receiving elements of the radio camera. Each antenna element drives a receiver and the several receivers deliver signals to a signal processor. The signal processor consists of three subsystems, each of which performs a distinct function. The first is a closed-loop, phase-synchronizing system (labeled I) which receives signals from the target or source and focuses the array back on that source. This is the *adaptive beamforming* process noted earlier. It is a retrodirective process for which several means of accomplishment are described in [47] and [39]. It may be accomplished by any number of methods. Three references which deal specifically with phase synchronization of the radio camera are [83], [87], and [94].

To form a beam in an arbitrary direction it is necessary only that all signals received from that direction be cophased, after which they may be summed. It is not necessary to know the locations of the receiving elements, provided that the circuits or signal processor have the ability to cophase the signals received by the antenna elements. Closed-loop phase shifters that consist of phaselock loops driven by the phase differences between the element signals and some reference element in the array can perform the task. So can off-line digital processing. The circuits may be analog, digital, or hybrid. Because the elements are widely separated on the airfield, their signals cannot be compared in phase at RF but must be heterodyned to some convenient lower frequency for transmission to a common location. A receiver, which includes a local oscillator and mixer, is incorporated with each element into an array module. The oscillators must run at a common frequency and their phases must be identical or they must have some known relations between them. A reference wave must be delivered from the radar to the signal-processing system to cohere the local oscillators. The reference wave may be inserted directly, as shown, by radio or cable or by an RF transmission path through a sidelobe of the transmitting antenna's radiation pattern. Alternately, the echo from the target can be used to phase-synchronize the receiver (Section 5.8). The advantage of direct synchronization is that the signal level and therefore the signal-to-noise ratio in the synchronizing signal can be controlled by the designer, whereas the same is not true in the target-derived echo, the virtue of which is that the system self-corrects for Doppler shift due to target motion because the echoes from the target contain the Doppler information. The direct reference is used in the Valley Forge experimental equipment in which the received signals are demodulated coherently to video and transmitted as in-phase and quadrature digital numbers to the signal processor. The technique is described in Chapter 5.

Adaptive beamforming has been accomplished successfully in dozens of experiments and the process is no longer in question. Figure 1.10 shows an

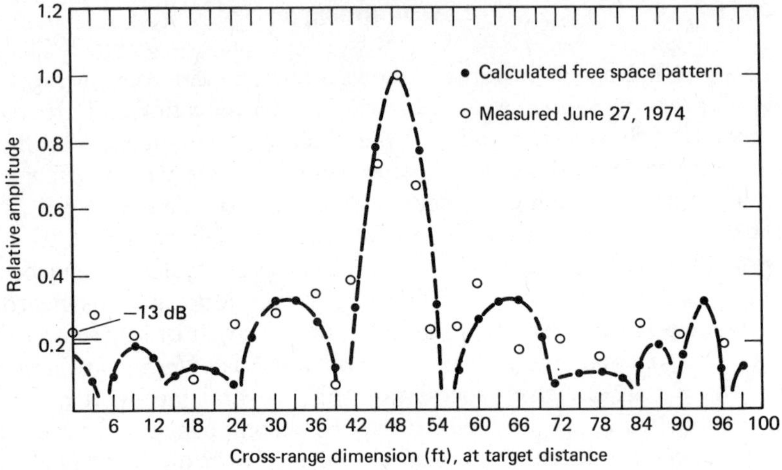

FIGURE 1.10. Beam pattern of self-cohering, 19-element, random 100-ft *L*-band receiving array compared to calculated free space pattern. Target is in near field (approximately 800 ft from array). (From B. D. Steinberg, *Principles of Aperture and Array System Design*, Wiley, 1976.)

early adaptive beamforming experiment. In that test, as well as in all subsequent experiments, gain, beamwidth, beam shape, and average sidelobe level consistently conform to theory.

Imaging

Beam scanning or imaging, which is the second function (labeled II) of the signal processor, also has been accomplished repeatedly and satisfactorily. Some knowledge of target and element coordinates is required to scan the beam, in contrast to the self-cohering process that requires no such knowledge. In the scanning or imaging process the beam is scanned over the target, or over some larger field of view that includes the target, to produce the image. Scanning is an open-loop procedure. It is because the operation is open-loop that target and array element-coordinate information is needed. The error-tolerance theory is given in Chapter 13 of [85] and is placed in the perspective of imaging in Chapter 7 of this book, in which tests of the theory are reported. The main effect of inaccuracies in the scanning procedure is to diminish the mainlobe gain. Phase errors in measurement, phasefront distortion due to multipath and scattering, and phase errors due to inaccurate knowledge of element positions are mainly responsible. Amplitude errors have little influence. Beamwidth is hardly affected because it is determined by aperture size and to a lesser extent by the specific relations between amplitude and phases within the aperture. The details of the side radiation pattern are altered by the errors but the statistics are unchanged because the sidelobes are those of a random array; the introduction of phase and amplitude errors does not increase the randomness nor does it change the statistics of the sidelobes. Errors therefore decrease the ratio between mainlobe and average sidelobe level and consequently reduce the contrast in the image. Many imaging experiments with milliradian and submilliradian systems have been conducted successfully. Several of the results are discussed in Chapters 2 and 5.

Image-Quality Enhancement

The third function (labeled III) is image-quality optimization or improvement. Figure 1.9 shows the function being performed by closed-loop feedback to the phase shifters in the scanner with the results of a quality measurement of the image. The process is called image feedback control. The more general term is image-quality enhancement. This subject deals with any algorithmic or signal-processing technique that improves the quality of the image before or after it has been formed by the scanning process. Any process that improves image contrast or dynamic range or resolution or removes or smooths artifacts in the image falls into this category. Certain open-loop techniques are discussed in Chapter 3 and feedback control is discussed in Chapter 4.

1.8. THE RANDOM ARRAY

It is not essential to the system organization that the elements of a radio camera be distributed randomly over the aperture. Its large size and the potential for flexure of the system will, however, often result in an element distribution with the properties of a random array. The primary properties of the radiation pattern of a one-dimensional array, shown in Figure 1.11, consist of a mainlobe, which almost always has well formed properties, and a side radiation pattern that expresses the random character of the array and can be described only statistically.

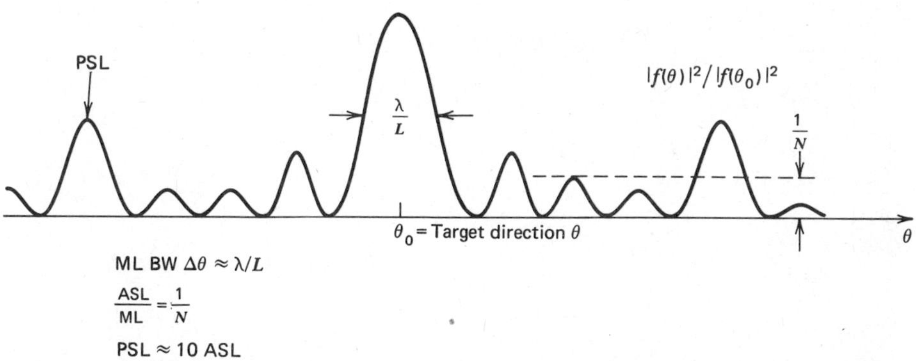

FIGURE 1.11. Properties of the radiation pattern of a random array. Grating lobes avoided by random distribution of element locations. λ = wavelength; L = size (length or width); N = number of elements.

The theory of the radiation pattern is given in Chapter 2 and a summary of the pertinent properties appears in Figure 1.11. These are the expected characteristics of the radiation pattern that result from the retrodirective, closed-loop process. A mainlobe is developed in the direction of the target: its beamwidth is approximately λ/L, where λ is the wavelength and L is the size of the receiving aperture. The average sidelobe power level (ASL) is N^{-1} in relation to the strength of the mainlobe and the peak sidelobe (PSL) (with probability close to unity) is approximately $10N^{-1}$. The statistics of the amplitude of the side radiation pattern are those of the envelope of band-limited Gaussian noise.

CHAPTER TWO

Image Properties

2.1. WAVELENGTH EFFECTS

The difference in wavelength between the optical/IR and microwave regions—a difference of three to six orders of magnitude—has several important effects on image properties. The first is in the scattering and reflecting properties. Optical waves are readily scattered because they are smaller than or comparable to the surface roughness of most objects. As a consequence, we, or a camera, can "see" a surface; that is, light from a source striking an object is scattered by the roughness of the surface (Figure 2.1). Some of the scattered light reaches our eyes or the camera lens and an image is formed on the retina or film. The image portrays the strength of the two-dimensional scattering from the surface illuminated by the radiation. Because the scattering is from the surface, it is easy to perceive its shape, texture, and color. This is true even if a surface is largely obscured by artifacts. A table or desk, for example, largely covered with books, papers, and ashtrays, nevertheless still appears as a table or desk.

Most Surfaces Appear Black

The same surface illuminated with microwaves would produce an entirely different image. Its irregularities are generally much smaller than the wavelength. Hence the surface would primarily reflect the microwave energy and only secondarily scatter it. Consequently the surface would appear dark to an observer unless he viewed it from the reflection angle. Only if the observer or detector intercepted a reflected ray would the surface brighten, in which case a strong signal would be received from a small portion of the surface, much like a specular optical reflection from a highly polished automobile surface. This

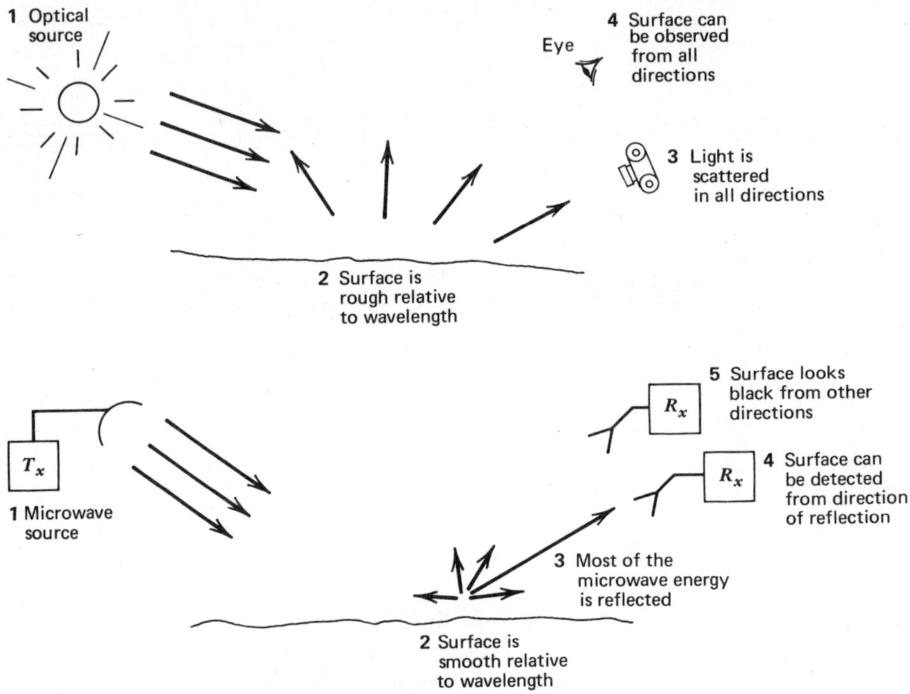

FIGURE 2.1. Short wavelength light is scattered; long wavelength microwave radiation is largely reflected.

strong highlight contributes nothing to shape information and only obscures weaker information-bearing echoes.

Therefore the surface would appear black. The other objects on the table or desk, assuming that they reflect or interact with the radiation, would scatter energy because they introduce discontinuities on the surface comparable to or larger than the wavelength. The metal back of a three-ring binder would form a two-cornered reflector with the surface of a metal desk and produce a strong signal in some direction.

Other objects as well would scatter or reflect significant energy. These signals, when imaged by the radio camera, would portray the articles on the desk but not the desk itself. What, then, would the desk contribute to the image? At best it would provide edge and corner scattering from the top surface, legs, and drawers. These are image elements that belong to the desk, whereas the others belong to the articles on the desk; consequently they are artifacts† of the image of the desk. A picture of a desk dominated by the artifacts would bear little resemblance to a desk. A second picture, taken at a later time when the desk-top materials had been rearranged, would look

†A second type of artifact, literally a false or nonexistent source or target appearing in the image, is due to high sidelobes. See Section 2.7 and Chapter 3.

different from the first and also unlike a desk. Thus radio camera microwave images of bodies with large, smooth (to microwaves) surfaces would look exceedingly different from the optical images of the same bodies.

Lower Information Content

A second effect of the longer wavelength is a drastic reduction in the information content of an image. An object immersed in a radiation field can be sensed or probed by that field with no finer resolution than the order of wavelength of the radiation. A few square centimeters of exposed desk surface viewed optically discloses to the eye or camera a huge amount of information concerning the color and texture of the surface. An area of similar size exposed to microwave radiation might yield one or a few independent samples of its scatter strength. The potential difference in information content per unit area is the square of the wavelength ratio, which is 10^6 to 10^{12}. Needless to say that even without the problem of loss of reflecting surface this loss of information content would cause radio camera images to be drastically different from optical images. The theory and application of Shannon's theory of information [75] is discussed in [28], [50], [51], [45], [57], [58], [77].

Specular Reflections Dominate

A third factor separates the natures of the images. The long wavelengths diminish scattering from smooth surfaces and increase specular reflections. Consequently a manmade structure, which usually consists largely of flat surfaces, will appear in a microwave image as a structure of mirrors. Highlights or specular reflections become the brightest parts of the image. Optical images, on the other hand, are largely devoid of specularities. Mirrors, windows, and certain shiny surfaces produce specularities, but they are small in number and rarely compete with the scattered optical energy from a scene. The converse, however, is true in many microwave scenes. The microwave image properties of scenes that contain large, smooth surfaces would be determined by the reflections from those surfaces. These microwave images would bear little resemblances to optical images.

Large Dynamic Range

A fourth difference results from the tendency toward specular reflections in microwave images. Reflected energy is confined to a narrow angular interval approximately the ratio (in radians) of the wavelength to the size of the reflector; scattered energy, on the other hand, is broadly dispersed in angle. The specular signal detected by an imaging system is stronger than the scattered signal from the same surface by the ratio of the angle through which the scattered energy is dispersed to the angular spread of the specular signal. As a consequence the specular signal can obscure the much weaker scattered energy that contains the bulk of the information regarding the shape and structure of the object being imaged. This is a dynamic range limitation. Every system can handle only a limited range of target strengths. The stronger the specular reflections, the higher the threshold for scattered signals below which they are ignored by the system. The dynamic range of radar targets varies from 50 to almost 90 dB [62], whereas it is rarely larger than 20 dB for most optical images, reaching 30 dB at most for sunlit snow [36].

For these reasons the microwave radio camera image is different from the optical image. This does not, however, mean that it is useless or uninteresting. Examples shown in Chapter 5 offer contrary evidence. Synthetic aperture radar experience has shown that high resolution microwave images yield exceedingly valuable information. The mapping of natural terrain is an example. Here the problems raised by lack of surface scattering do not pertain. Furthermore, complicated manmade structures image satisfactorily. Hence we may conclude that radio camera images will be different from optical images but we cannot conclude that they will not be useful images. Utility depends entirely on the nature of the scenes to be imaged and the purpose for which the imaging is done.

2.2. WHY THE RANDOM ARRAY?

Nothing is particularly attractive about a random array except the ease of its mathematical description [59], [85]. In fact, there is always, (i.e., with probability one) a distribution of element locations better than any set chosen at random. In addition, the random array suffers from a serious sidelobe problem that is totally avoided with conventional array design. Nevertheless, it is likely that most radio cameras will be designed and built with random arrays.

The reasons stem entirely from the sad logic that follows from the requirement for large size. In an array as large as hundreds of meters to tens of kilometers the difficulty of placing elements accurately or of surveying their locations accurately, following placement, forces the array design to be random. Conventional design requires control or measurement of element placement to an accuracy of one-tenth the wavelength or less (see Section 2.4). When the placement accuracy degrades significantly from this tolerance the sidelobe properties decay to those of the random array. This is almost independent of the original distribution of elements. Thus largeness itself will tend to force the statistics of the radiation pattern into those of the random array.

Another concomitant of large size is high cost unless the array is drastically thinned. An array of 10^5 wavelengths in linear dimension, or about 10^{10} square wavelengths in area, is needed for a resolving power of 10^{-5} rad or two arc seconds, equivalent to that of a 2-in. optical telescope. A phased array of that size would require at least 10^{10} antenna elements and RF circuits. This is an outrageous number of parts. The array must be thinned or made sparse by six or seven orders of magnitude to bring the cost within realistic bounds. Thinning by removal of regularly spaced elements would result in grating lobes. Therefore in the distribution of the remaining elements all periodicities or regularities must be eliminated [85, Chapter 7]. A designer has two alternate routes for ensuring an aperiodic distribution: deterministically, based on a variety of algorithms, or by random placement. In either case the rigorous control of the sidelobes that conventional array synthesis permits is lost. With respect to the worst sidelobes it has been shown that random placement is no worse on the average than the best deterministic aperiodic algorithms reported in the literature [82]. Figure 2.2 shows measured cumulative distribution functions of the peak sidelobes of a group of computer-simulated random arrays and a collection of algorithmic aperiodic arrays. The abscissa is the ratio of the PSL of a sample array normalized to the expected value of the PSL of the random array designed with the same resources and parameters (e.g., size, number of elements, wavelength, and beamsteering angle). The scale is a decibel scale and the data are plotted on a Gaussian probability graph. Both experimental distributions are approximately linear, which suggests log normal probability density functions for the ratios of the peak-to-average sidelobe

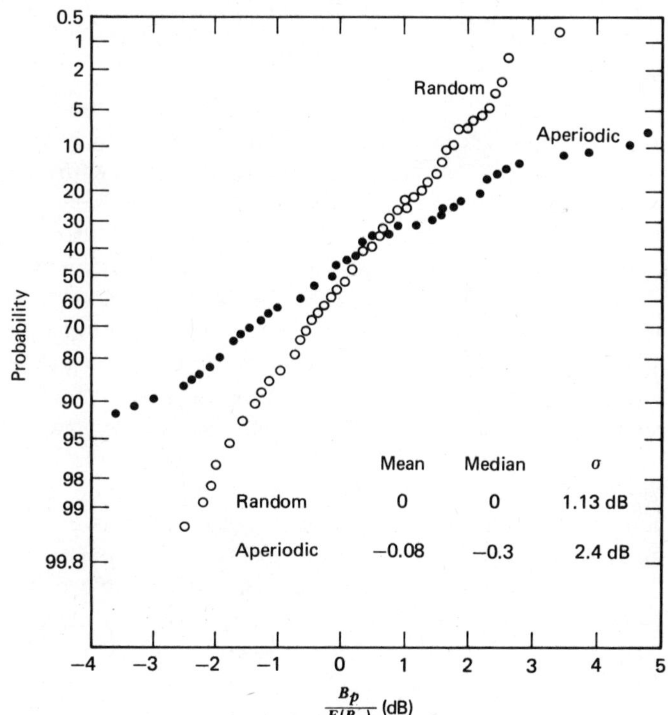

FIGURE 2.2. Cumulative distribution functions of peak sidelobes of 170 random arrays and 70 aperiodic arrays. (From B. D. Steinberg, *IEEE Trans. Antennas and Propagation*, AP-21, May 1973.)

levels. The means and medians of the distributions are indistinguishable (within a few tenths of a decibel). Although there is no advantage in choosing a random array over an algorithmic aperiodic design, there is, on the average, no penalty.

Other reasons for interest in the random array when the large system is nonrigid or inherently time-varying include two factors that tend to drive the design in that direction. First, the designer is likely to be denied the necessary knowledge to predict system performance from conventional, deterministic array theory. Second, even if the element locations were known at all times, their motions would cause the radiation pattern properties to degrade to those of the random array.

Because of these reasons, it is recommended that radio camera design be based on the assumption that the array properties are as bad as those of the random array. If the array proves to be nonrandom with superior sidelobe properties, so much the better.

2.3. PROPERTIES OF THE RADIATION PATTERN OF THE RANDOM ARRAY

Resolution

The properties of the radiation pattern of the random array that are most relevant in the design of an imaging system are the beamwidth and the sidelobe level. The beamwidth sets the limit to the resolving power of the imaging system, the sidelobe level limits image contrast and dynamic range. Subsequent sections in this chapter, to which this section is introductory, deal quantitatively with these image attributes. The facts are illustrated in Figure 2.3. Three sources or targets are shown at the top of the figure and below is the one-dimensional image that results from scanning the scene. The targets at bearings θ_1 and θ_2 are resolvable as separate targets, provided that their angular separation is at least the order of one beamwidth. The exact value is determined by three factors: (1) the relative target amplitudes, (2) the relative target phases, and (3) the resolution criterion of the observer. The greater the clarity demanded by the observer in distinguishing between two targets, the larger the required separation for resolvability. Also, the larger the ratio of the source strengths, the greater the angular separation needed for their responses to be separately identifiable in the image. Nonetheless, the resolving power is always proportional to beamwidth and is usually the order of the beamwidth [84], [85, Chapter 10].

The target at bearing θ_3 is several beamwidths distant from the other targets. Its detectability is determined not by position but by its strength in relation to the size of the sidelobe responses of the system to the larger targets. In the case shown it would not be identifiable as a distinct target.

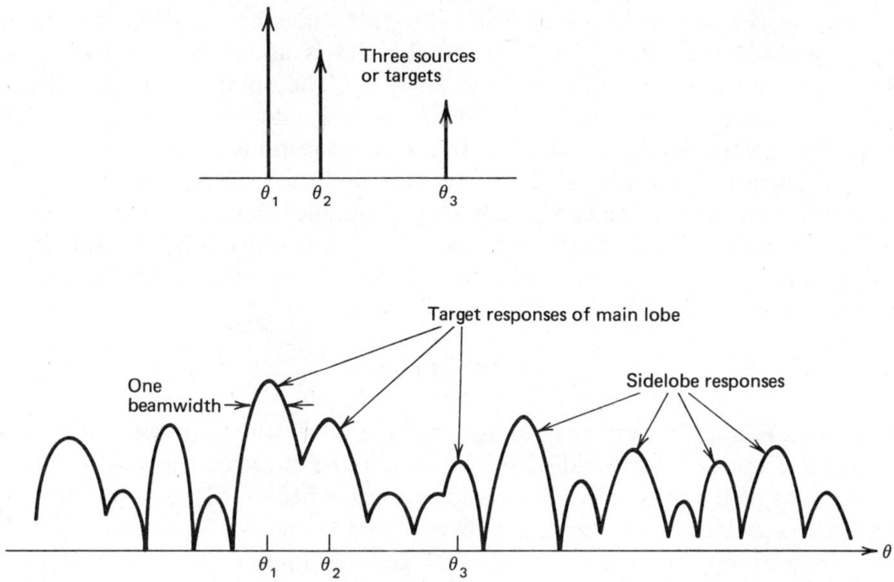

FIGURE 2.3. Image (in one dimension) obtained by scanning a three-target scene with a random array.

Contrast

The statistics of the side radiation pattern have a dual impact on the image. The first effect is to limit the image contrast. The side radiation pattern of the random array has the statistical properties of the envelope of narrowband Gaussian noise. The sketch in Figure 2.3 is typical of the sidelobe response of an array with high sidelobes; the average sidelobe level (ASL) represents the background or gray level against which the image is displayed. It is determined primarily by the number N of antenna elements in the array and is approximately a fixed intensity, that is, power, ratio N^{-1} with respect to the largest target response. This ratio establishes the maximum image contrast. Smaller targets appear with lesser contrast. Hence the number of antenna elements is a dominant factor in image contrast.

Dynamic Range

The second effect of the side radiation pattern of the random array is the limit that it sets to the dynamic range of the sources or targets. The dynamic range is the ratio of the largest to the smallest target that the system can handle. It is set by the PSL of the array. The mainlobe response to the smallest target that can be identified as belonging to the scene and not to the sidelobe response of the array to another target must exceed the peak sidelobe response of the largest target. The mainlobe response of the target at θ_3 in Figure 2.3 failed to exceed the largest sidelobe response. The peak sidelobe of a linear random array typically exceeds the ASL by 10 dB and is about 3 dB higher for a two-dimensional array (Section 2.4). Therefore the smallest distinguishable target response for the one-dimensional array should be more than 10 dB above the average background. The largest target response will not exceed the average sidelobe power level by more than a factor of N, the number of elements in the array. Hence the maximum dynamic range for targets is about $N/10$ for a one-dimensional array and $N/20$ for a two-dimensional array. Image contrast and dynamic range are discussed in greater detail in Section 2.5.

Depth of Field

Two other factors influence the image characteristics, although their effects are secondary: the depth of field of the array and the effect of bandwidth on the dimensions of the focal zone. Because the array is necessarily large to have the desired resolving power, the targets being imaged may be in its near field. In this circumstance the focal interval in range is no longer infinite but becomes proportional to the wavelength multiplied by the square of the ratio of the target distance to the array size (Section 2.6). This ratio is the *f*-number of the array, analogous to the ratio of focal length to lens diameter in a camera or telescope. When the depth of field is sufficiently small, resolving power in range can be achieved even with long pulses or continuous-wave radiation. Near-field imaging systems can make use of the depth of field to obtain the desired range resolution of the system while easing some of the bandwidth requirements of the transmitter and receiver. Wide bandwidth, on the other hand, sharpens the focal zone of a large array and improves its resolving power. It also reduces the sidelobe level and thereby improves dynamic range. Bandwidth effects are discussed in Chapter 3.

2.4. RANDOM ARRAY THEORY

The Radiation Pattern

This section is a brief review of random array theory.† Figure 2.4 shows a one-dimensional or linear random array. The antenna elements are located at random positions $x_1, x_2, \ldots, x_N$; their excitations are $I_1, I_2, \ldots, I_N$. The location x_n of the nth element is a random variable with a probability density function (pdf) $w(x)$. The excitation of the array can be represented by

$$i(x) = \frac{1}{N} \sum_{n=1}^{N} I_n \delta(x - x_n) \tag{2.1}$$

†For a fuller treatment see [59] and Chapters 8 and 9 of [85].

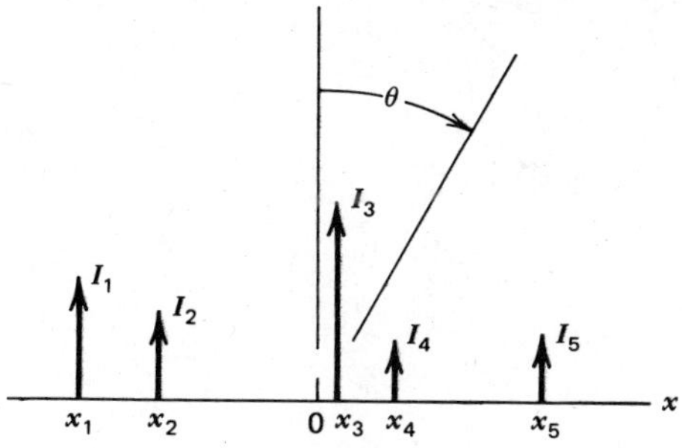

FIGURE 2.4. One-dimensional or linear random array.

where $\delta(\cdot)$ is the Dirac delta function. The far-field radiation pattern (the near-field pattern is discussed in Section 2.6) is the Fourier transform of $i(x)$, which is given by

$$f(u) = \mathcal{F}\{i(x)\} = \frac{1}{N}\sum_{n=1}^{N} I_n \exp(jkx_n u) \qquad k = \frac{2\pi}{\lambda} \tag{2.2}$$

The variable u in (2.2) is a reduced angular variable related to the real angle θ by the trignometric relation $u = \sin\theta$; θ is measured from the normal to the array. To steer the beam to the direction $\theta_0 = \sin^{-1}u_0$ the array is linearly phase-weighted by the factor $\exp(-jkx_n u_0)$ and the radiation pattern becomes

$$f(u) = \frac{1}{N}\sum^{N} I_n \exp[jkx_n(u - u_0)] \tag{2.3}$$

At the beamsteering angle u_0 all components in the sum are cophased and the value of the radiation pattern is

$$f(u_0) = \frac{1}{N}\sum^{N} I_n = \bar{I} \tag{2.4}$$

where the overbar represents the average value. The power density on the axis of the mainlobe is proportional to the square of (2.4).

Amplitude and Phase Errors

Random amplitude and phase errors in the weights alter the radiation pattern (2.3) to

$$f(u) = \frac{1}{N}\sum^{N} (I_n + \delta I_n)\exp\{j[kx_n(u - u_0) + \delta\phi_n]\} \tag{2.5}$$

At the beamsteering angle

$$f(u_0) = \frac{1}{N}\sum^{N}(I_n + \delta I_n)\exp(j\delta\phi_n) \tag{2.6}$$

and its expected value is

$$Ef(u_0) = \frac{1}{N}\sum^{N}(I_n + E[\delta I_n])E[\exp(j\delta\phi_n)] \tag{2.7}$$

assuming that the amplitude errors and phase errors are independent. In the common case in which the error statistics are independent of position in the array

$$Ef(u_0) = (\bar{I} + E[\delta I])E[e^{j\delta\phi}] \tag{2.8}$$

Most small errors are unbiased and their pdfs are usually even functions. With these additional assumptions

$$Ef(u_0) = \bar{I}E[\cos(\delta\phi)] \tag{2.9}$$

which leads to a particularly simple result for a Gaussian distribution of phase errors with phase-error variance equal to σ_ϕ^2:

$$Ef(u_0) = \bar{I}e^{-\sigma_\phi^2/2} \tag{2.10}$$

The ratio of (2.10) to (2.4) is the expected value of the ratio of the mainlobe amplitude $A(u_0)$ to the design value $A_0(u_0)$:

$$E\left[\frac{A(u_0)}{A_0(u_0)}\right] = e^{-\delta_\phi^2/2} \tag{2.11}$$

The power at the beamsteering angle from (2.6) is

$$\begin{aligned} p(u_0) &= f(u_0)f^*(u_0) \\ &= \frac{1}{N^2}\sum_n^N\sum_m^N(I_n + \delta I_n)(I_m + \delta I_m)^* e^{j(\delta\phi_n - \delta\phi_m)} \end{aligned} \tag{2.12}$$

where the asterisk indicates the complex conjugate. The expected value is

$$Ep(u_0) = \frac{1}{N^2} \sum \sum (I_n I_m^* + I_n E\delta I_m^* + I_m^* E\delta I_n^* + E\delta I_n \delta I_m^*) E e^{j(\delta\phi_n - \delta\phi_m)} \tag{2.13}$$

By invoking the assumptions introduced into (2.7) through (2.9) the expected mainlobe power becomes

$$Ep(u_0) = \frac{1}{N^2}\left\{ E^2[\cos(\delta\phi)] \sum_{n \neq m}^{N} \sum^{N} I_n I_m^* + \sum_{n}^{N} E|\delta I_n|^2 + \sum_{n}^{N} |I_n|^2 \right\}$$

$$= \overline{I_n I_m^*}\left(1 - \frac{1}{N}\right) E^2[\cos(\delta\phi)] + \frac{\sigma_I^2 + \overline{|I|^2}}{N} \tag{2.14}$$

where the coefficient of the first term is the average value of the terms in the current covariance matrix and σ_I^2 is the amplitude variance.

The ratio of (2.14) to the square of (2.4) is the expected mainlobe power normalized to the design value. The second term is generally a small contributor and can be ignored. In the important case of uniform excitation of the array ($I_n = 1$, all n), a large number of antenna elements, and a phase-error distribution that is Gaussian with variance σ_ϕ^2, the normalized expected mainlobe power gain G becomes

$$E\left[\frac{G}{G_0}\right] = e^{-\sigma_\phi^2} \tag{2.15}$$

which is the square of the expected value of the normalized mainlobe amplitude.† Based on (2.15), the loss in decibels becomes

$$\Delta G(\text{dB}) = 10\sigma_\phi^2 \log e \simeq 4.3\sigma_\phi^2 \tag{2.16}$$

In various parts of this book a 1-dB loss in gain is taken as an acceptable tolerance, which implies that the rms phase error can be as large as $\frac{1}{2}$ rad or about 30°.

Equations (2.15) and (2.16) are excellent approximations to the expected loss even when the phase-error distribution deviates from the Gaussian. Digital phase shifters introduce phase quantization errors that are distributed uniformly over the phase-quantum interval $\Delta\phi$. If M is the number of levels of

†This equation is correct for Gaussian phase errors. The more general expression is $E[G/G_0] = |E[\exp(j\delta\phi)|^2$, where $\delta\phi$ is any random phase error [85, Chapter 13].

phase quantization, the interval $\Delta\phi = 2\pi/M$ and the pdf of phase error equals $M/2\pi$ within the interval and is zero outside. Application of this pdf to the earlier expressions results in the following formula for the expected normalized gain of the mainlobe:

$$E\left[\frac{G}{G_0}\right] = \left[\frac{\sin(\Delta\phi/2)}{\Delta\phi/2}\right]^2 = \left[\frac{\sin(\pi/M)}{\pi/M}\right]^2 \tag{2.17}$$

The variance of a uniform distribution of width $\Delta\phi$ is $(\Delta\phi)^2/12$. When substituted into (2.17) the expected value of the mainlobe power gain for the quantized digital phase shifter may be written

$$E\left[\frac{G}{G_0}\right] = \left(\frac{\sin\sqrt{3}\,\sigma_\phi}{\sqrt{3}\,\sigma_\phi}\right)^2 \tag{2.18}$$

Equations (2.18) and (2.15) are compared in Figure 2.5. It is evident for small losses that (2.15), therefore (2.16), is an excellent approximation even for a pdf so different from the Gaussian as the uniform.

A major source of phase error in a large array is due to errors in element position. Let σ_x be the rms value of the position error in the x-direction. The tolerance on element position may be determined from (2.16) to be the order of $\lambda/10$, for then $\sigma_\phi = k\sigma_x = \pi/5$, which, according to (2.15), results in a loss of mainlobe gain of 1.7 dB. An element position tolerance approximating $\lambda/4\pi$ constrains the loss to 1 dB.

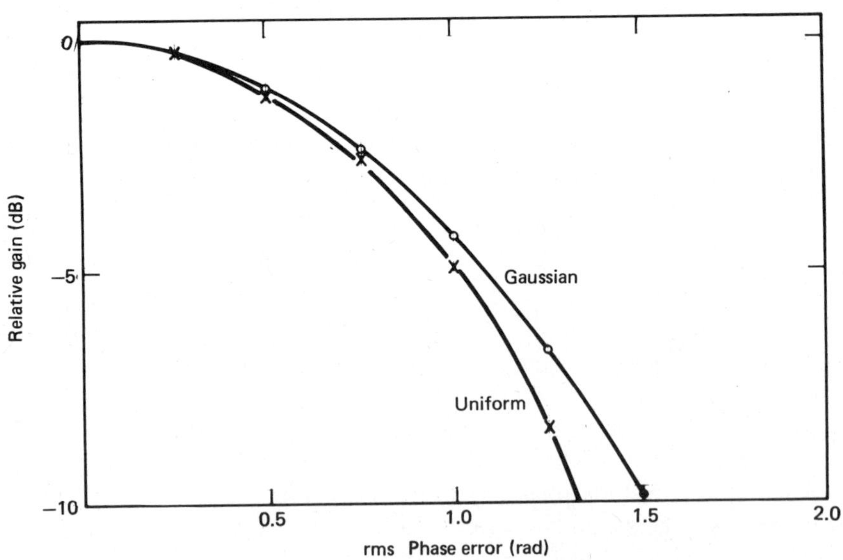

FIGURE 2.5. Degradation in mainlobe gain due to random phase errors. Gaussian and uniform error distributions.

In the sidelobe region ($|u - u_0| >$ beamwidth) the phases in the sums of (2.2) and (2.3) are random and uniformly distributed; hence the expected value is zero. The average power is not zero however; it is the sum of the power contributions of each of the elements:

$$\text{average power in sidelobe region} = \frac{1}{N^2}\sum_{n}^{N} I_n^2 = \frac{\overline{I^2}}{N} \tag{2.19}$$

Thus the ratio of the average sidelobe power level to the error-free mainlobe power [the square of (2.4)] is

$$\frac{\text{ASL}}{\text{ML}} = \frac{\left(\overline{I^2}/\bar{I}^2\right)}{N} \tag{2.20}$$

When the elements are excited equally the numerator is unity and

$$\frac{\text{ASL}}{\text{ML}} = \frac{1}{N} \qquad \text{uniform excitation} \tag{2.21}$$

which is smaller than (2.20) by 1 to $1\frac{1}{2}$ dB for many current distributions. This decrease represents a lowering of the background level of the image, or an increase in the contrast, by this amount.

Effect on Image Contrast

When phase errors are present the mainlobe gain is decreased according to (2.15) or (2.16). The average sidelobe power level, on the other hand, does not change because the phases in the sums are already random and further random errors do not alter their statistics. Hence the ratio (2.20) of ASL to ML increases by (2.16) or by the reciprocal of (2.15). The effect in the image is to increase the apparent background level by that amount or, equivalently, to decrease the average image contrast by the same quantity.

Effect on Dynamic Range

The introduction of phase errors in the array changes the details of the side radiation pattern but not the statistics of the sidelobes. The location of the peak sidelobe may move, but the probability of achieving a sidelobe of some arbitrary level above the average does not change. Thus the sidelobe response of the system to a large target will not be reduced by phase errors, whereas the mainlobe response to all targets will be reduced according to (2.15) or (2.16). Hence the dynamic range is reduced by exactly that amount.

The Peak Sidelobe Level (PSL)

Although the average sidelobe level can be derived by these simple considerations, the peak sidelobe level is somewhat of an anomaly. It is not a simple statistic of the sidelobe distribution. It can, however, be derived in a useful fashion by assigning a probability of occurrence or a confidence level to the concept of *the* peak sidelobe [81], [85]; that is, we can define the peak sidelobe as the value that will not be exceeded with some given probability. In other words, a threshold is calculated for which there is some assigned probability or level of confidence that no sidelobe will exceed it. That threshold is called *the* peak sidelobe for that confidence level. Its calculation requires a knowledge of the pdf of sidelobe amplitude. From (2.2) and (2.3) it is easy to show that the pdf of the amplitude of $f(u)$, except in the neighborhood of the beamsteering angle, is Rayleigh; that is, letting $A = |f(u)|$, the pdf of A is

$$w(A) = \frac{2A}{N}\exp\left(\frac{-A^2}{N}\right) \tag{2.22}$$

The probability that an arbitrary sample of the radiation pattern will not exceed some value A_0 is

$$P(A < A_0) = 1 - \int_{A_0}^{\infty} w(A)\,dA = 1 - e^{-A_0^2/N} \triangleq 1 - e^{-B} \tag{2.23}$$

and the probability that none of n independent samples will exceed this arbitrary level A_0 is

$$P(n \text{ independent samples} < A_0) = (1 - e^{-B})^n \triangleq \beta \tag{2.24}$$

where β is the confidence level and n is the number of independent samples of the side radiation pattern, called the array parameter. It can be estimated by dividing the extent of the visible region of the radiation pattern by the Nyquist sampling interval, which, in turn, is about equal to the beamwidth λ/L. In [85, Chapter 8] it is shown that n is reasonably well approximated by

$$n = \frac{L}{\lambda}(1 + |u_0|) \tag{2.25}$$

for a line array and by its square for a two-dimensional array.

By solving (2.24) for B in terms of the array parameter n and the confidence level β

$$B = -\ln(1 - \beta^{1/n}) \simeq \ln n - \ln(1 - \beta) \tag{2.26}$$

In (2.26) B may be interpreted as the sidelobe power level in relation to the average power level that, with probability β, will not be exceeded when the array is characterized by the array parameter n. In the same sense B/N is the peak sidelobe power level in relation to the mainlobe. The approximation in (2.26) is correct within 10% for $\beta \geq 0.5$; B is a slightly biased estimate of the peak sidelobe. A small additive correction leads to the estimate

$$B_P = B + 1 + \frac{2}{B} \tag{2.27}$$

which more closely matches computer simulation data for $B_P \geq 5$ [81], [85]; B_P is plotted against n in Figure 2.6 for a range of values of β. In [20] this estimate is found to be conservative (too high) by a few percent.

The term β is a system tolerance, whereas B_P/N is a system specification. The cost of the array and the signal processor is generally monotonic with β. The sensitivity of the PSL to β, however, is small, as revealed in the following example. Let a $10^4\lambda$ linear, random array designed for near-broadside ($\theta_0 \approx 0$) operation be required to meet a PSL specification of -20 dB; $n \approx 10^4$, $\ln n = 9.2$, and $B_P/N = 0.01$. Compare the design requirements for $\beta = 0.5$ and 0.9. The first value represents *expected* array performance; the second represents fairly high confidence that the specification will be met; $-\ln(1 - \beta) = 0.7$ and 2.3, respectively for these values of β. The corresponding values of B are 9.9 and 11.5 and for B_P they are 11.1 and 12.7. The required numbers of elements are 1110 and 1270 for the two. Thus only 14% more elements are needed to satisfy the PSL specification with a confidence of 90% as compared to a confidence level of 50%.

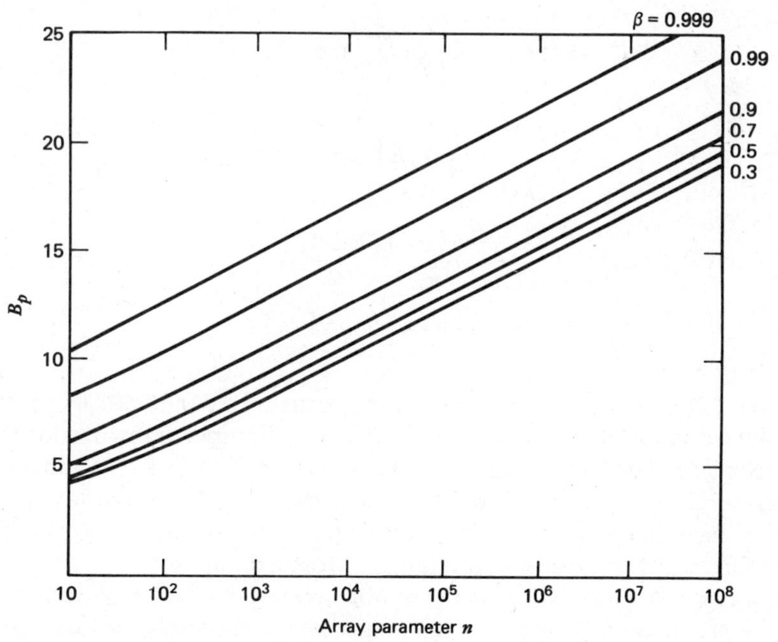

FIGURE 2.6. Ratio of peak-to-average sidelobe level of thinned random arrays.

Two-Dimensional Array

Many imaging arrays will be two-dimensional to produce high angular resolution in two dimensions. In this case, the increment in the required number of elements is even smaller. The array parameter is approximately the square of (2.25) for the two-dimensional array. Consequently, the values of B_P rise to 20.2 and 22.2, and the required numbers of elements become 2020 and 2220, which differ by less than 10%. Thus an additional system cost of the order of 10% increases the likelihood from 50% to 90% that the system will meet the peak sidelobe specification.

The low sensitivity of PSL to β might have been surmised from the data of Figure 2.2, which showed the distribution of the PSL to be very compact, i.e., the experimental standard deviation shown there was 1.13 dB.

Average Radiation Pattern

The ensemble average radiation pattern can be calculated from (2.2). For the important case of uniform excitation it is evaluated by substituting $I_n = 1$, interchanging the order of expectation and summation, recognizing that the x_n are identically distributed variates and factoring out the common expected value, and replacing the expected value of the function with the integral of that function weighted by the pdf $w(x)$ of the random variable. These steps are indicated on successive lines leading to (2.28).

$$\begin{aligned} Ef(u) &= E\left[\frac{1}{N}\sum_{n=1}^{N}\exp(jkx_n u)\right] \\ &= \frac{1}{N}\sum_n E\left[\exp(jkx_n u)\right] \\ &= E\left[\exp(jkxu)\right] \\ &= \int w(x)\exp(jkxu)\,dx \end{aligned} \tag{2.28}$$

Equation 2.28 is recognized as the Fourier transform of the pdf $w(x)$, that is, the expected value of the radiation pattern is the Fourier transform of the pdf of the element distribution; $w(x)$ therefore plays the role of a current density. Furthermore, $w(x)$ can be identified with some desired current density $i_0(x)$, which, in turn, is the Fourier transform of the radiation pattern $f_0(u)$ that the random array is designed to approximate. Provided only that $w(x) = i_0(x)$ (to within a constant multiplier), the ensemble average radiation pattern (2.28) *is* $f_0(u)$. In short, the properties of the expected or ensemble average radiation pattern can be deduced directly from the pdf of the element distribution exactly as the properties of a conventionally designed antenna array are deduced from the current density across the aperture. Beamwidth Δu approximately equals λ/L rad. This value, given in (2.29), is usually within 10 to 20% of the exact value, almost independent of the pdf of element distribution:

$$\Delta u \simeq \frac{\lambda}{L} \tag{2.29}$$

Equation (2.29) is the far-field beamwidth of an unfocused array. It also is the near-field beamwidth of a focused array at the focal range; L in that equation is the projected array size normal to the steering direction.

Deviations of actual beamwidths from the nominal value given in (2.29) are shown in Table 2.1, which lists the 3-dB beamwidths for several aperture excitations. These values are the beamwidths of the expected radiation patterns of random arrays whose pdfs of element location correspond to the excitations listed. Aperture excitation functions are tapered or shaped (as in the table) for the purpose of reducing the sidelobe level. The third column lists the level of the largest sidelobe in relation to the mainlobe. The large reduction in sidelobe level effected by aperture tapering more than justifies the slight increase in beamwidth that results.

Aperture Taper Is Rarely Warranted in Random Arrays

When the array is large, random, and highly thinned, however, this conclusion is no longer correct. The second column (beamwidth) does follow the pdf $w(x)$ according to the shapes in the table. The third column (PSL) does not; PSL is determined by the aperture taper only to the extent that it influences the array parameter n, which is proportional to the reciprocal of the beamwidth, hence deviates only slightly from (2.25) over a wide range of pdfs. The effect on the PSL is further weakened because of the logarithmic relation (2.26) between n and B. Consider, for example, a 100-λ random array steered to $\theta_0 = 26°$. At the 90% confidence level the ratio of PSL to ASL is 8.51 when $w(x)$ is uniform, whereas it is 8.18 when the pdf is triangular. This means that the PSL drops 0.2 dB and the beamwidth increases by 44%. It is hardly a worthwhile exchange. It can be concluded therefore that large changes in $w(x)$ induce minor variations

TABLE 2.1. Three-decibel Beamwidths and Highest Sidelobe for Several Different Aperture Excitations

Aperture	Beamwidth	Peak Sidelobe
Uniform	$0.88\lambda/L$	−13.4 dB
Circular $\sqrt{1-x^2}$	1.02	−17.5
Parabolic $(1\text{-}x^2)$	1.16	−22.0
Triangular	1.27	−26.8

in PSL for the large, highly thinned random array and generally are not worth it. The best pdf is a compromise between those with the smallest beamwidths and those that are easiest to construct, which often is the uniform distribution.

There are two exceptions to this general conclusion. First, a distribution that produces a poor mainlobe pattern must be avoided. An extreme example is an element distribution in which most of the elements are clustered at the two ends of the array. Its pattern will be multimodal, similar to that of the interferometer.

The second exception occurs when the number of elements is so large that the PSL due to the random character of the array is *small* compared with the first few sidelobes of the deterministic portion, that is, $f_0 f_0^*$; for example, the calculated value of the PSL of a 10^4-element, linear random array would be about -30 dB in relation to the mainlobe. However, if $w(x)$ were uniform, the first sidelobe would be -13.4 dB and subsequent lobes would decrease at the rate of 6 dB per octave. Hence a large portion of the visible region that surrounds the mainlobe would be dominated by the deterministic pattern. In this case a more appropriate pdf is the triangular for which the first sidelobe is -26.8 dB and subsequent lobes fall off at the rate of 12 dB per octave. Further suppression of the deterministic sidelobe pattern by control of $w(x)$ is unwarranted in most cases because all but the region of the first sidelobe is already dominated by the random properties of the array.

Average Power Pattern

Although the ensemble average radiation pattern (2.28) can be identified with the Fourier transform of the pdf of element location, the same is not true for any given sample pattern *except* in the neighborhood of the mainlobe. The sidelobe properties are dominated by the number of elements rather than by the pdf. This fact is evident from an examination of the expected or ensemble average power pattern. The power pattern is the product of $f(u)$ and its complex conjugate $f^*(u)$:

$$\begin{aligned} p(u) &= f(u) f^*(u) \\ &= \frac{1}{N^2} \sum_n^N \sum_m^N \exp[jk(x_n - x_m)u] \end{aligned} \tag{2.30}$$

when the elements are equally excited and its expected value is

$$\begin{aligned} Ep(u) &= \frac{1}{N^2}\left[N + Ee^{jkxu} Ee^{-jkxu}(N^2 - N)\right] \\ &= \left(1 - \frac{1}{N}\right) f_0(u) f_0^*(u) + \frac{1}{N} \end{aligned} \tag{2.31}$$

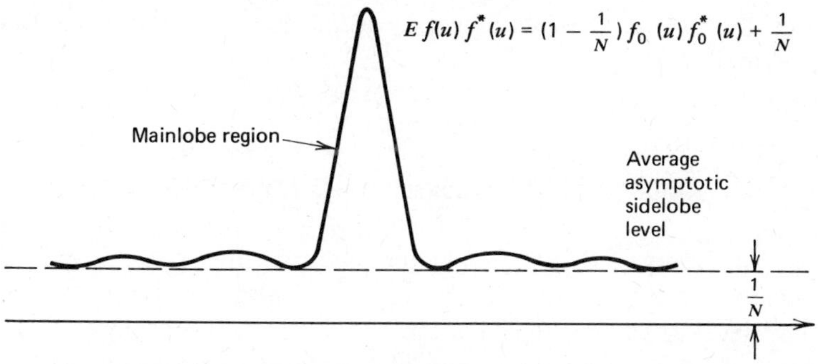

FIGURE 2.7. Ensemble average power pattern of the random array.

The ensemble average power pattern has two terms. The first is identical in shape to the desired or underlying power pattern $f_0 f_0^*$ and is slightly reduced in strength by the factor $1 - 1/N$. The second is independent of angle. It represents a constant average level of $1/N$. At the beamsteering angle $f_0 f_0^* = 1$ and $Ep(u_0) = (1 - 1/N) + 1/N = 1$. Thus the average sidelobe is a factor of N below the peak value of the ensemble average power pattern. This is the same as the result obtained earlier in (2.21). When the elements are not equally excited the coefficient changes according to (2.20).

Figure 2.7 shows the ensemble average power pattern when the elements are equally excited. It is dominated by the Fourier transform of the pdf of element location $w(x)$ in the region of the beamsteering angle. When that function, which is the design power pattern, becomes small compared with the asymptotic average level the latter constant function becomes dominant. Any particular sample function of the random process represented by this ensemble average pattern would exhibit its own specific characteristics. In the mainlobe region, however, it would not be likely to deviate much from the ensemble average pattern because the variance of the function is small. The mainlobe and perhaps the first few sidelobes follow the design pattern until that function is reduced in squared amplitude to the average sidelobe level $1/N$, beyond which it becomes obscured by the random character of the sidelobes. In actuality, the deterministic pattern characteristic is obscured before it drops to $1/N$ because, as noted earlier, the peak sidelobe power is typically 10 times larger than the average sidelobe power level.

The variance of the radiation pattern is

$$\begin{aligned} \operatorname{var} f(u) &= E|f(u) - Ef(u)|^2 \\ &= E|f|^2 - |Ef|^2 \\ &= Ep - f_0 f_0^* \\ &= \frac{1 - f_0 f_0^*}{N} \end{aligned} \tag{2.32}$$

which goes to zero at the beamsteering angle [as expected from (2.4)] and asymptotes to $1/N$.

2.5. BACKGROUND LEVEL AND DYNAMIC RANGE

The peak and average sidelobe level influence two important properties of an imaging system: the background, or gray, level and the dynamic range. The background level in the image limits the contrast, which is defined in this book as the ratio of the system response of the largest target or source to the background level. The dynamic range is the ratio of the largest target to the smallest that can be detected or perceived unambiguously in the presence of the many sidelobe responses in the system. No single pair of numbers suffices to characterize the required contrast and dynamic range of target response for all systems. Each imaging system has its own specifications. In few systems, however, would a range of target strengths of less than 20 to 30 dB be acceptable.

Contrast

Contrast is determined by the average sidelobe response of the array and dynamic range is determined by the peak sidelobe. Consider a one-dimensional scene that consists of sources of strength s_k located at angles u_k. An example of three sources appears in the upper portion of Figure 2.3. These sources can be modeled as a set of weighted delta functions and the scene of which they consist can be written

$$s(u) = \sum_{k=1}^{K} s_k \delta(u - u_k) \tag{2.33}$$

The average background power contribution from each source is s_k^2/N and the total average sidelobe power level due to all sources is

$$\text{ASL} = \frac{1}{N} \sum s_k^2 \tag{2.34}$$

This value, which is the sum of the average background contributions from all sources, represents the background level in the image. The contrast in the image between the largest target $s_{\max}$ and the background is

$$\text{contrast} = \frac{N s_{\max}^2}{\sum s_k^2} \tag{2.35}$$

TABLE 2.2. Target Dynamic Range

Confidence Level β	Antenna Size L/λ	
	100	1000
0.9	$0.125N$	$0.096N$
0.99	$0.096N$	$0.072N$

Dynamic Range

The dynamic range is determined by the smallest target that can be distinguished from the sidelobe responses of the larger targets. The peak sidelobe level is the largest sidelobe response in the radiation pattern to the largest target or source. Unlike (2.35), it is a single component of the system response rather than a collection or sum. It is given by (2.25), (2.26), and (2.27) and plotted in Figure 2.6, from which it may be calculated or seen that

$$\text{PSL} \sim \frac{10}{N} s_{\max}^2 \tag{2.36}$$

for arrays up to about 1000 wavelengths. Therefore the dynamic range for a one-dimensional array of that size is approximately

$$\text{dynamic range} = \frac{s_{\max}^2}{\text{PSL}} \approx \frac{N}{10} \tag{2.37}$$

For a 10^5-λ array the estimate is $\approx N/15$. Because the peak sidelobe is nearly 3 dB higher in a two-dimensional array, the dynamic range is approximately $N/20$ or $N/30$.

Table 2.2 shows the calculated dynamic range based on (2.37) at levels of confidence equal to 90 and 99% for antennas of 100 and 1000 wavelengths, respectively. It is evident that approximately 1000 elements are needed for the system to achieve a dynamic range of 20 dB or 10^4 elements to achieve a 30-dB target-strength ratio.

2.6. FOCAL PROPERTIES

The Near Field

The analysis and discussion in Section 2.4 were based on far-field theory, but because of the large size of the array of the radio camera targets may be in its near field. It is necessary, then, to consider the properties of the radiation pattern when the array is focused in the near field. In this region the radiation pattern is no longer well approximated by the Fourier transform (2.2) of the current excitation (or the aperture weighting function).

The correct relation between the field strength $f(u, R)$ at a point that has coordinates u, R, due to an aperture excitation $i(x)$, is given by the diffraction integral

$$f(u, R) = \int i(x)\exp[-jkr(x, u, R)]\, dx \tag{2.38}$$

where $r(x, u, R)$ is the distance between a point in the aperture and the field point. At distances $R \gtrsim L^2/\lambda$ the distance dependence vanishes and this integral simplifies to the Fourier transform, which was expressed by (2.2) in terms of the sampled aperture. At shorter ranges the integral is better approximated by the Fresnel transform. That integral, in turn, is replaced by sum (2.39) because the array is discrete:

$$f(u, R) = \frac{1}{N}\sum^{N} I_n \exp j\left(kx_n u - \frac{k'x_n^2}{2R}\right) \tag{2.39}$$

Equation 2.39 accounts explicitly for the range dependence. In (2.39) $k' = 2\pi\cos^2\theta/\lambda$ and θ = scan angle from the normal to the array.

Focusing the beam to u_0, R_0 requires that linear and quadratic phase-weights be given to the element currents so that the phase-weighted sum will maximize at u_0, R_0:

$$\begin{aligned} f(u, R) &= \frac{1}{N}\sum^{N}\left(I_n \exp j\left(-kx_n u_0 + \frac{k'x_n^2}{2R_0}\right)\right)\exp j\left(kx_n u - \frac{k'x_n^2}{2R}\right) \\ &= \frac{1}{N}\sum^{N} I_n \exp j\left[kx_n(u - u_0) - \frac{k'x_n^2}{2}\left(\frac{1}{R} - \frac{1}{R_0}\right)\right] \end{aligned} \tag{2.40}$$

The linear phase term $-kx_n u_0$ steers the beam to u_0, whereas the quadratic term $k'x_n^2/2R_0$ moves the focal distance from infinity to R_0. The radiation pattern at the focal range $R = R_0$ is unchanged from (2.2); hence all the statistical properties discussed in Section 2.4 characterize a focused array at its focal distance. What remains to be added are the range-dependent properties of the main beam of an array focused in its near field, the effects of phase errors on the focal zone, and their implications with respect to imaging.

Cross-Range Dimension

The most important focal properties of the main beam are its width and cross section, both transverse to the propagation direction (called the beamwidth) and in the direction of propagation (called the depth of field). The nominal transverse beamwidth may be expressed in angular units (rad), as in (2.29), or as the cross-range dimension of the beam at the focal distance R, given by

$$\Delta S \simeq \frac{R\lambda}{L} \tag{2.41}$$

A more accurate measure is obtained by multiplying (2.29) or (2.41) by the appropriate coefficient given in the middle column of Table 2.1. When a large array is synthesized by SAR techniques the beamwidth is smaller by a factor of 2 (Section 2.4). When the same antenna is used for transmission and reception, which is common practice in radar, the "two-way" beamwidth is $\sqrt{2}$ times smaller.

Depth of Field

The depth of field is determined from (2.40). It is the length (range dimension) of the focal zone of the radiation pattern at the focal angle $u = u_0$. By defining it as the interval in range ΔR about the focal range R_0 for which $f(u, R)$ reduces by a factor of $\sqrt{2}$ (-3 dB) the depth of field is given by [85, Chapter 3]

$$\Delta R \simeq 7\lambda\left(\frac{R}{L}\right)^2 \tag{2.42}$$

when ΔR is small compared with R. The focal zone is approximately elliptical, with its minor and major axes given by (2.41) and (2.42). The geometry is sketched in Figure 2.8. ΔR also is reduced by a factor of 2 in a synthetic aperture array.

The focal properties in Figure 2.10 in Section 2.7 are shown in an experimental two-dimensional diffraction pattern of a random, thinned, linear, synthetic aperture array. The contours in the center of the figure show the fall-off of amplitude in range and angle. The contour separation is 10% of peak amplitude. The 3-dB contour falls on the third contour. The observed depth of field and transverse angular dimension are 8.5 m and 0.028°, respectively. On the basis of the parameters of the array ($L = 27$ m and $\lambda = 3$ cm) and the distance to the focal point ($R = 240$ m) the nominal values predicted from (2.29) and (2.42) are 8.5 m and 1.1 mr = 0.032°. Because the probability density function of the element position in that experiment was uniform, the cross-range estimate can be improved by reducing the theoretical beamwidth to $0.88\ \lambda/L$, as indicated in Table 2.1. This reduces the predicted cross-range dimension to 0.028°, which matches the experimental value.

Effects of Phase Errors

In addition to inducing a loss in gain (2.16), phase errors across the array shift the focal point. The linear component of the phase variation missteers the

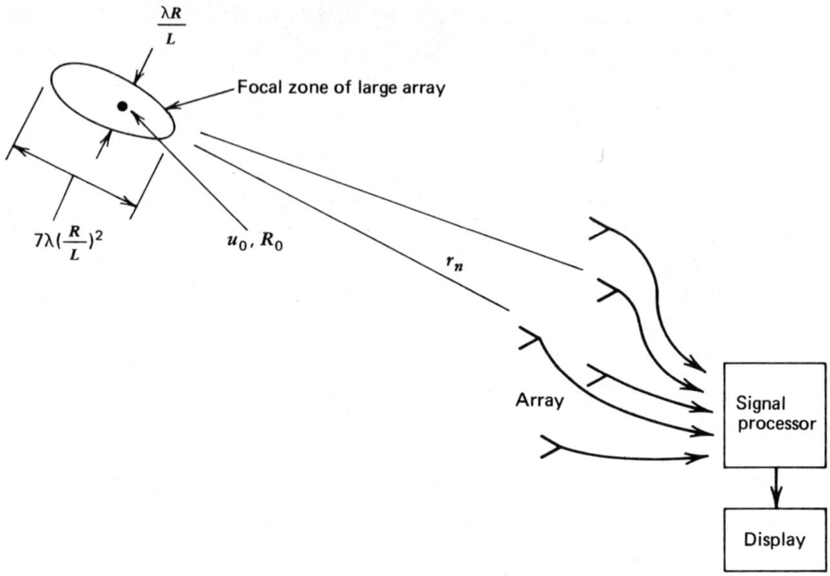

FIGURE 2.8. Dimensions of the focal zone of a large, random array. > = antenna element; L = size of array; R = nominal distance to array; r_n = distance to ith element.

beam (known as the beam-pointing error), whereas the quadratic component changes the focal length of the array. These two effects are readily seen by rewriting (2.40) in terms of the continuous linear variable x and adding an error phase $\delta\phi(x)$ to the argument:

$$f(u, R) = \frac{1}{L}\int_L i(x)\exp\left\{ j\left[kx(u - u_0) - \frac{k'x^2}{2}\left(\frac{1}{R} - \frac{1}{R_0}\right) + \delta\phi(x)\right]\right\} dx \tag{2.43}$$

Now expand $\delta\phi(x)$ in a polynomial in x. Also replace the range term with $-(R - R_0)/R_0^2$, making the assumption that $R \sim R_0$ (which implies small depth of field). The radiation pattern then becomes

$$f(u, R) = \frac{1}{L}\int_L i(x)\exp\left\{ j\left[kx(u - u_0) - \frac{k'x^2}{2R_0^2}(R - R_0) + \alpha_0 + \alpha_1 x + \alpha_2 x^2 + \cdots \right]\right\} dx \tag{2.44}$$

The beam-pointing error is the angular deviation from u_0 for which the sum of

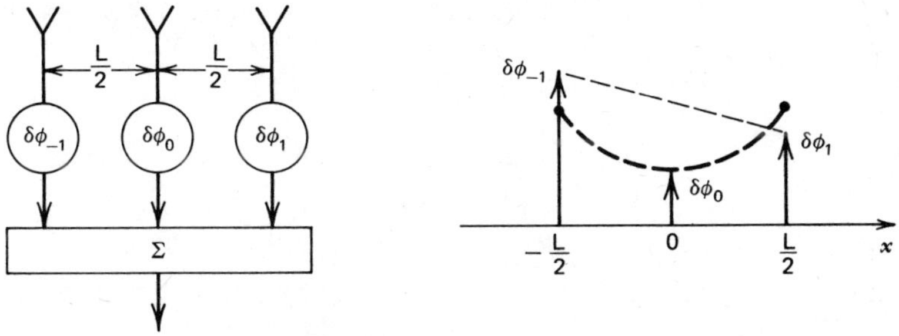

FIGURE 2.9. Three-element interferometer array. Linear and quadratic phase errors lead to beam-pointing error and focal-length error.

the linear terms is zero:

$$u - u_0 = \frac{\alpha_1}{k} \tag{2.45}$$

Similarly, the focal-length error is the range deviation from R_0 for which the sum of the quadratic terms is zero; it is

$$R - R_0 = -\frac{2R_0^2\alpha_2}{k'} \tag{2.46}$$

Figure 2.9 illustrates these errors with a simple example of a three-element array in which the elements are spaced by $L/2$ and their phase errors are $\delta\phi_{-1}$, $\delta\phi_0$, and $\delta\phi_1$. These errors are assumed to have the same statistics and to be independent. For this example the error slope is calculated from any pair of independent errors, such as the outer pair illustrated. Then $\alpha_1 = (\delta\phi_1 - \delta\phi_{-1})/L$ and

$$u - u_0 = \frac{\delta\phi_{-1} - \delta\phi_1}{kL} \tag{2.47}$$

Its mean value is zero because the phase errors have common statistics. The rms pointing error is the standard deviation of (2.47) which is

$$\sigma_u = \frac{\sqrt{2}\,\sigma_\phi}{kL} = \frac{\sqrt{2}\,\sigma_\phi\lambda}{2\pi L} \simeq \frac{\sigma_\phi \Delta u}{\sqrt{2}\,\pi} \tag{2.48}$$

where σ_ϕ is the rms phase error and the beamwidth Δu is identified with λ/L. The rms fractional pointing error (in units of beamwidth) is

$$\frac{\sigma_u}{\Delta u} = \frac{\sigma_\phi}{\sqrt{2}\,\pi} \tag{2.49}$$

In an N-element array $N - 1$ independent pairs of phase measurements may be made. Averaging these observations reduces (2.49) by $\sqrt{N-1}$. Assuming that N is large, the general result is

$$\frac{\sigma_u}{\Delta u} \propto \frac{\sigma_\phi}{\sqrt{N}} \tag{2.50}$$

where the proportionality constant is determined by the particular distribution of element locations. The constant is generally the order of unity.

The focal-length error can be estimated similarly. The quadratic coefficient of phase error is estimated from the dashed curve in Figure 2.9, which leads to the relation $\alpha_2(L/2)^2 = (\delta\phi_1 - 2\delta\phi_0 + \delta\phi_{-1})/2$. When this expression is combined with (2.46) the focal length error is

$$R - R_0 = -\frac{4}{k'}\left(\frac{R_0}{L}\right)^2(\delta\phi_1 - 2\delta\phi_0 + \delta\phi_{-1}) \tag{2.51}$$

Its mean value also is zero and the rms error is

$$\sigma_R = \frac{4\sqrt{6}}{k'}\left(\frac{R_0}{L}\right)^2\sigma_\phi = \frac{4\sqrt{6}\,\lambda\sigma_\phi}{2\pi\cos^2\theta}\left(\frac{R_0}{L}\right)^2 \tag{2.52}$$

The depth of field ΔR is given in (2.42). The ratio of (2.52) to (2.42) is the fractional error in the focal length; it evaluates to

$$\frac{\sigma_R}{\Delta R} = \frac{0.7\sigma_\phi}{\pi\cos^2\theta} \tag{2.53}$$

which is the value in (2.49) when the beam is normal to the array. Thus the fractional beam-pointing errors in the two dimensions are nearly identical. Similarly,

$$\frac{\sigma_R}{\Delta R} \propto \frac{\sigma_\phi}{\sqrt{N}} \tag{2.54}$$

for a large array.

The proportionality factors for (2.50) and (2.54) can be calculated by observing the effect of phase errors on the power pattern. By adding a phase error $\delta\phi_n$ to the nth weight and defining

$$\psi = kx(u - u_0) - \frac{k'x^2}{2}\left(\frac{1}{R} - \frac{1}{R_0}\right) + \delta\phi \tag{2.55}$$

we obtain the power pattern

$$p(u, R) = f(u, R)f^*(u, R)$$

$$= \frac{1}{N^2} \sum_n^N \sum_m^N I_n I_m^* \exp j(\psi_n - \psi_m) \tag{2.56}$$

$$= \frac{\sum^N I_n I_n^*}{N^2} + \frac{1}{N^2} \sum^N_{n \neq m} \sum^N I_n I_m^* \cos(\psi_n - \psi_m) \tag{2.57}$$

The displacement in the focal point from u_0, R_0 is found by differentiating (2.57) with respect to u and R and setting the derivatives to zero at $R = R_0$ and $u = u_0$, respectively. Thus

$$\left.\frac{\partial p}{\partial u}\right|_{R=R_0} = 0 = -\frac{k}{N^2} \sum_{n \neq m} \sum I_n I_m^* (x_n - x_m) \sin(\psi_n - \psi_m) \tag{2.58}$$

Provided that the phase errors are small, which is a necessity for a properly operating imaging system, the sine term can be replaced by its argument. Doing so leads to

$$k(u - u_0) = -\frac{\sum_{n \neq m} \sum I_n I_m^* (x_n - x_m)(\delta\phi_n - \delta\phi_m)}{\sum_{n \neq m} \sum I_n I_m^* (x_n - x_m)^2} \tag{2.59}$$

the expected value of which is zero. The second moment is

$$E[k(u - u_0)]^2$$

$$= \frac{E \sum_{n \neq m} \sum \sum_{r \neq p} \sum I_n I_m^* I_r I_p^* (x_n - x_m)(x_r - x_p)(\delta\phi_n - \delta\phi_m)(\delta\phi_r - \delta\phi_p)}{\left[\sum_{n \neq m} \sum I_n I_m^* (x_n - x_m)^2\right]^2} \tag{2.60}$$

The numerator terms are zero, except when r or $p = n$ or m. The four nonzero terms sum to

$$4\sigma_\phi^2 \left(\sum_n I_n\right)^2 \sum_n I_n^2 x_n^2 \tag{2.61}$$

The denominator is the square of

$$\sum_{n \neq m} \sum I_n I_m^* (x_n - x_m)^2 = \sum_{n \neq m} \sum I_n I_m^* (x_n^2 - 2x_n x_m + x_m^2) \tag{2.62}$$

When the element excitation is real and symmetric about the origin and the element position distribution is an odd function the right hand side of (2.62) reduces to

$$2\sum_n I_n \sum_n I_n x_n^2 - 2\Big(\sum_n I_n x_n\Big)^2 = 2\sum_n I_n \sum_n I_n x_n^2 \tag{2.63}$$

and the second moment becomes

$$E[k(u - u_0)]^2 = \frac{\sum I_n^2 x_n^2}{\left(\sum I_n x_n^2\right)^2} \sigma_\phi^2 \tag{2.64}$$

This equation evaluates to

$$\frac{\sigma_u}{\Delta u} = \left(\frac{3\overline{I^2}}{N\bar{I}^2}\right)^{1/2} \frac{\sigma_\phi}{\pi} \tag{2.65}$$

for a random array with elements distributed according to a uniform probability density function over its aperture. In the important case in which the element weights are equal

$$\frac{\sigma_u}{\Delta u} = \sqrt{\frac{3}{N}} \frac{\sigma_\phi}{\pi} \tag{2.66}$$

which is also the solution for a large, uniformly excited, periodic array [85, Chapter 3].

By setting the derivative of (2.57) with respect to range to zero

$$\left.\frac{\partial p}{\partial R}\right|_{u=u_0} = 0 = -\frac{k'}{2R_0^2} \sum^N_{n \neq m} \sum^N I_n I_m^* (x_n^2 - x_m^2) \sin(\psi_n - \psi_n) \tag{2.67}$$

which leads to

$$R - R_0 = -\frac{2R_0}{k'} \frac{\sum\limits_{n \neq m} \sum I_n I_m^* (x_n^2 - x_m^2)(\delta\phi_n - \delta\phi_m)}{\sum\limits_{n \neq m} \sum I_n I_m^* (x_n^2 - x_m^2)^2} \tag{2.68}$$

The expected value of the deviation in range is zero and the fractional rms deviation, under the same conditions, evaluates to

$$\frac{\sigma_R}{\Delta R} = \frac{2\sigma_\phi}{\pi \cos\theta}\sqrt{\frac{2}{N}} \tag{2.69}$$

Equations 2.66 and 2.69 describe the rms displacement of a target from its true coordinates in an image, normalized to the dimensions of the focal zone. Both are small quantities and generally can be ignored; for example, it was shown earlier that σ_ϕ would not be permitted to exceed about $\frac{1}{2}$ rad; otherwise the loss in image contrast could become excessive. It was also shown that N is likely to be in the 10^3 to 10^4 range, although it is possible that by using the techniques discussed in Chapter 3 some designs will be satisfactory with as few as 100 elements. Assuming that the number of elements is as small as 100, (2.66) and (2.69) are, at most, a few percent. Hence the movement of the focal point always will be a small amount.

2.7. ASYMMETRIES IN THE TWO-DIMENSIONAL DIFFRACTION PATTERN

A point source at distance R appears in the image as a blob with the approximate dimensions of the focal zone. As the beam is scanned to the left and right of the source, spurious responses due to the sidelobes can and do occur. A similar phenomenon occurs as the array is focused at shorter and greater distances than R. The sidelobe problem is therefore two-dimensional; that is, a target located at R_0, u_0 will produce a maximum response in the image plane near these coordinates and artifacts at other locations in the image plane.

The easiest mode of visualization of this two-dimensional problem is as a contour plot in u and R of the response of the system to a point target. As shown in the analysis that follows, certain unexpected asymmetries exist which lead to a shearing or sloping quality in the image. This fact is demonstrated in Figure 2.10, which is a two-dimensional contour map of a point source made with a radio camera equipped with a random antenna array. The source is a corner reflector located 0.1° to the right of the array normal and at a distance of 240 m from the array. Because a corner reflector is designed to act as a point source, Figure 2.10 is an approximation of the two-dimensional diffraction pattern of the aperture. The shear or tilt-distortion in the image is evident.

The analysis proceeds by writing the near-field radiation pattern of a current distribution $i(x)$; it is given approximately by†

$$f(u, R) = \int_{-L/2}^{L/2} i(x)\exp\left[jkxu - jk'\frac{x^2}{2R}\right] dx \tag{2.70}$$

† The analysis and experimental data that follow are based on [92].

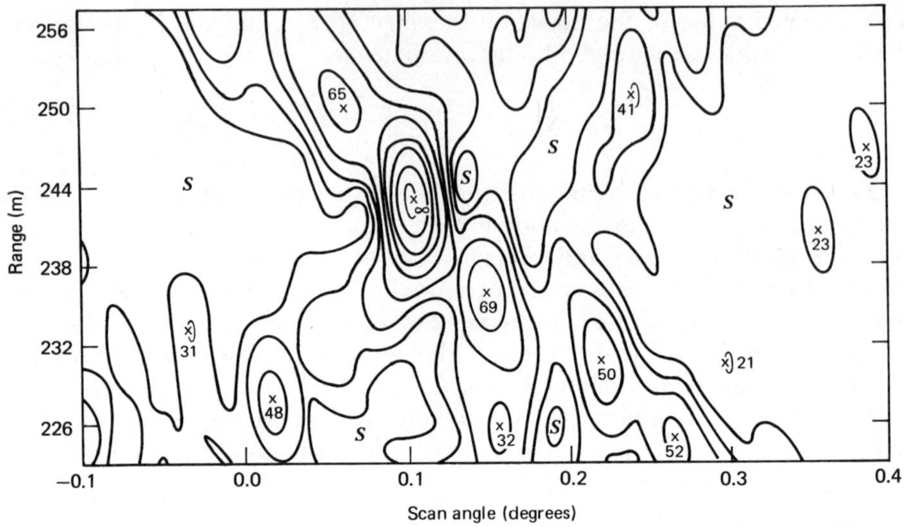

FIGURE 2.10. Two-dimensional diffraction pattern of array near focus at Valley Forge Research Center. Twisting of pattern is evident. The difference between successive contour lines is 10 amplitude units; *S* designates areas of less than 20 amplitude units. (From B. D. Steinberg and A. K. Luthra, *IEEE Trans. Antennas and Propagation*, AP-29, No. 4, July 1981.)

which is the continuous-current distribution analog to the expression for the sampled aperture given by (2.39). The direction in which the beam is steered is taken as zero for convenience. To bring the focal distance in from infinity to an arbitrary distance R_0 the phase of the current $i(x)$ is advanced by $k'x^2/R_0$. The radiation pattern at any distance R then becomes

$$f(u, R) = \int_{-L/2}^{L/2} i(x)\exp\left[jkxu - jk'\frac{x^2}{2R} + jk'\frac{x^2}{2R_0}\right] dx \tag{2.71}$$

Now let $R = R_0 + r$, where r is a displacement from the focus. In terms of the distance r the radiation pattern is

$$f(u, r) = \int i(x)\exp\left[jkxu - jk'\frac{x^2}{2}\left(\frac{1}{R_0 + r} - \frac{1}{R_0}\right)\right] dx$$

$$\approx \int i(x)\exp\left[jkxu + jk'\frac{x^2 r}{2R_0^2}\right] dx \qquad r \ll R_0 \tag{2.72}$$

For a positive increment, that is, $R > R_0$,

$$f(u, |r|) = \int i(x)\exp\left[jkxu + jk'\frac{x^2|r|}{2R_0^2}\right] dx \tag{2.73}$$

whereas for a negative increment

$$f(u, -|r|) = \int i(x)\exp\left[jkxu - jk'\frac{x^2|r|}{2R_0^2}\right] dx \tag{2.74}$$

The complex conjugate of (2.74) is

$$f^*(u, -|r|) = \int i^*(x)\exp\left[-jkxu + jk'\frac{x^2|r|}{2R_0^2}\right] dx \tag{2.75}$$

which for real $i(x)$ becomes

$$f^*(u, -|r|) = \int i(x)\exp\left[-jkxu + jk'\frac{x^2|r|}{2R_0^2}\right] dx \tag{2.76}$$

Now consider $f^*(-u, -|r|)$:

$$f^*(-u, -|r|) = \int i(x)\exp\left[jkxu + jk'\frac{x^2|r|}{2R_0^2}\right] dx \tag{2.77}$$

which is identical to (2.76). Therefore

$$f(u, |r|) = f^*(-u, -|r|) \tag{2.78}$$

and

$$|f(u, |r|)| = |f(-u, -|r|)| \tag{2.79}$$

Equation 2.79 shows that the amplitude of the two-dimensional (range, angle) radiation pattern (which is the image of a point source) will exhibit point symmetry around the origin in the u, r plane. It can also be seen that if $u_0 \neq 0$ and the variation in k' due to the change in scan angle is neglected the radiation pattern will be symmetric around $u_0, 0$. Further, it can be deduced that at the focal range ($r = 0$) the power pattern should be symmetric around u_0, that is, $|f(u - u_0)|^2 = |f(-(u - u_0))|^2$, and the image of a point source should be symmetric around u_0. In addition, it can be seen that in general

$$|f(u, |r|)| \neq |f(u, -|r|)| \tag{2.80}$$

$$|f(u, |r|)| \neq |f(-u, |r|)| \tag{2.81}$$

and that the diffraction pattern will not be symmetric with respect to the coordinate axes but will be with respect to the point of peak amplitude.

Equation 2.79 explains the tendency toward point symmetry in Figure 2.10 and (2.80) and (2.81) explain the nonsymmetric property of the image with respect to the axes.

These inequalities disappear when the current excitation is even; that is, $i(x) = i(-x)$. Beginning with (2.73) and replacing u with $-u$, we have

$$f(-u, |r|) = \int_{-L/2}^{L/2} i(x)\exp\left(-jkxu + jk'\frac{x^2}{2R_0^2}|r|\right) dx$$

$$= -\int_{L/2}^{-L/2} i(-y)\exp\left(jkyu + jk'\frac{y^2}{2R_0^2}|r|\right) dy, \qquad y = -x$$

$$= \int_{-L/2}^{L/2} i(-y)\exp\left(jkyu + jk'\frac{y^2}{2R_0^2}|r|\right) dy$$

$$= \int_{-L/2}^{L/2} i(y)\exp\left(jkyu + jk'\frac{y^2}{2R_0^2}|r|\right) dy, \qquad i(y) = i(-y)$$

$$= f(u, |r|) \tag{2.82}$$

In combination with (2.79) this equation implies that

$$|f(u, |r|)| = |f(-u, |r|)| = |f(u, -|r|)| = |f(-u, -|r|)| \tag{2.83}$$

when $i(x) = i(-x)$.

Equation 2.83 shows that the two-dimensional picture will exhibit axial symmetry as well as symmetry around the point of peak amplitude but will not exhibit shear when the current excitation is even. Thus periodic phased arrays as well as aperiodic arrays in which the elements are located symmetrically around center will exhibit axial symmetry in their diffraction patterns.

Asymmetry is introduced by the third moment of the aperture distribution, which may be found by examining the second derivative of $f(u, r)$ at the origin in the u, r plane; that is, at the location of the source or target. The first derivative is zero when the aperture weights or currents are real; this is common practice in phased-array design. The second derivative indicates the curvature in the neighborhood of the peak of the function. An even aperture excitation leads to axially symmetric curvature, whereas a nonzero third moment couples the two variables and destroys axial symmetry.

The analysis is made with the realistic assumption that the aperture excitation is real. Further, it is convenient to assume that the aperture distribution is normalized such that

$$\int i(x)\, dx = 1 \tag{2.84}$$

and that the origin in the aperture is chosen so that the first moment

$$\int x i(x)\, dx = 0 \tag{2.85}$$

Higher moments are indicated by

$$\mu_n = \int x^n i(x)\, dx \tag{2.86}$$

The radiation pattern near the focus is given by (2.72). By substituting

$$\nu = ku \qquad \rho = \frac{k'r}{2R_0^2} \tag{2.87}$$

$$f(\nu, \rho) = \int i(x) \exp[j(x\nu + x^2\rho)]\, dx \tag{2.88}$$

The power pattern

$$p(\nu, \rho) = ff^* = \int\int i(x) i(x') \exp\{j[(x - x')\nu + (x^2 - x'^2)\rho]\}\, dx\, dx' \tag{2.89}$$

can be expanded in a Taylor series at the origin:

$$p(\nu, \rho) = p(0,0) + \nu\frac{\partial p}{\partial \nu} + \rho\frac{\partial p}{\partial \rho} + \frac{1}{2}\nu^2\frac{\partial^2 p}{\partial \nu^2} + \nu\rho\frac{\partial^2 p}{\partial \nu\, \partial \rho} + \frac{1}{2}\rho^2\frac{\partial^2 p}{\partial \rho^2} + \cdots \tag{2.90}$$

in which the partial derivatives are evaluated at the origin. The first partial derivatives are

$$\frac{\partial p}{\partial \nu} = \int\int j(x - x') i(x) i(x') \exp\{j[(x - x')\nu + (x^2 - x'^2)\rho]\}\, dx\, dx' \tag{2.91}$$

$$\frac{\partial p}{\partial \rho} = \int\int j(x^2 - x'^2) i(x) i(x') \exp\{j[(x - x')\nu + (x^2 - x'^2)\rho]\}\, dx\, dx' \tag{2.92}$$

which are zero at the origin. The second partial derivatives are

$$\frac{\partial^2 p}{\partial \nu^2} = -\iint (x - x')^2 i(x) i(x') \exp\{j[(x - x')\nu + (x^2 - x'^2)\rho]\}\, dx\, dx' \tag{2.93}$$

$$\frac{\partial^2 p}{\partial \nu\, \partial \rho} = -\iint (x - x')(x^2 - x'^2) i(x) i(x') \times \exp\{j[(x - x')\nu + (x^2 - x'^2)\rho]\}\, dx\, dx' \tag{2.94}$$

$$\frac{\partial^2 p}{\partial \rho^2} = -\iint (x^2 - x'^2)^2 i(x) i(x') \exp\{j[(x - x')\nu + (x^2 - x'^2)\rho]\}\, dx\, dx' \tag{2.95}$$

which evaluate to

$$\begin{aligned} \left.\frac{\partial^2 p}{\partial \nu^2}\right|_{\nu=\rho=0} &= -\iint (x - x')^2 i(x) i(x')\, dx\, dx' \\ &= -\iint (x^2 - 2xx' + x'^2) i(x) i(x')\, dx\, dx' \\ &= -\int (x^2 - 2x\mu + \mu_2) i(x)\, dx \\ &= -2\mu_2 \end{aligned} \tag{2.96}$$

$$\left.\frac{\partial^2 p}{\partial \nu\, \partial \rho}\right|_{\nu=\rho=0} = -2\mu_3 \tag{2.97}$$

$$\left.\frac{\partial^2 p}{\partial \rho^2}\right|_{\nu=\rho=0} = -2(\mu_4 - \mu_2^2) \tag{2.98}$$

Thus the intensity of the power pattern at a small distance from the origin is

$$p(\nu, \rho) \simeq p(0,0) - \mu_2 \nu^2 - 2\mu_3 \nu\rho - (\mu_4 - \mu_2^2)\rho^2 \tag{2.99}$$

The second and fourth terms contribute to axial symmetry because positive and negative increments in ν and ρ result in identical changes in p. The third term contributes to point symmetry in the product $\nu\rho$; positive or negative increments in both variables result in the same change in p and an opposite change occurs when the increments in ν and ρ are of opposite sign. This contribution is responsible for the skewed appearance of the image of a point reflector in Figure 2.10. The distortion to the image disappears when the third moment goes to zero, which occurs when the aperture excitation is even.

Equating (2.99) to a constant results in contours of constant intensity: μ_2 and μ_4 (and all even moments) are positive; $\mu_4 \geq \mu_2^2$, for otherwise the pattern would *increase* with ρ when $\nu = 0$.† Hence contours of constant intensity are ellipses for small decrements in p.

Two modifications are required to apply the theory to the contour plot of Figure 2.10. First, amplitude $|f|$ is plotted rather than $p = |f|^2$. Because $p(0,0) = 1$ and the decrement given in (2.99) is small,

$$|f(\nu,\rho)| \simeq 1 - \tfrac{1}{2}\mu_2\nu^2 - \mu_3\nu\rho - \tfrac{1}{2}(\mu_4 - \mu_2^2)\rho^2 \tag{2.100}$$

Second, the origin in Figure 2.10 is not at the center of the array; that is, (2.88) ≠ 0. When a displacement d is taken into account such that

$$\int x i(x)\, dx = d \tag{2.101}$$

(2.99) changes to

$$\begin{aligned} p(\nu,\rho) \simeq p(0,0) &- \mu_2\nu^2 - 2(\mu_3 + 2\mu_2 d)\nu\rho \\ &- (\mu_4 - \mu_2^2 + 4\mu_2 d^2 + 4\mu_3 d)\rho^2 \end{aligned} \tag{2.102}$$

where μ_n, $n = 2, 3, 4$, now are the central moments given by

$$\mu_n = \int (x-d)^n i(x)\, dx$$

The coefficient of the third term in (2.102) includes the displacement of the origin of the coordinate system from the center of gravity of the array.

The theory can be tested with the experimental array used to obtain the two-dimensional diffraction pattern in Figure 2.10. The second, third, and

†Also shown by Schwartz's inequality:

$$\left|\int_a^b f(t)g(t)\,dt\right| \leq \left[\int_a^b |f(t)|^2\,dt\right]^{1/2}\left[\int_a^b |g(t)|^2\,dt\right]^{1/2}$$

Let $f(t) = x^2\sqrt{i(x)}$, $g(t) = \sqrt{i(x)}$, $dt = dx$. Then

$$\left|\int x^2 i(x)\,dx\right| \leq \left[\int x^4 i(x)\,dx\right]^{1/2}\left[\int i(x)\,dx\right]^{1/2}$$

and $\mu_4 \geq \mu_2^2$ because $\int i(x)\,dx = 1$.

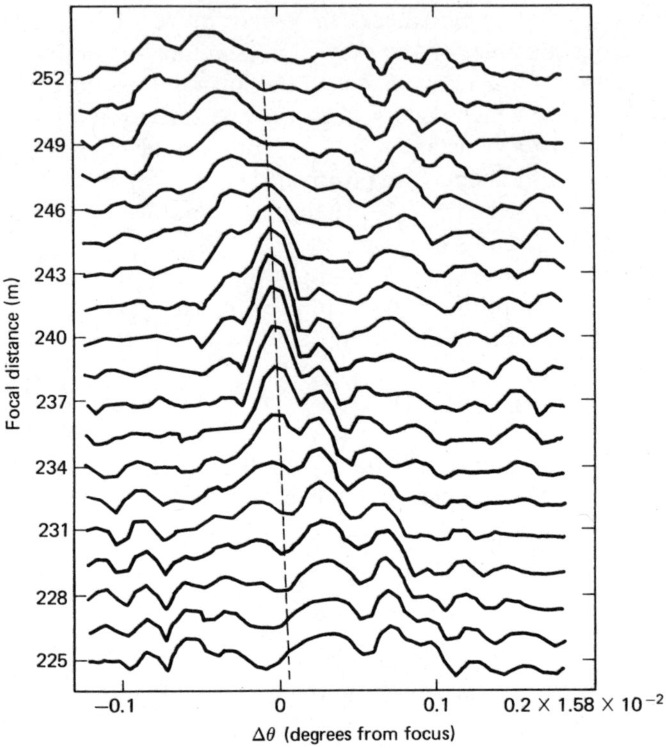

FIGURE 2.11. Angle-scanned slices of the amplitude of the radiation pattern in Figure 2.10. Dashed line shows the theoretical slope at the focus. The instrument is successively focused at 1.5-m intervals. (From B. D. Steinberg and A. K. Luthra, *IEEE Trans. Antennas and Propagation*, AP-29, No. 4, July 1981.)

fourth central moments of that array, calculated from measured element locations, are

$$\mu_2 = 56.93 \text{ m}^2$$

$$\mu_3 = -52.03 \text{ m}^3$$

$$\mu_4 = 7263 \text{ m}^4$$

and $d = 0.86$ m. Assuming that ρ remains constant, the derivative of the power pattern (2.102) becomes

$$\left.\frac{\partial p}{\partial \nu}\right|_{\rho=\text{constant}} = -2\mu_2\nu - 2(\mu_3 + 2\mu_2 d)\rho \qquad (2.103)$$

Setting (2.103) equal to zero and making the necessary substitutions, we obtain the slope at the focal point:

$$\frac{\Delta_{\max}}{r} = \frac{-(\mu_3 + 2\mu_2 d)}{\mu_2} \frac{k'}{2kR_0^2} \tag{2.104}$$

The polarity of the coefficient in (2.104) is consistent with the observed direction of inclination of the contours of the mainlobe. The numerical agreement is also satisfactory. The value of the slope calculated from these quantities is more easily compared with the data in the presentation of Figure 2.11, which shows angle-scanned slices of the diffraction pattern when the system was focused at successive 1.5-m intervals. The calculated slope is the dashed line. The agreement with the experimentally observed shear is good.

2.8. EXAMPLES

In this section two brief examples are given of one-dimensional angle scans or images. Further and more detailed imaging experiments are reported in Chapter 5. Figure 2.12 shows the result of an X-band ($\lambda = 3$ cm) radio-camera imaging experiment with a low-power (5-W) pulsed radar transmitter and a corner reflector for the phase synchronizer. The size of the reflector is 0.46 m and its radar cross section is calculated as 56 m^2. The estimated clutter cross section is 2 m^2. The receiving array is shown in Figure 5.7; it is 27 m long and 1.2 m wide and 100 element positions are located randomly within it. Uniform probability density functions of element position were chosen for the length and width dimensions of the array. The experiment was conducted on a time-shared basis in which a single radar receiver was moved successively from position to position to deliver a radar echo trace to a microprocessor from each receiver position. The signal processing operations are described in Chapter 5. The image is one-dimensional, deflection-modulated. The target consists of two additional reflectors, each 0.61 m, drawn to scale below the image. The target is 240 m from the array and subtends an angle of 14 mrad. The reference reflector is 41 m closer. Drawn also in Figure 2.12 is the calculated response of the array in free space, *had it self-organized perfectly*. It is evident that the experiment was exceedingly successful.

The theoretical beamwidth for this experiment, based on diffraction theory, is $0.88\lambda/L\cos\theta$. The coefficient corresponds to the particular probability density function of element location used (Section 2.4); θ, the target angle from the array normal, is 25°. This expression evaluates to 1.1 mrad, which is close to the measured beamwidth.

Figure 2.13 is the image of the same reflectors when the equipment was operated in the synthetic aperture mode: the low-power transmitter and receiver were moved from position to position for each radar transmission and reception. Because of the doubling of the wavenumber due to the synthetic

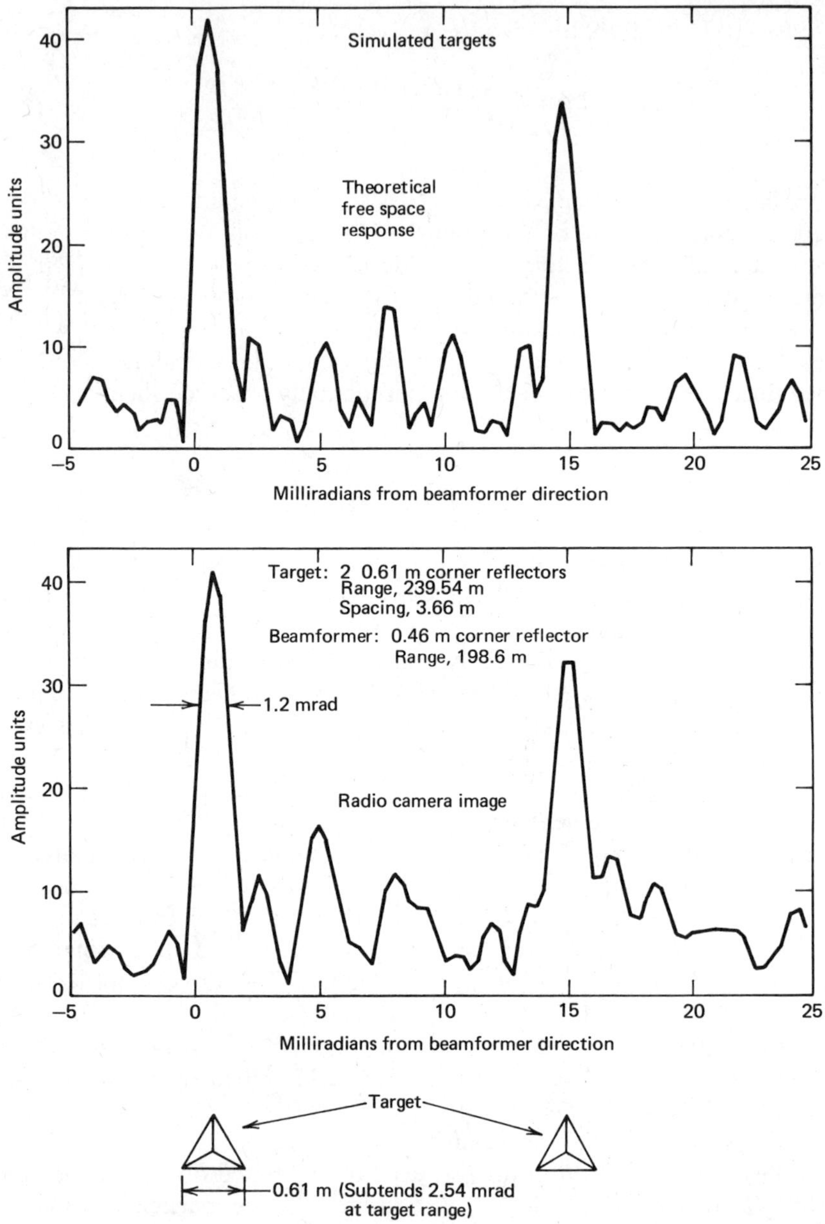

FIGURE 2.12. One-dimensional image of two corner-reflector targets with another corner reflector for adaptive beamforming. Theoretical response in free space shown for comparison. (From B. D. Steinberg, *IEEE Trans. Antennas and Propagation*, AP-29, No. 5, September 1981.)

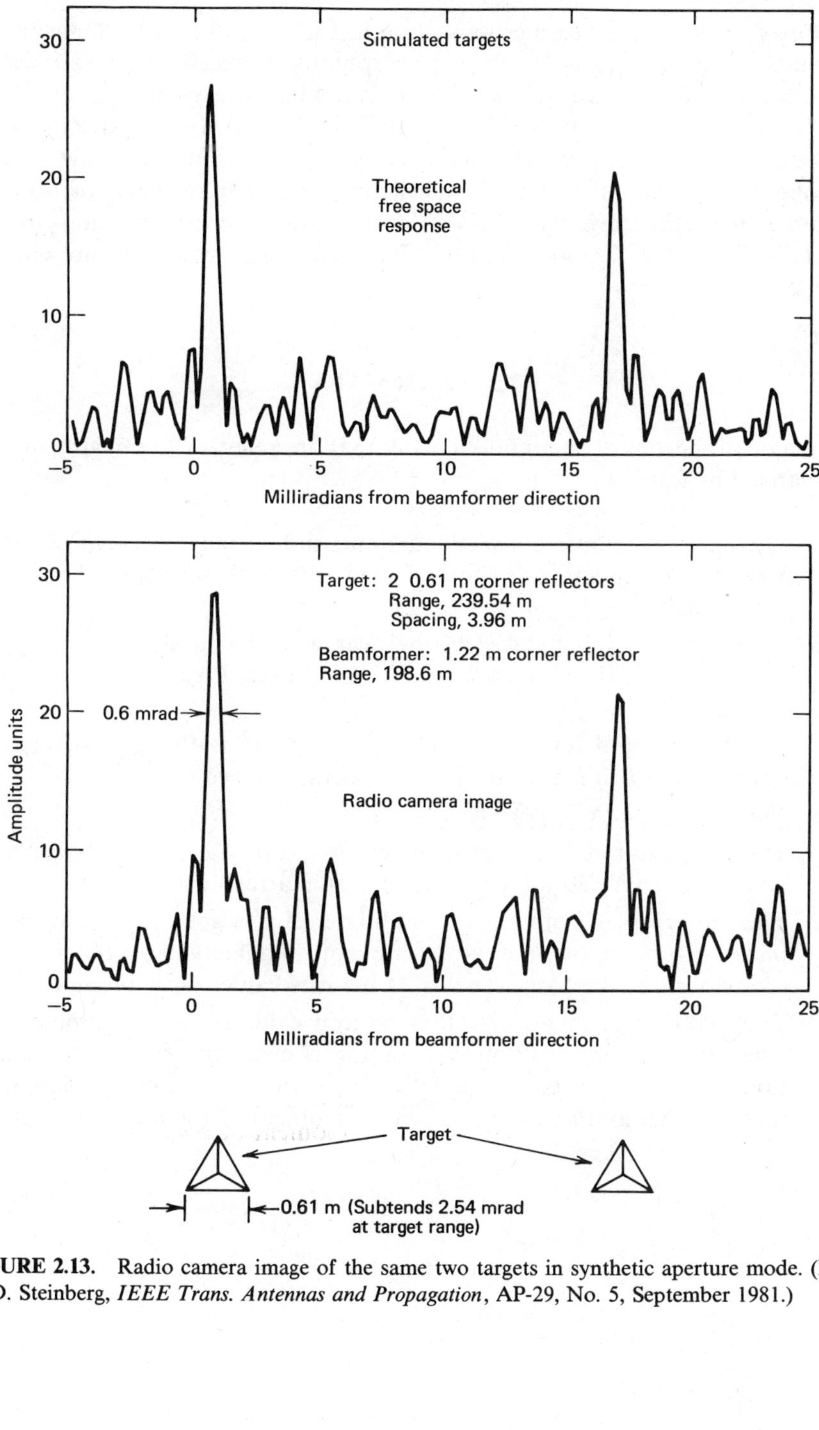

FIGURE 2.13. Radio camera image of the same two targets in synthetic aperture mode. (From B. D. Steinberg, *IEEE Trans. Antennas and Propagation*, AP-29, No. 5, September 1981.)

aperture operation, the beam cross section in Figure 2.13 is halved. Again, the comparison with the calculated free space response is excellent. The separation of the corner reflectors was increased to 16 mrad in this experiment.

In both experiments the sidelobe properties conform to the theory of the random array [59], [85], which predicts that the contribution to the average sidelobe power level is N^{-1} times the mainlobe power response of a target. Thus the theoretical average sidelobe level for the two nearly equal-strength targets in Figures 2.12 and 2.13 is -17 dB; the measured levels are close to this value.

2.9. SUMMARY

The major properties of the random array that are relevant to imaging may be summarized as follows:

1. The far-field beam cross-sectional dimension at range R is $R\lambda/L$, where λ is wavelength and L is the projected array size normal to the steering direction.
2. When the array is focused in the near field the focal zone is approximately elliptical, with minor and major axes $R\lambda/L$ and $7\lambda(R/L)^2$, respectively.
3. The background level is $(1/N)\Sigma s_k^2$, where s_k^2 is the intensity (power units) of the kth source in the scene being imaged.
4. The contrast is $Ns_{\max}^2/\Sigma s_k^2$.
5. The dynamic range is approximately $N/10$ to $N/15$ for a linear array and $N/20$ to $N/30$ for a two-dimensional array.
6. The rms errors in both image coordinates of a target due to independent phase errors across the array are approximately $\sigma_\phi/\sqrt{N}$ times the dimensions of the focal zone; σ_ϕ is the rms value of the phase errors.
7. The two-dimensional diffraction pattern exhibits axial symmetry only when the element position distribution is even. In general, the diffraction pattern is skewed, the direction being determined by the sign of the first moment and by the third central moment of the element locations.

CHAPTER THREE

Diversity Techniques

3.1. IMAGE ARTIFACT REDUCTION

In Chapter 2 we determined that image artifacts can form in two ways. The first, discussed in Section 2.1, is due to reflections from bodies on or near the target being imaged, as in the desk-top example. The second source of trouble is intrinsic to the array itself. A natural concomitant of a high angular resolution microwave imaging system is high sidelobes, which implies the likely presence of false targets in the images. This chapter is devoted to the reduction or suppression of image artifacts due to the side radiation pattern of the array.

The reasoning developed in the earlier chapters is as follows: high angular resolution requires an aperture thousands to hundreds of thousands of wavelengths in size. Because typical wavelengths are 1 to 30 cm, apertures must measure hundreds of meters to tens of kilometers. These apertures must be highly thinned or their costs will be excessive. Apertures of such size are unlikely to be mechanically and electrically rigid. The last two facts imply that the side radiation pattern is most likely to be that of a random array. Although the random array preserves the desired resolving power, its sidelobe properties are poor. The average sidelobe power level is N^{-1}, where N is the number of antenna elements, and the peak sidelobe will usually be 10 or more dB higher. If the design is constrained by cost to allow no more than 1000 elements, the peak sidelobe is likely to be as high or higher than -20 dB. If the display is designed to accommodate a dynamic range of targets of 20 dB or more in the scene to be imaged, it is highly likely that sidelobe responses of the larger targets will produce false targets or artifacts in the image. It is important therefore to develop techniques for reduction of the larger sidelobes.

A straightforward approach increases the number of antenna elements. Each doubling reduces the average sidelobe and peak sidelobe levels by 3 dB. In every design problem, however, there is some maximum number beyond which the cost is excessive in relation to the benefit achieved. The question addressed in this chapter is whether other resources are available to the system designer for increasing the target dynamic range without increasing the number of elements.

Experience in the use of diversity techniques in several other fields suggests that diversity combining may also be useful here [12]. Diversity techniques are proven tools for the enhancement of stable components of signals or images in a stochastic background. Required is some parameter (space, time, frequency, or polarization) in which the undesired signal component becomes decorrelated by variation of the parameter. Space diversity on reception was introduced early in HF radio communication to reduce the effect of random signal fading [4]. Angle diversity has proved to be useful in tropospheric scatter communication [66], as have frequency and polarization diversity [74]. Even the integration of successive radar echoes in a background of noise or time-varying clutter is a temporal form of diversity [5]. More recently RF phase diversity has become indispensable in obtaining stable maximum entropy images [30]. As shown later, second-order diversity, when applied to sparse, random arrays, reduces the peak sidelobe by 2–$2\frac{1}{2}$ dB, which is close to the 3 dB that would be expected from doubling the number of elements. Thus significant payoff can be expected by the introduction of diversity techniques.

A second difference between diversity combining and increasing the number of elements is that diversity combining only marginally reduces the average sidelobe power level, whereas the number of elements reduces it in direct proportion to numbers. Thus diversity combining of the images improves the dynamic range of the system but not the image contrast.

3.2. DIVERSITY COMBINING APPLIED TO RANDOM ARRAYS

Any technique that decorrelates the side radiation pattern of a random antenna array is potentially useful. Specifically, if $f_1(u)$ is the radiation pattern of a random array and a technique is applied (see later in this chapter) that changes the radiation pattern to $f_2(u)$ the technique is a useful one if the cross-correlation between the patterns is close to unity in the neighborhood of the mainlobe and close to zero elsewhere. Given this condition, it is useful to display the sum of the moduli of the images, for then the radiation patterns enhance one another in the mainlobe region and the sidelobe peaks and valleys tend to average toward the mean sidelobe level.

The improvement may be estimated from the statistics of the pattern in the sidelobe region. The pdf of the side radiation pattern of a random array is

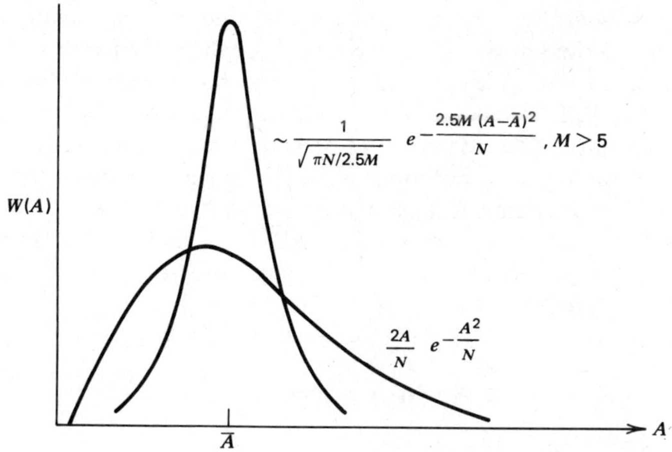

FIGURE 3.1. Pdfs of random-array sidelobes with and without averaging of moduli of patterns.

given in (2.22); it is the Rayleigh distribution

$$w(A) = \frac{2A}{N}\exp\left(-\frac{A^2}{N}\right) \tag{3.1}$$

where A is an arbitrary sample of the amplitude of the side radiation pattern and N is the number of elements. The mean square value $\overline{A^2}$, which is the average sidelobe power level, is N; N is also the ratio of the mainlobe power gain to the average sidelobe power because the power on the axis in the mainlobe is N^2. The average of the Rayleigh distribution $\bar{A} = \sqrt{\pi N}/2$. Hence the variance

$$\sigma = \overline{A^2} - \bar{A}^2 = N\left(\frac{1-\pi}{4}\right) \tag{3.2}$$

After M-fold averaging of independent side radiation patterns the variance reduces to

$$\sigma_M^2 = \left(\frac{N}{M}\right)\left(1 - \frac{\pi}{4}\right) \simeq \frac{N}{5M} \tag{3.3}$$

and the average sidelobe power drops a small amount to

$$\overline{A^2}(M) = \frac{\pi N}{4} + \frac{N(1-\pi/4)}{M} \tag{3.4}$$

Although the reduction in the variance expressed by (3.3) is exact, the corresponding reduction in PSL cannot be calculated easily. By using two different techniques, however, adequate knowledge of its behavior is obtained. First, because of the known (Rayleigh) shape of the pdf of random samples of the amplitude of the side radiation pattern, the effect of smoothing two samples on the PSL is calculated. Next it is recognized that the pdf of the average of several independent variates tends toward the Gaussian distribution. Thus averaging or superimposing many images changes the amplitude distribution from Rayleigh to approximately Gaussian. The average values of the two distributions are the same ($\bar{A} = \sqrt{\pi N}/2$), whereas the variances differ by the factor M equal to the number of images superimposed. These facts are illustrated in Figure 3.1.

3.3. DIVERSITY TOOLS

The key to the potential improvement lies in the independence of the side radiation patterns. Because the far-field pattern of a one-dimensional array is

$$f(u) = \sum_{n}^{N} \exp[jkx_n(u - u_0)] \tag{3.5}$$

it is evident that the number of diversity tools is limited by the small number of variables and parameters. Two of the most useful tools are element position diversity (randomly varying the element locations) and frequency (or wavelength or wavenumber) diversity. Both require that significant changes be made in the system design to acquire the multiple images derived from patterns that have independent or nearly independent sidelobes.

The more general formulation of this far-field radiation pattern includes complex weights:

$$f(u) = \sum_{n}^{N} I_n \exp[jkx_n(u - u_0)] \tag{3.6}$$

The additional two variables (amplitude and phase of I_n) suggest the potential for additional diversity capability. Randomly varying the phases decorrelates the sidelobes but also destroys the mainlobe; it certainly cannot be used as a diversity tool. Randomly varying the amplitudes $|I_n|$ from array to array does decorrelate the side radiation pattern somewhat but not very efficiently. Little merit is found in amplitude-weight diversity.

Another parameter in the equation for the radiation pattern (3.6) is beamforming angle u_0. It could be supposed that phase synchronizing the array in different directions would result in radiation patterns with decorrelated sidelobe patterns. Unfortunately that is not true. The shape of the radiation pattern does not change with beamforming direction. Because the side radiation pattern is independent of beamforming angle, the sidelobes do not decorrelate when u_0 is changed.

In addition to knowing the tools that are available for diversity, two separate questions are important to the designer. First, how valuable is it to change the system design, which always is a costly process, from one without diversity to one that permits the introduction of diversity techniques? Thus the designer has a basic decision to make whether to include diversity in the system. If the answer to this question indicates sufficient value to alter the system to accommodate diversity, the second is how much diversity should be included? These two questions are answered separately in the sections that follow immediately.

3.4. DUAL DIVERSITY

The answer to the first question is found by determining the ratio of the peak sidelobe at some given level of confidence without diversity (i.e., with diversity of order one) and with dual diversity (diversity of order two). The first-order diversity solution was developed in Section 2.4. A similar technique is discussed here for a dual diversity peak sidelobe estimator.

Given two random arrays of the same length L with N elements each, the averaged side magnitude pattern is $A(u) = \frac{1}{2}[A_1(u) + A_2(u)]$, where $A_1(u)$ and $A_2(u)$ are the component magnitude patterns. The probability that a single sample of the random process $A(u)$ will not exceed a certain level A_0 is given by

$$\mathrm{P}\left(\frac{A_1}{2} + \frac{A_2}{2} \leq A_0\right) = \int_0^{A_0}\int_0^{A_0 - y_1} w(y_1, y_2)\, dy_1\, dy_2 \tag{3.7}$$

where $y_1 = A_1/2$, $y_2 = A_2/2$, and $w(y_1, y_2)$ is the joint pdf of the two random variables y_1 and y_2. Because the element locations of each array are assumed to be chosen independently, y_1 and y_2 are independent random variables. Furthermore, each of them is Rayleigh distributed with a mean square value $N/4$. Thus (3.7) becomes

$$\begin{aligned}\mathrm{P}(A < A_0) &= \int_0^{A_0}\int_0^{A_0 - y_1} \frac{2y_1}{N/4}\exp\left(-\frac{y_1^2}{N/4}\right)\cdot\frac{2y_2}{N/4}\exp\left(-\frac{y_2^2}{N/4}\right) dy_1\, dy_2 \\ &= 1 - \exp\left(-\frac{4A_0^2}{N}\right) - \sqrt{\frac{\pi}{2}}\left(\frac{2A_0}{\sqrt{N}}\right)\exp\left(-\frac{2A_0^2}{N}\right)\mathrm{erf}\left(\frac{A_0}{\sqrt{N/2}}\right)\end{aligned} \tag{3.8}$$

For a random array steered to broadside the number of independent samples of the side magnitude pattern is $n = L/\lambda$. It is clear that this number will remain the same after averaging. As before, the confidence level β is defined as the probability that none of these independent samples will exceed the level A_0. In other words,

$$\beta = [\mathrm{P}(A < A_0)]^n \tag{3.9}$$

or

$$1 - \beta^{1/n} = e^{-4B_2} + \sqrt{2\pi B_2}\, e^{-2B_2}\operatorname{erf}\left(\sqrt{2B_2}\right) \tag{3.10}$$

where $B_2 = A_0^2/N$. This quantity is analogous to the quantity B defined in Section 2.4, which in the context of diversity we call B_1; B_2 represents a downward-biased estimator for the peak sidelobe power level at a level of confidence β, normalized to the number of array elements N. Note that N no longer represents the average power level of the side radiation pattern. The average power level has dropped to $(\pi/8 + 1/2)N$ due to averaging. Although B_2 is smaller than B_1, both are the same order of magnitude. In most practical situations in which high resolution is required and a high confidence level is specified values of n and β are such that $B_2 > 2$; for example, for $L/\lambda = 1000$ and $\beta = 0.9$, $B_2 = 5.46 < B_1$. Thus $\operatorname{erf}(\sqrt{2B_2}) \simeq 1$ and (3.10) can be simplified to

$$B_2 \simeq \frac{1}{2}\ln\left(\frac{\sqrt{2\pi}}{1 - \beta^{1/n}}\right) + \frac{1}{4}\ln B_2 \tag{3.11}$$

Equation 3.11 can be solved numerically for B_2. Table 3.1 shows B_1 and B_2, and the ratio B_1/B_2 in decibels, as a function of array length for a value of $\beta = 0.9$.

The results in Table 3.1 show that a reduction of the order of $2\text{–}2\frac{1}{2}$ dB in the peak sidelobe response due to dual incoherent element position diversity is possible. This is significant because the improvement in peak sidelobe level due to doubling the number of elements in a single array is 3 dB. Further insight into the significance of (3.11) can be gained by observing that because $B_1 = -\ln(1 - \beta^{1/n})$, Section 2.4, (3.11) can be rewritten as

$$B_2 = \tfrac{1}{2}B_1 + \tfrac{1}{2}\ln\sqrt{2\pi} + \tfrac{1}{4}\ln B_2 \tag{3.12}$$

TABLE 3.1. Improvement in Peak Sidelobe Level, in Decibels, Due to Dual Diversity ($\beta = 0.9$): B_1 is the PSL in the Absence of Diversity. B_2 is the PSL with Dual Diversity.

L/λ	10	10^2	10^3	10^4	10^5
B_1	4.56	6.86	9.16	11.46	13.76
B_2	3.01	4.25	5.46	6.66	7.86
$10\log B_1/B_2$	1.8	2.08	2.24	2.36	2.44

Because the last term is always the order of 0.5, (3.12) becomes

$$B_2 \simeq \tfrac{1}{2}B_1 + 1 \tag{3.13}$$

which indicates that the improvement due to dual diversity approaches 3 dB as B_1 increases because of an increase in L/λ, β, or both.

Experimental verification is made later in Figure 3.2. The first two curves in that figure correspond to first- and second-order diversity. The reduction in the sidelobe level is evident.

3.5. ASYMPTOTIC IMPROVEMENT: DOWNWARD-BIASED ESTIMATOR

The second question is addressed by determining the asymptotic reduction in peak sidelobe level as the order of diversity increases. Two biased estimators are derived. One is downward biased, the other, upward biased. The estimators prove to be closely spaced and experimental simulation results fall between them. Both estimators are sound for $M \geq 5$. The downward-biased estimator is derived in this section.

After M-fold averaging of independent side radiation patterns the amplitude distribution changes from Rayleigh (3.1) to approximately Gaussian ($M \geq 5$), as shown in Figure 3.1. The average values of the two distributions are the same ($\bar{A} = \sqrt{\pi N}/2$), whereas the variance becomes

$$\sigma_M^2 = \frac{\sigma^2}{M} = \frac{N}{M}\left(1 - \frac{\pi}{4}\right) \tag{3.3}$$

Hence the resulting pdf takes the approximate form

$$w_M(A) \simeq \frac{1}{\sqrt{2\pi}\,\sigma_M} \exp\left[-\frac{(A - \bar{A})^2}{2\sigma_M^2}\right], \qquad A \geq 0 \tag{3.14}$$

where the subscript denotes the M-fold averaged magnitude of the side radiation pattern.

Following the argument beginning with (2.23), define A_0 as the peak sidelobe, subject to some prespecified confidence level. Let α be defined as

$$\alpha = \mathrm{P}(A > A_0) = \int_{A_0}^{\infty} w_M(A)\, dA = \int_{A_0}^{\infty} \frac{1}{\sqrt{2\pi}\,\sigma_M} \exp\left[-\frac{(A - \bar{A})^2}{2\sigma_M^2}\right] dA$$

$$= \frac{1}{2}\operatorname{erfc}\left(\frac{A_0 - \bar{A}}{\sqrt{2\sigma_M^2}}\right) \tag{3.15}$$

This equation presents the probability that an arbitrary sample of the average side radiation pattern will exceed the specified peak sidelobe A_0. The complement $1 - \alpha$ is the probability that such a sample will be less than A. As in (2.24), let β be the probability that none of the n independent samples will exceed A_0. The result is a downward-biased estimator because with unity probability the largest samples of the side radiation pattern will fall somewhat below the crest of the largest sidelobe. For n independent samples

$$\beta = (1 - \alpha)^n = \left[1 - \frac{1}{2}\operatorname{erfc}\left(\frac{A_0 - \bar{A}}{\sqrt{2\sigma_M^2}}\right)\right]^n \tag{3.16}$$

For the common case of high confidence level $\alpha \ll 1$ and

$$\operatorname{erfc}(x) \cong \frac{e^{-x^2}}{x\sqrt{\pi}} \quad \text{for} \quad x \gg 1 \tag{3.17}$$

from which

$$\alpha \cong \frac{e^{-b}}{2\sqrt{\pi}\, b^{1/2}} \tag{3.18}$$

where

$$b = \frac{(A_0 - \bar{A})^2}{2\sigma_M^2} \tag{3.19}$$

hence

$$\beta = \left[1 - \frac{e^{-b}}{2\sqrt{\pi}\, b^{1/2}}\right]^n \tag{3.20}$$

Solving for b,

$$b + \tfrac{1}{2}\ln b = -\ln 2\sqrt{\pi}\left(1 - \beta^{1/n}\right) \tag{3.21}$$

The number of independent samples n of the side radiation pattern for a single random array steered to broadside is $n = L/\lambda$ (2.25). This number is unaffected by M-fold averaging of independent side radiation patterns. For given values of β and L/λ the right-hand side of the equation above can be calculated, after which (3.21) can be solved numerically for b. From (3.3) and (3.19)

$$b = \frac{M\left(A_0^2 - 2\bar{A}A_0 + \bar{A}^2\right)}{2(1 - \pi/4)N} \tag{3.22}$$

Solving for A_0^2/N

$$B_L(M) \triangleq \frac{A_0^2}{N} = \frac{\pi}{4} + \frac{\sqrt{\pi(2-\pi/2)b}}{\sqrt{M}} + \frac{(2-\pi/2)b}{M}, \qquad M \geq 5 \quad (3.23)$$

This equation is the desired result; B_L corresponds to B in (2.26) and B_1 in (3.12). It is a downward-biased estimator of the peak sidelobe power level normalized to the average sidelobe level in the absence of diversity.

3.6. ASYMPTOTIC IMPROVEMENT: UPWARD-BIASED ESTIMATOR

This estimator was derived in [91]. It is based on level-crossing theory [9], first used for this type of problem by Sodin [79] and later by DonVito and Kassam [20]. Except for a few beamwidths centered around the main beam, the side radiation pattern of a thinned random array can be considered a stationary random process. After M-fold incoherent averaging of independent side radiation patterns the amplitude distribution changes from Rayleigh to approximately Gaussian of mean $\bar{A} = \sqrt{\pi N}/2$ and variance $\sigma^2(N/M)(1-\pi/4)$. In an interval of length l the expected number of upcrossings of the level A_0 by the random process $A(u)$ is given by

$$E\left[U_{A_0}\right] = l\int_0^\infty |A'| w(A_0, A')\, dA' \qquad (3.24)$$

where $w(A, A')$ is the joint pdf of the random variables $A(u)$ and $A' = dA(u)/du$. It can be shown that $A(u)$ and $A'(u)$ are uncorrelated [104]. Furthermore, they are independent because each is Gaussian. Thus we can write

$$\begin{aligned} w(A, A') &= w_M(A) \cdot w_{M'}(A') \\ &= \frac{1}{2\pi\sigma_M\sigma_{M'}} \exp\left[-\left(\frac{(A-\bar{A})^2}{2\sigma_M^2} + \frac{A'^2}{2\sigma_{M'}^2}\right)\right] \end{aligned} \qquad (3.25)$$

where $\sigma_{M'}^2$ is the variance of the Gaussian random variable A'. Substituting (3.25) in (3.24) and performing the integration, we have

$$E\{U_{A_0}\} = \frac{l}{2\pi} \cdot \frac{\sigma_{M'}}{\sigma_M} \exp\left[-\frac{(A_0-\bar{A})^2}{2\sigma_m^2}\right] \qquad (3.26)$$

Substituting $\sigma_M^2 = (N/M)(1-\pi/4)$ in (3.26) and solving for A_0^2/N, we get

$$B_H \triangleq \frac{A_0^2}{N} = \frac{\pi}{4} + \frac{\sqrt{\pi(2-\pi/2)b'}}{\sqrt{M}} + \frac{(2-\pi/2)b'}{M} \qquad M \geq 5 \quad (3.27)$$

where

$$b' = \ln\left[\frac{l}{2\pi(1-\beta)} \cdot \frac{\sigma_{M'}}{\sigma_M}\right] \tag{3.28}$$

and

$$\beta \triangleq 1 - E\left[U_{A_0}\right] \tag{3.29}$$

Because we are interested in the peak of the process, we consider only levels A_0 that are high with respect to the expected value of $A(u)$. It can be shown that for such high levels the expected number of local maxima above A_0 approaches the expected number of upcrossings of A_0 [20], [70]. Also the expected value $E[U_{A_0}]$ of the number of upcrossings in the sidelobe region is an upperbound on the probability $\Pr\{U_{A_0} \geq 1\}$ of at least one upcrossing of A_0 in the same region [20]. To show this let P_k be the probability of exactly k upcrossings of A. It is clear that $E[U_{A_0}] = \sum_{k=0}^{\infty} kP_k$ and $\Pr\{U_{A_0} \geq 1\} = \sum_{k=1}^{\infty} P_k$. Therefore, $\Pr\{U_{A_0} \geq 1\} \leq E[U_{A_0}]$ and $E[U_{A_0}]$ is an upperbound on $\Pr\{U_{A_0} \geq 1\}$. We expect that this bound becomes tight for small values of $E[U_{A_0}]$ because the probability of obtaining more than one upcrossing in the sidelobe region is negligible compared with the probability of obtaining zero or one upcrossing. It is also clear that $\beta \geq \Pr\{U_{A_0} = 0\}$, which is the probability that the process will remain below the level A_0, and we expect the two sides of the inequality to become close to each other for values of β close to unity.

From (3.27) and (3.28) we see that B_H is a monotonically increasing function of β. If we define $\beta' = \Pr\{U_{A_0} = 0\}$, then the corresponding normalized level B'_H will satisfy $B'_H \leq B_H$ because $\beta' \leq \beta$. Thus B_H is always an upward-biased estimate of B'_H.

Next we turn to the calculation of the ratio σ'_M/σ_M in (3.28). The variance of $A'(u)$ is $\sigma^2_{M'} = -R''_a(0)$, where $R''_a(u) = d^2R_a(u)/du^2$ is the second derivative of the autocorrelation function of $A(u)$ [66], and because differentiation is a linear operation it is clear that $\sigma^{2\prime} = K^2\sigma^2$, where K^2 is a constant of proportionality. Because a closed form expression for $R_a(u)$ is not available for incoherent averaging, simulation experiments were conducted to determine the constant K for an assumed uniform distribution of element positions. A value of $K = 2.77\ L/\lambda$ was found.

Equation (3.27) is the desired result. It gives an upward biased estimate of the normalized level A_0^2/N that with probability β will not be exceeded.

3.7. ELEMENT POSITION DIVERSITY

The theory developed in the preceding three sections pertains to those forms of diversity combining in which the sidelobes of the M patterns are decorrelated.

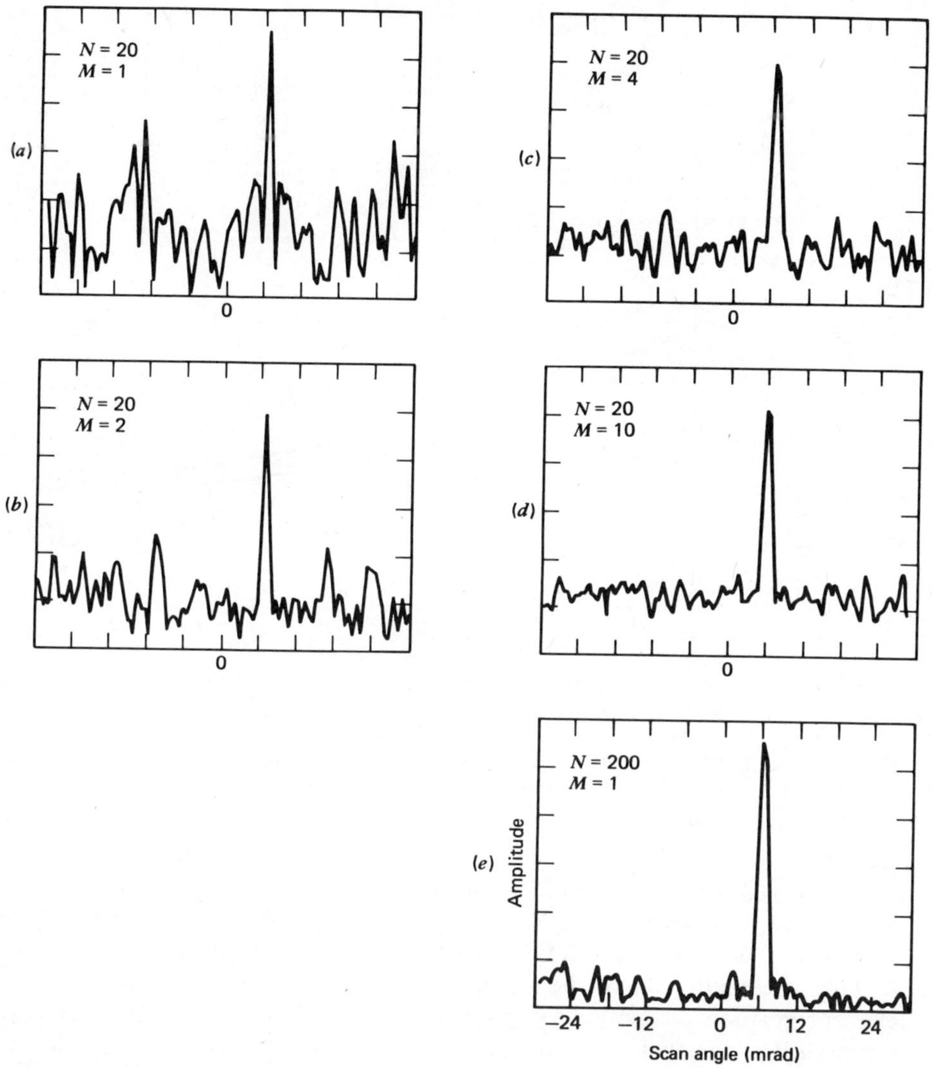

FIGURE 3.2. Element position diversity experiment; 28 m array, 3-cm wavelength. Cement block target at distance of 146 m. (*a–d*) Twenty elements each, diversity orders 1, 2, 4, and 10; (*e*) 200 elements, diversity order one.

Among the several types described in Section 3.3 element position diversity corresponds most closely to this case.

Figure 3.2 shows the results of an element position diversity experiment. Ten random arrays were formed with the same equipment used in the experiments reported in Section 2.8. Each array consisted of 20 elements and a single cement-block target with its 6 × 14 in. surface facing the array at a distance of 146 m was imaged: (*a*) is the scanned image of a single array

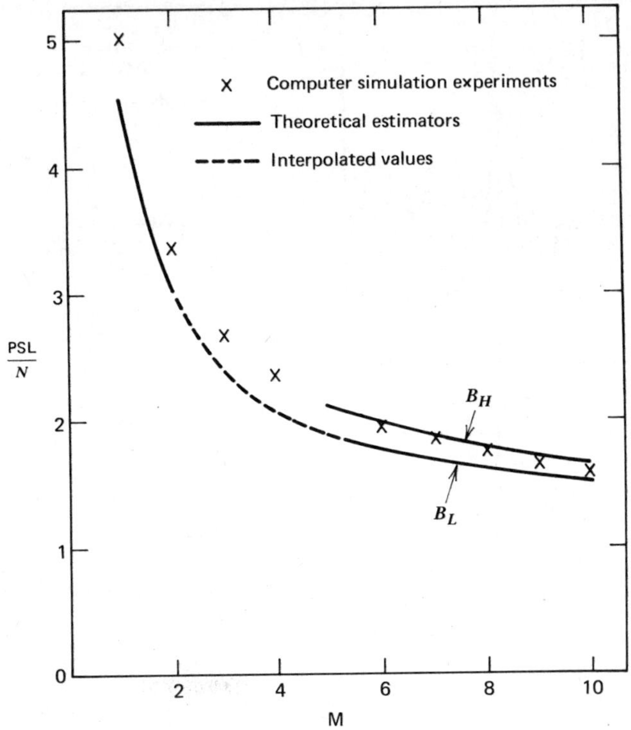

FIGURE 3.3. Peak sidelobe power level normalized to the number of array elements as a function of the order of diversity: $N = 20$, $\beta = 0.9$.

($M = 1$); (b) shows the average of two images ($M = 2$); (c) and (d) are $M = 4$ and 10, respectively; (e) shows the effect of coherent combining of the signals from all 200 elements, that is, a single, 200-element array.

Two observations may be made. First, in comparing (a) through (d), element position diversity does produce the expected results: the peak sidelobes shrink, the valleys fill in, and the average level shows little change. Second, (d) and (e) are images based on the same 200 elements. Their comparison reveals the expected phenomena: the average in (e) drops in relation to (d) and the variances of the side radiation patterns are similar.

Figure 3.3 shows the results of computer simulation experiments of element position diversity. One thousand random arrays were formed, each of length $L = 100\lambda$ and containing 20 elements. Their radiation patterns were computed in the interval [0.9, 1.0] in u-space and the peak sidelobe of each was measured.

The pattern amplitudes were then averaged by twos, threes, fours, etc., and the peak sidelobes of each combination was measured. For each value of M 100 experiments were conducted. Empirical cumulative distribution functions of the peak sidelobe level were formed from each set of 100 data points. The

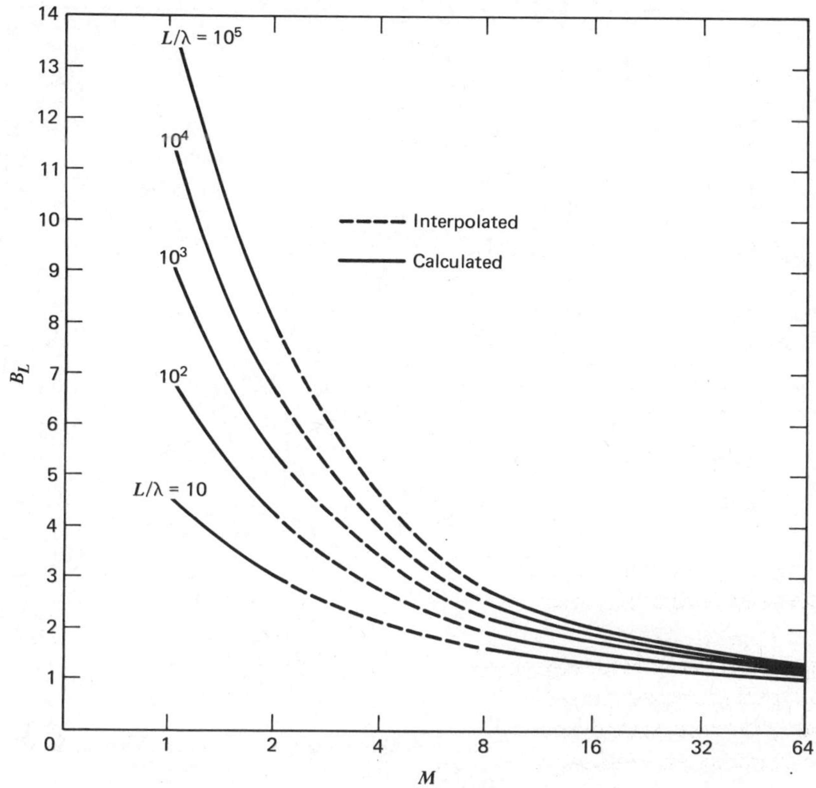

FIGURE 3.4. Downward-biased estimator for the peak sidelobe: $N = 20$, $\beta = 0.9$.

points in Figure 3.3 for $M \geq 5$ represent the normalized level that exceeded the peak sidelobe level for 90% ($\beta = 0.9$) of the experiments.

The solid curves connect computed values for $M = 1$ and 2, both downward-biased, and curves for the two asymptotic estimators (upward- and downward-biased). The agreement is good. The dashed curve is an interpolation between the two portions of the lower solid curve. The solid curves represent theoretical estimates given by (2.26), (3.11), (3.23), and (3.27), calculated for $\beta = 0.9$. In (3.28) an interval length $l = 0.1$ was assumed, whereas in (3.21) $n = 0.1\ L/\lambda$ was taken. It is seen that all the experimental values fall between B_H and B_L, in good agreement with the theories that led to both.

Figure 3.4 plots $B_L(M)$ over a wide range of diversity order and for a wide range of array sizes L/λ, which is essentially the array parameter n. Figure 3.5 plots $10 \log[B_L(M)/B_L(2M)]$ versus M with L/λ as a parameter. The confidence level $\beta = 0.9$. The ordinate is the reduction in PSL, in decibels, caused by a doubling of the order of diversity. The abscissa is the order of diversity. It

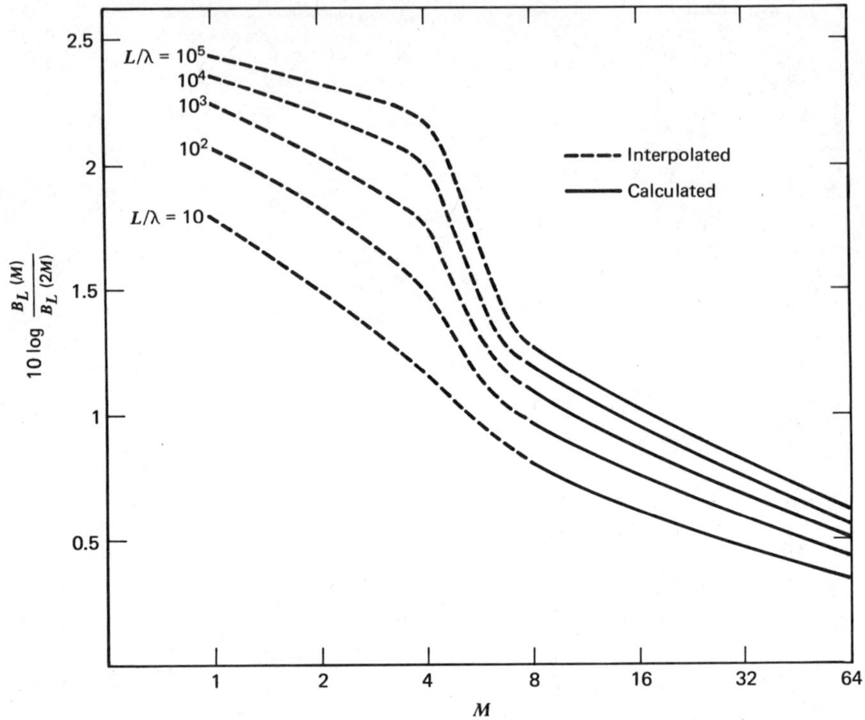

FIGURE 3.5. PSL reduction in dB due to doubling the diversity order: $N = 20$, $\beta = 0.9$.

is evident that the larger the array, the larger the potential gain due to diversity. The reason is that B is the maximum reduction in PSL that can be obtained and B grows with n according to

$$B \simeq \ln n - \ln(1 - \beta) \approx \ln n \qquad \text{large } n \tag{3.30}$$

as shown in Chapter 2.

3.8. FREQUENCY DIVERSITY

Other forms of diversity provide less than complete decorrelation of the sidelobes. Nevertheless, they can reduce the high sidelobes significantly, provided that sufficient partial decorrelation results from them. Large signal bandwidth is one form of diversity.

Both coherent and noncoherent combining of the frequency components of a wideband source can be used. Coherent combining results when a wide bandwidth signal is radiated. In coherent frequency diversity the frequency components coexist simultaneously, and the radiation pattern that results involves all the interference phenomena between the constituents. In noncoherent combining the signal constituents are separately radiated, each producing its own radiation pattern, and the moduli of the patterns are averaged. Frequency hopping is an example of noncoherent frequency diversity. Notwithstanding the intrinsic difference between them, their sidelobe reduction effects are essentially the same.

Coherent Combining

The effect of coherent combining follows from (2.26) or (3.30). The peak-to-average sidelobe level ratio of a one-dimensional random array is approximately $\ln(L/\lambda)$, plus a small additive term; L/λ is the order of the number of sidelobes when the radiation is monochromatic and the elements are isotropic. Except for the first few, the statistics of the sidelobes are the same everywhere in the pattern; L/λ is approximately the number of independent locations in the side radiation pattern at which the peak sidelobe can occur.

Bandwidth reduces the number of likely locations of the peak. In Figure 3.6 the radiation patterns of a random array are sketched at two different wavelengths. Their shapes are identical but their abscissa scales differ; both are centered at $u = u_0$, the beamforming angle, but differ in horizontal scale by the factor λ_1/λ_2. The wavelength-dependent factor is evident from the basic formulation of the radiation pattern given in (2.2) in which the wavenumber $k = 2\pi/\lambda$. Hence $f(u)$ more properly should be written $f_\lambda(u)$ or $f(u, \lambda)$; however, the λ-dependence is normally ignored because radiation is usually sufficiently narrowband to create a negligible effect.

Because the abscissa scale of the radiation pattern is linear with frequency, it is evident that for a fixed set of element locations the radiation patterns of an array at two different frequencies become progressively different at angles away from the main lobe. The decorrelation is illustrated in Figure 3.6. The patterns are highly correlated in the near sidelobe region and become progressively decorrelated beyond. If radiation patterns with these characteristics are averaged, the near portion will remain unchanged, whereas in the distant

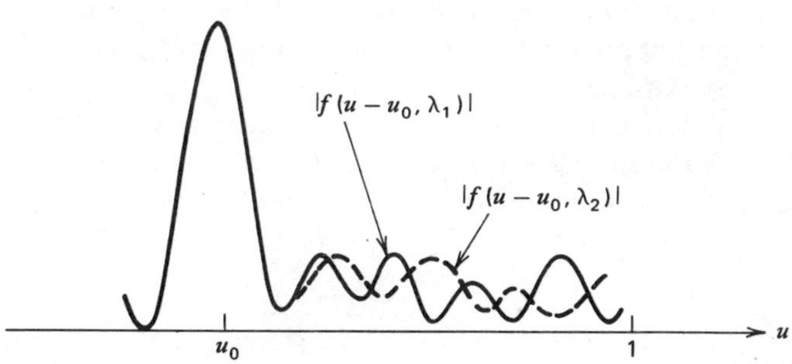

FIGURE 3.6. The scale of the radiation pattern is a function of wavelength.

sidelobe region peaks will reduce and the valleys will fill in. After averaging, therefore, the largest sidelobe will most likely fall in the near-sidelobe region. By reducing the extent of the visible region in which the largest sidelobe is likely to occur the effective array parameter n is also reduced. Because the peak sidelobe is approximately proportional to $\ln n$, the sidelobe level can be expected to be reduced.

The diversity theory developed in earlier sections no longer applies in the calculation of the peak sidelobe because the near-sidelobe region in which the peak is likely to be found is not beneficially affected by frequency diversity. Thus the formula for $M = 1$ remains relevant; however, the array parameter must be reinterpreted to correspond to the reduced portion of the visible region in which little or no sidelobe smoothing takes place.

To estimate the extent of the reduction in the array parameter it is necessary only to calculate the number of lobes of the radiation pattern that occur before the two patterns in Figure 3.6 have slipped by the order of $\frac{1}{2}$ to 1 lobe. Let λ_1 and λ_2 be components in the spectrum of the signal passing through the array. Further assume that these components are close to the edges of the signal spectrum. Hence $\Delta\lambda = \lambda_1 - \lambda_2$ is the width of the band. The radiation patterns at these wavelengths, sketched in Figure 3.6, are nearly identical in the mainlobe region and drift apart progressively with scan angle. Because sidelobes are nominally spaced by one beamwidth, the pth sidelobe at $\lambda = \lambda_1$ occurs approximately at angle $p\lambda_1/L$. At the same angle the sidelobe number at $\lambda = \lambda_2$ is $(p\lambda_1/L)/(\lambda_2/L) = p\lambda_1/\lambda_2$.

The angle u_p at which the number of lobes at λ_1 and λ_2 differ by, say, 1/2 to 1 defines the transition between the portions of the radiation patterns that are nearly the same ($u < u_p$) and the portions that differ significantly ($u > u_p$). The approximate number of lobes in this region is found by equating $p + \frac{1}{2}$ or $p + 1$ to $p\lambda_1/\lambda_2$, from which $p \approx Q/2$ or Q, where Q is the reciprocal of the fractional bandwidth.

The number p is the number of sidelobes left relatively unchanged by widening the spectral width of the radiation, whereas the remainder of the radiation pattern tends to be smoothed toward the average sidelobe level. The array parameter n reduces from approximately L/λ [see (2.25)] to approximately Q. Hence, from (2.26) and (2.27) or (3.30) when $Q < L/\lambda$, the power ratio of the peak sidelobe to the average sidelobe level, minus a small additive constant, becomes approximately $\ln Q$ rather than $\ln(L/\lambda)$. Thus Q replaces L/λ as the array parameter in Figure 2.6 when $Q < L/\lambda$.

The decorrelation between the side radiation patterns at two frequencies and the effect on the sidelobe level also can be calculated from the cross-correlation coefficient between the two patterns at these frequencies:

$$\rho_{12} = \frac{Ef_1 f_2^*}{\left[E|f_1|^2 E|f_2|^2\right]^{1/2}} \tag{3.31}$$

where ρ_{12}, f_1, and f_2 are functions of u. The numerator

$$\text{num} = E\sum_n^N \sum_m^N e^{j\omega_0 x_n u/c} e^{-j(\omega_0+\Delta\omega)x_m u/c}$$

$$= \sum_n^N Ee^{-j\Delta\omega x_n u/c} + \sum_{n \neq m}^N \sum^N Ee^{j\omega_0 x_n u/c} Ee^{-j(\omega_0+\Delta\omega)x_m u/c}$$

$$= NEe^{-j\Delta\omega x u/c} + (N^2 - N)Ee^{j\omega_0 x u/c} Ee^{-j(\omega_0+\Delta\omega)x u/c}$$

$f_0 = \omega_0/2\pi$ is either one of the frequencies and $\Delta f = \Delta\omega/2\pi$ is the frequency separation. The denominator

$$\text{den} = \left[E|f_1|^2 E|f_2|^2\right]^{1/2} = E|f|^2$$

$$= E\sum_n^N \sum_m^N e^{j\omega x_n u/c} e^{-j\omega x_m u/c}$$

$$= N + \sum_{n \neq m}^N \sum^N Ee^{j\omega x_n u/c} Ee^{-j\omega x_m u/c}$$

$$= N + (N^2 - N)Ee^{j\omega x u/c} Ee^{-j\omega x u/c}$$

Let x be distributed uniformly in the interval $[-L/2, L/2]$, which is customary for most random arrays. Then

$$Ee^{-j\Delta\omega x u/c} = \frac{1}{L}\int_{-L/2}^{L/2} e^{-j\Delta\omega x u/c}\, dx = \frac{\sin(\Delta\omega L u/2c)}{\Delta\omega L u/2c}$$

Hence

$$\text{num} = N\frac{\sin(\Delta\omega Lu/2c)}{\Delta\omega Lu/2c} + (N^2 - N)\frac{\sin(\omega_0 Lu/2c)}{\omega_0 Lu/2c} \cdot \frac{\sin[(\omega_0 + \Delta\omega)Lu/2c]}{(\omega_0 + \Delta\omega)Lu/2c} \tag{3.32}$$

and

$$\text{den} = N + (N^2 - N)\left[\frac{\sin(\omega Lu/2c)}{\omega Lu/2c}\right]^2 \tag{3.33}$$

For a low-density array ($N/(L/\lambda) \ll 1$) and a high $Q(\omega_0 \gg \Delta\omega)$ the second terms in (3.32) and (3.33) are much smaller than the first for values of u a few beamwidths away from the mainlobe. Then

$$\text{num} \simeq N \frac{\sin(\Delta\omega Lu/2c)}{\Delta\omega Lu/2c}$$

$$\text{den} \simeq N$$

$$\rho_{12} \simeq \frac{\sin(\Delta\omega Lu/2c)}{\Delta\omega Lu/2c} \tag{3.34}$$

Thus the patterns become decorrelated when

$$\frac{\Delta\omega Lu}{2c} = \pi \tag{3.35}$$

for then $\rho_{12} = 0$. Equation (3.35) can be written in terms of the frequency separation required to achieve decorrelation at and beyond a given value of u:

$$\Delta f = \frac{\Delta\omega}{2\pi} = \frac{c}{Lu} = \frac{f\lambda}{Lu} \tag{3.36}$$

or, in terms of the Q of the microwave source,

$$Q = \frac{f}{\Delta f} = \frac{Lu}{\lambda} \tag{3.37}$$

In (3.37) u is the angle beyond which the sidelobe patterns are decorrelated and

$$\frac{u}{1 + |u_0|} \qquad u_0 = \text{beamsteering angle}$$

is the fraction of the visible region in which the sidelobes are *not* decorrelated. Because the array parameter with monochromatic radiation is

$$n = \frac{L}{\lambda}(1 + |u_0|) \tag{3.38}$$

the effective or reduced array parameter

$$n' = n \frac{u}{1 + |u_0|} = \frac{Lu}{\lambda} = Q \tag{3.39}$$

and the power ratio of the peak sidelobe to the average level, minus the same small additive constant mentioned earlier, becomes approximately ln Q. Thus both calculations lead to the same result.

These calculations pertain to two signal components at opposite edges of the signal band. Interior frequency components contribute less to the decorrelation of the sidelobes but the combined effect remains proportional to Q.

Figure 3.7 shows the effect of bandwidth on a thinned random array. The radiation patterns are those of a 200-element, linear random array, 800 wavelengths long, for three different signal bandwidths. In (a) the signal is monochromatic, where the Q in (b) is 11 and in (c), 2.8. The ragged sidelobe behavior of (a) nearly disappears when the Q is lowered to that of (c). The reduction in the sidelobe envelope is evident in the composite plot shown in (d).

Large bandwidth is helpful in reducing the peak sidelobe, especially for the very large apertures needed for high angular resolution imaging. The payoff, however, is exceedingly slow because of the logarithmic relation (2.26) between PSL and the array parameter n; for example, the expected value of PSL/ASL of a narrowband radio camera in which the aperture size $L = 10^5\lambda$ is 13.4. If the signal bandwidth is increased to 1% of the carrier frequency ($Q = 100$), the expected value of the peak drops to 6.7, a reduction of 3 dB. A further broadening of the bandwidth by a factor of 10 reduces PSL by only another 1.6 dB. The generation of microwave power with so low a value of Q is difficult to achieve. Furthermore, time delays would be needed to replace phase shifters. Consequently signal bandwidth plays a limited role in microwave radio camera design.

Noncoherent Combining

The alternative to coherent utilization of the entire spectrum of the signal is noncoherent frequency-diversity combining, which is generally a more practical technique because wideband frequency-hopping generators are more common than broadband generators of modulated microwave waveforms. The available waveform is a succession of narrowband signals (such as RF pulses) at frequencies $f_1, f_2, \ldots,$ and f_M. The radio camera phase-synchronizes and images on each pulse; it then superimposes the images. Provided that the frequency spacing is sufficient to decorrelate the sidelobes, the PSL/ASL should reduce according to the formulation in the earlier sections. The expressions there were reasonably accurate when the diversity order M is large enough for the sum of M Rayleigh variates to approximate a Gaussian variate, which takes place for $M \geq 5$.

M is the number of images with statistically independent sidelobe patterns. The frequency separation Δf which results in independence in the sidelobe region may be calculated from the cross-correlation coefficient, as for the wideband signal waveform. The cross correlation, however, must be calculated between the power patterns at the two frequencies rather than between the

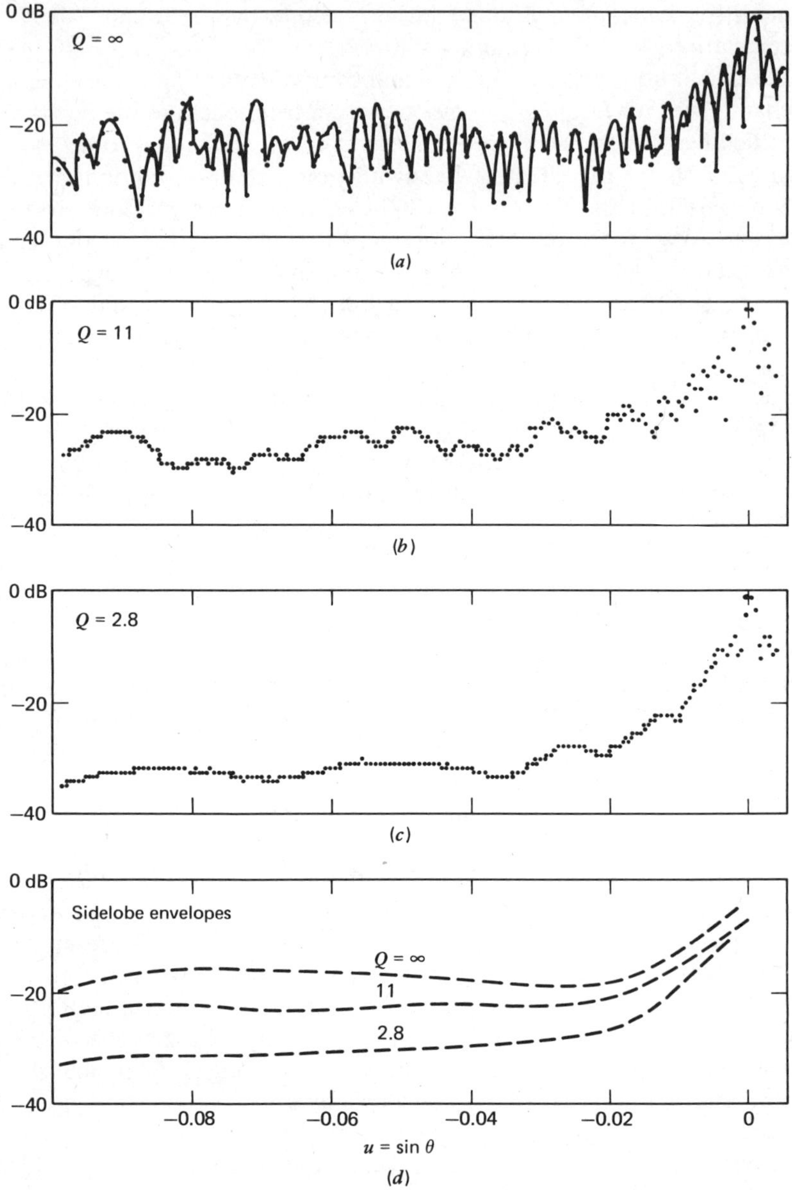

FIGURE 3.7. Effect of bandwidth on side radiation pattern; 800 wavelength random array; 200 elements.

complex radiation patterns. Decorrelation occurs at the same frequency separation as coherent frequency diversity.

Writing the radiation pattern as the sum of two quadrature, frequency-dependent components $a(u) + jb(u)$, where $a(u) = \Sigma^N \cos kx_i u$ and $b(u) = \Sigma^N \sin kx_i u$, we obtain the power pattern

$$p(u) = a^2(u) + b^2(u) \tag{3.40}$$

The correlation function between patterns with different wavenumbers (dropping the angular variable) is

$$\begin{aligned}\mathcal{R}_{12}(k_1, k_2) &= E[p(k_1) \cdot p(k_2)] \\ &= E[[a^2(k_1) + b^2(k_1)][a^2(k_2) + b^2(k_2)]] \\ &= E[a^2(k_1)a^2(k_2)] \\ &\quad + E[a^2(k_1)b^2(k_2)] + E[b^2(k_1)a^2(k_2)] \\ &\quad + E[b^2(k_1)b^2(k_2)]\end{aligned} \tag{3.41}$$

Because a and b are Gaussian, the fourth moment theorem

$$\begin{aligned}E[x_1x_2x_3x_4] &= E[x_1x_2]E[x_3x_4] \\ &\quad + E[x_1x_3]E[x_2x_4] + E[x_1x_4]E[x_2x_3] - 2\mu_1\mu_2\mu_3\mu_4\end{aligned} \tag{3.42}$$

may be applied. The first term in (3.41) is

$$\begin{aligned}E[a^2(k_1)a^2(k_2)] &= E[a^2(k_1)]E[a^2(k_2)] \\ &\quad + 2E^2[a(k_1)a(k_2)] - 2E^2[a(\cos k_1)]E^2[a(k_2)] \\ &\simeq R_a(k_1, k_1)R_a(k_2, k_2) + 2R_a^2(k_1, k_2)\end{aligned} \tag{3.43a}$$

because $E[a]$ typically approaches zero in a large, thinned array for values of u greater than few beamwidths. Similarly,

$$\begin{aligned}E[a^2(k_1)b^2(k_2)] &\simeq E[a^2(k_1)]E[b^2(k_2)] + 2E^2[a(k_1)b(k_2)] - 0 \\ &\simeq R_a(k_1, k_1)R_b(k_2, k_2) + 2R_{ab}^2(k_1, k_2) \\ E[b^2(k_1)a^2(k_2)] &\simeq E[b^2(k_1)]E[a^2(k_2)] + 2E^2[b(k_1)a(k_2)] - 0 \\ &\simeq R_a(k_2, k_2)R_b(k_1, k_1) + 2R_{ba}^2(k_1, k_2) \\ E[b^2(k_1)b^2(k_2)] &\simeq R_b(k_1, k_1)R_b(k_2, k_2) + 2R_b^2(k_1, k_2)\end{aligned} \tag{3.43b}$$

Assuming a uniform distribution of element locations in the interval $[-L/2, L/2]$,

$$R_a(k_1, k_2) = E[a(k_1)a(k_2)] = E\left[\left(\sum_i^N \cos k_1 x_i u\right)\left(\sum_j^N \cos k_2 x_j u\right)\right]$$

$$= \frac{N}{2}\operatorname{sinc}\left(\frac{\Delta f L u}{c}\right) + \frac{N}{2}\operatorname{sinc}\left(\frac{[f_1 + f_2]Lu}{c}\right)$$

$$+ (N^2 - N)\operatorname{sinc}\left(\frac{f_1 L u}{c}\right)\operatorname{sinc}\left(\frac{f_2 L u}{c}\right) \tag{3.44}$$

The second term is always smaller than the first, except when $Q = f/\Delta f \approx 1$. This term may be dropped for realistic microwave conditions. The third term may also be dropped when $Q \gg 1$ and when the density of elements is low $(N \ll L/\lambda)$. Under these conditions

$$R_a(k_1, k_2) \simeq \frac{N}{2}\operatorname{sinc}\left(\frac{\Delta f L u}{c}\right) \tag{3.45}$$

and is stationary. Also

$$R_b(k_1, k_2) = E[b(k_1)b(k_2)] = E\left[\left(\sum_i^N \sin k_1 x_i u\right)\left(\sum_j^N \sin k_2 x_j u\right)\right]$$

$$\simeq \frac{N}{2}\operatorname{sinc}\left(\frac{\Delta f L u}{c}\right) = R_a(k_1, k_2) \tag{3.46}$$

$$R_{ab}(k_1, k_2) = E[a(k_1)b(k_2)] = E\left[\left(\sum_i^N \cos k_1 x_i u\right)\left(\sum_j^N \sin k_2 x_j u\right)\right] = 0 \tag{3.47}$$

$$R_{ba}(k_1, k_2) = 0 \tag{3.48}$$

Therefore

$$\mathcal{R}_{12}(f_1, f_2) = \mathcal{R}_{12}(\Delta f) = 4R_a^2(0) + 4R_a^2(\Delta f) \tag{3.49}$$

Substitution from (3.45) results in

$$\mathcal{R}_{12}(\Delta f) = N^2\left[1 + \operatorname{sinc}^2\left(\frac{\Delta f L u}{c}\right)\right] \tag{3.50}$$

The constant additive term is the square of the nonzero mean of the power pattern and plays no part in the decorrelation of the pattern with frequency. Removing it and normalizing the correlation function to unity at the origin leads to

$$\rho_{12}(\Delta\omega) = \left[\frac{\sin(\Delta\omega Lu/2c)}{\Delta\omega Lu/2c}\right]^2 \tag{3.51}$$

which is the square of (3.34). Hence the power patterns become decorrelated at the same frequency separation that decorrelates the complex radiation patterns; namely, that value given by (3.36).

The equivalence between coherent combining (wideband waveform) when a square law detector is used and noncoherent combining (frequency hopping) of spectral components is also evident in waveform analysis. Following the approach taken in [60], let $v(t)$ be the waveform radiated from a source at $u = u_0$. The wave received at the nth element is $v(t + t_n)$, where $t_n = t_n(u_0)$ is the travel time from the source. Let $T_n = T_n(u)$ be the steering delay for the nth element for the arbitrary steering direction u. The array output following square law detection and integration is

$$p(u; u_0) = \int \sum_m^N \sum_n^N v(t + t_m - T_m) v^*(t + t_n - T_n)\, dt \tag{3.52}$$

Let $V(\omega) = \mathcal{F}\{v(t)\}$ be the Fourier spectrum of $v(t)$. By Parseval's theorem (3.52) equals

$$p(u; u_0) = \frac{1}{2\pi} \int \sum_m \sum_n |V(\omega)|^2 \exp[j\omega(t_m - T_m - t_n + T_n)]\, d\omega \tag{3.53}$$

The exponential term is recognized as the wideband equivalent of the monochromatic function $\exp[jk(x_n - x_m)(u - u_0)]$; as in the monochromatic theory the double sum of that function is the power pattern (2.30). Thus

$$p(u; u_0) = \frac{1}{2\pi} \int |V(\omega)|^2 |f(u - u_0; \omega)|^2\, d\omega \tag{3.54}$$

in which $|f|^2$ is the power pattern at frequency ω when the array is steered to u_0.

Now let

$$V(\omega) = V_0 \sum_{g=-G/2}^{G/2} \delta(\omega - \omega_0 - g\,\Delta\omega) \tag{3.55}$$

The radiation consists of $G + 1$ spectral lines spaced by $\Delta\omega$ which occupy a

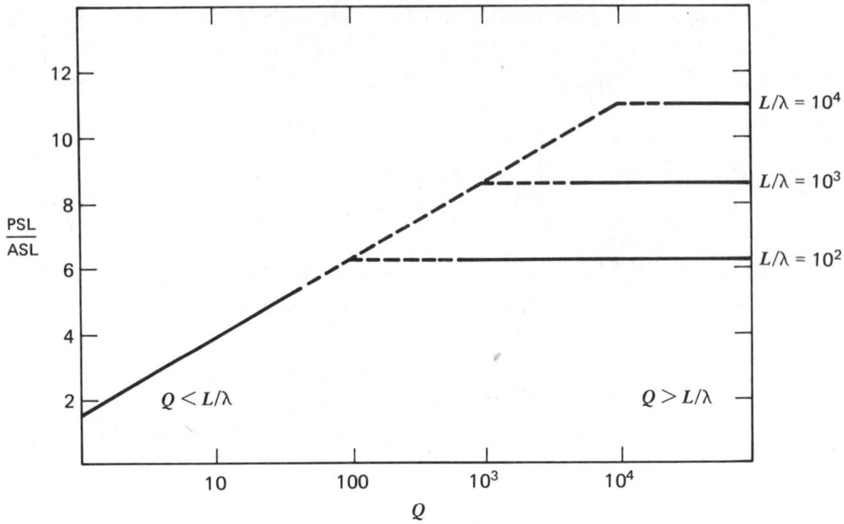

FIGURE 3.8. Expected asymptotic peak sidelobe behavior as a function of Q (reciprocal fractional signal bandwidth) and array size (L/λ).

band $W = G\Delta\omega$. Then

$$p(u; u_0) \propto V_0^2 \sum_{g=-G/2}^{G/2} |f(u - u_0; \omega_0 + g\,\Delta\omega)|^2 \tag{3.56}$$

Equation 3.56 is the sum of $G + 1$ power patterns, which is exactly the same output obtained when the spectral lines are radiated sequentially. Thus simultaneous transmission (wideband signal) followed by square law detection and integration and sequential transmission (frequency hopping) result in the same system output.

Because both coherent and noncoherent frequency-diversity combining exhibit similar decorrelation properties, a single test of the theory suffices. Figure 3.8 shows the predicted asymptotic behavior as a function of Q. The ordinate is the ratio of the peak-to-average sidelobe level. The abscissa is logarithmic in Q. When $Q < L/\lambda$ the curve should rise linearly with $\ln Q$ and be independent of L/λ. In the high-Q region the curves should be indepedent of Q and should be spaced $\ln(L_i/L_j)$.

The results of many frequency-hopping simulation experiments for $L/\lambda =$ 50 and 100 are shown in Figure 3.9. The expected asymptotic behaviors are evident. The slope of the portion linear with $\ln Q$ in the low-Q region is 1.06 in close agreement with theory, which predicts a slope approximately equal to unity. In the high-Q region the asymptotic spacing, based on linear least-mean-square fits to the data points, is 0.69, which equals the predicted spacing

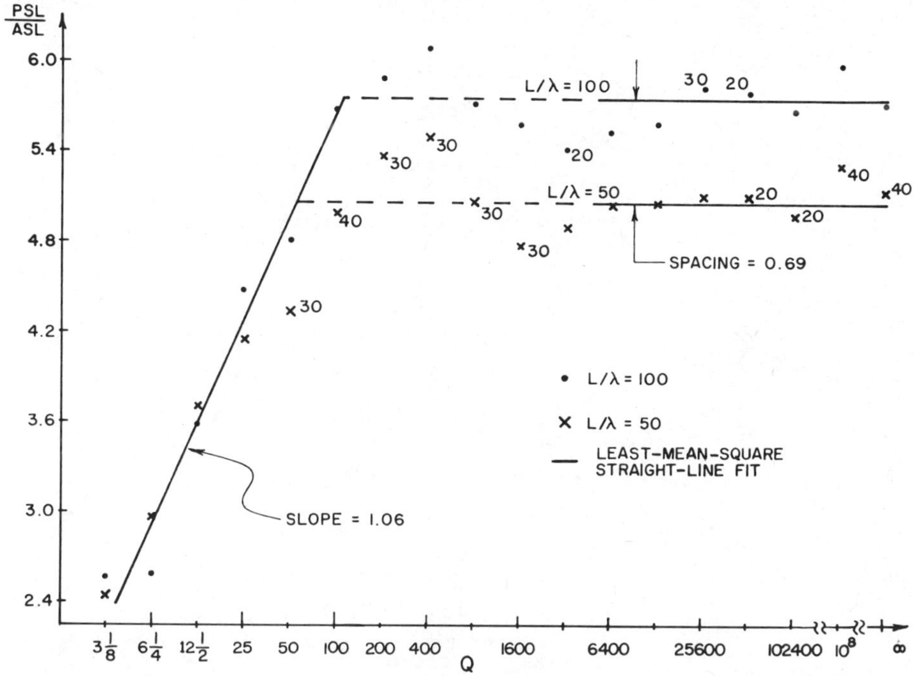

FIGURE 3.9. PSL/ASL versus Q in computer-simulation of frequency-hopping experiment. Each datum point is the average of 10 samples (unmarked) or 20 to 40 samples (as indicated).

of ln(100/50). The central region exhibits an overshoot–undershoot characteristic that cannot be predicted from the asymptotic theory.

3.9. THIRD MOMENT DIVERSITY

Third moment diversity attacks sidelobes in the two-dimensional image. It is applicable to radio camera design in which the distance dimension is obtained by range focusing the array rather than by propagation time delay. The effect of the third moment of the element-position distribution was derived in Section 2.7 in which it was shown that two-dimensional axial symmetry in the sidelobes (with respect to the focal point u_0, R_0) was obtained only when the third central moment of element position is zero. Otherwise a shearing or sloping quality develops in the diffraction pattern of the aperture. Figure 2.10 illustrates this experimental result.

Varying the third moment (by redistribution of the locations of the parts of the array) produces diffraction patterns with common mainlobes and resolution properties but independent sidelobes, as demonstrated in the simulation in Figure 3.10. Several 20-element, linear random arrays were simulated by random number generation and their two-dimensional diffraction patterns

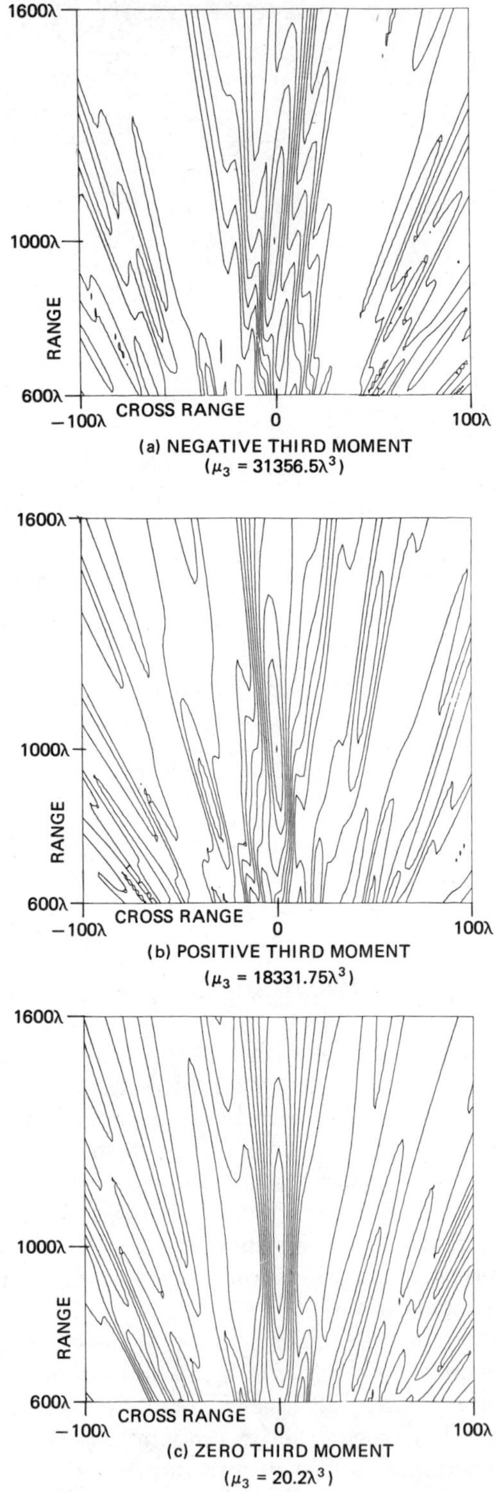

FIGURE 3.10. Sidelobe reduction by third-moment diversity.

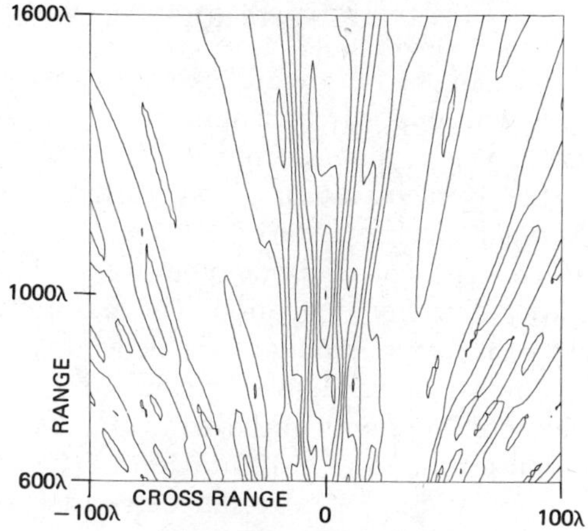

(d) SECOND ORDER DIVERSITY.
AVERAGE OF (a) AND (b)

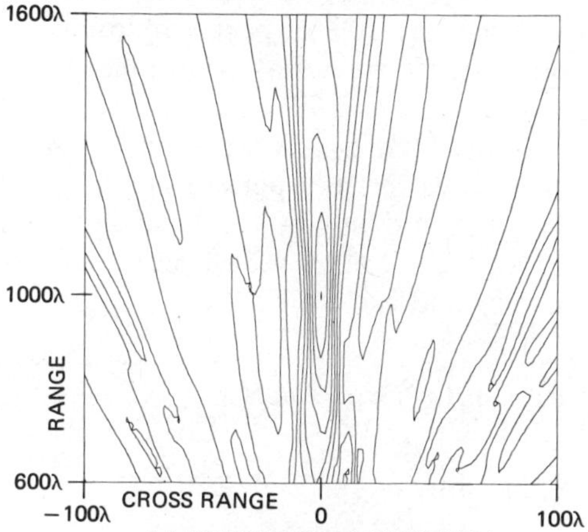

(e) FOURTH ORDER DIVERSITY.
AVERAGE OF (a), (b), (c),
AND PATTERN FOR $\mu_3 = 12259.15\lambda^3$.

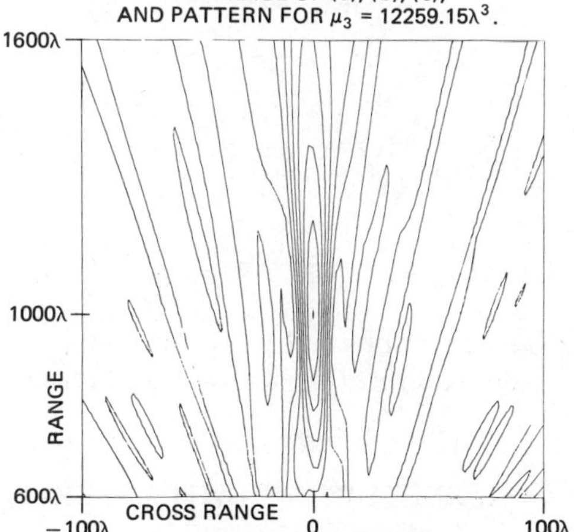

(f) EIGHT ORDER DIVERSITY.
AVERAGE OF (e) AND PATTERNS FOR
$\mu_3 = 6330.85\lambda^3, -3796.1\lambda^3$
$7090.4\lambda^3$, AND $-81617.75\lambda^3$

FIGURE 3.10. (*Continued*).

were calculated. Parts (a) and (b) in Figure 3.10 are two-dimensional contour patterns for two arrays with opposite first moments which show the expected opposing slopes. Part (c) is a pattern from an array with nearly zero first moment. Part (d) is the average pattern of two of the simulated arrays, (a) and (b). The improvement is evident. Part (e) shows fourth-order diversity to which the patterns of (a), (b), (c), and one other were added. Part (f) is the result of an eighth-order diversity simulation. The side radiation pattern has improved, evidenced by the larger amount of visible white area.

The lengths of the 20-element random arrays are 100λ. The point source is located normal to the center of gravity of the array at a distance of 1000λ, and the abscissa in each figure runs from −100 to 100λ. The ordinate runs from 600 to 1600λ. The presentations are in rectangular rather than polar coordinates, as used in Figure 2.10. The contours represent constant amplitudes of the two-dimensional radiation pattern of the array when it is focused on the source. The contour intervals are 10% of peak amplitude.

When the first moment of the array is displaced from the origin the coefficient of the coupling term in (2.99) increases from $2\mu_3$ to $2(\mu_3 + 2\mu_2 d)$ [see (2.102)], suggesting that the two-dimensional diffraction pattern varies with d. This is not true, however; the pattern is a physical property of the aperture excitation function and not of the array location, as is seen by a rectangular-coordinate analysis. Let x, y be a point in the near field of our array located in the x'-axis and excited by $i(x')$:

$$f(x, y) = \int i(x') \exp\left[-jk\sqrt{(x' - x)^2 + y^2}\right] dx'$$

When the array is displaced by d

$$f(x, y) = \int i(x' - d) \exp\left[-jk\sqrt{(x' - x)^2 + y^2}\right] dx'$$

Substituting $z = x' - d$

$$f(x, y) = \int i(z) \exp\left\{-jk\sqrt{[z - (x - d)]^2 + y^2}\right\} dz$$

$$= f[(x - d), y]$$

which is the original pattern displaced by the movement of the array. In other words the relation of the pattern to the array is unchanged.

3.10. AMPLITUDE WEIGHT DIVERSITY

When (3.5) is rewritten with real weight I_n, as in (3.6), it is evident that the amplitude of the weights is an additional variable that may be useful for diversity combining of images. Two arrays formed with two different sets of weights with the same statistics will have nearly identical mainlobes and identical sidelobe statistics (Section 2.4). The details of their side radiation patterns, however, may be expected to differ, in which case diversity combining might be used successfully in image formation. Although the patterns do differ, they do not become independent. The cross correlation between them remains high, however.

Let

$$f = \sum_{n}^{N} I_n e^{j\psi_n} \tag{3.57}$$

be the complex amplitude of the radiation pattern at an arbitrary angle away from the mainlobe. The $\psi_n = kx_n u$ are random variables that are assumed to be uniformly distributed in the interval from zero to 2π. Let two arrays be distinguished by subscripts 1 and 2 and let the weights be independently selected so that

$$EI_{n1}I_{n2} = EI_{n1}EI_{n2} = (EI)^2 = \bar{I}^2 \tag{3.58}$$

The normalized cross correlation between radiation patterns is

$$\rho_{12} = \frac{Ef_1 f_2^*}{\left[E|f_1|^2 E|f_2|^2\right]^{1/2}} \tag{3.59}$$

The numerator

$$Ef_1 f_2^* = E\sum_{n}^{N}\sum_{m}^{N} I_{n1}I_{m2}e^{j(\psi_n - \psi_m)} = E\sum_{n}^{N} I_{n1}I_{n2} = N\bar{I}^2 \tag{3.60}$$

The denominator

$$\left[E|f_1|^2 E|f_2|^2\right]^{1/2} = Eff^* = E\sum_{n}^{N}\sum_{m}^{N} I_n I_m e^{j(\psi_n - \psi_m)} = E\sum_{n}^{N} I_n^2 = N\overline{I^2}. \tag{3.61}$$

The ratio

$$\rho_{12} = \frac{\bar{I}^2}{\overline{I^2}} \tag{3.62}$$

is large for most distributions; for example, $\rho = \frac{3}{4}$ when the I_n are distributed uniformly from zero to some maximum value, $\pi/4$ when the pdf is Rayleigh,

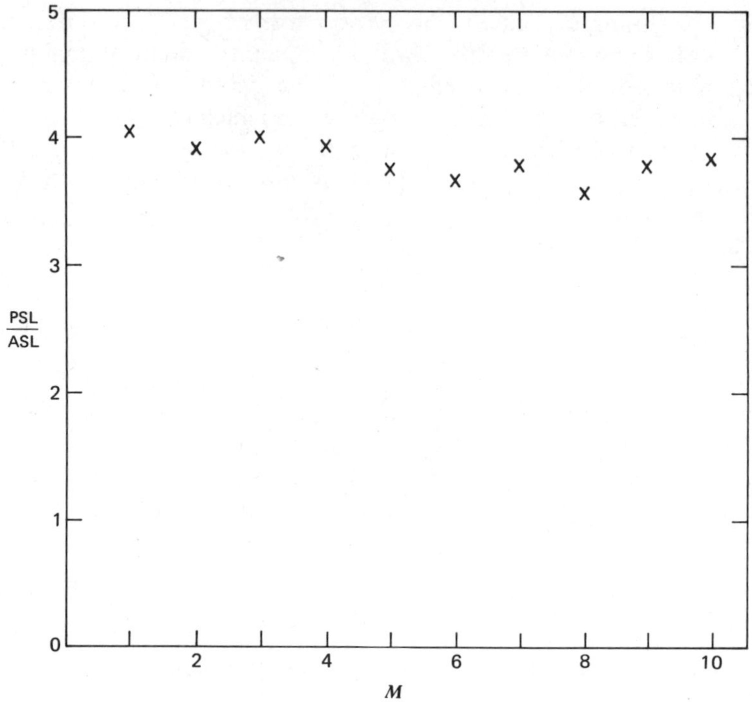

FIGURE 3.11. Noncoherent amplitude weight diversity. Peak-to-average sidelobe power ratio as a function of the order of diversity: $N = 20$.

and approximately 4/5 for a truncated Gaussian distribution with mean equal to twice the standard deviation.

Because (3.59) involves the products of complex radiation patterns, the correlation coefficient (3.62) corresponds to coherent amplitude weight diversity. The derivation is analogous to that which led to (3.34) for coherent frequency diversity. Coherent amplitude weight diversity is equivalent to adding or averaging sets of weight. Let

$$f_m(u) = \sum_n a_{nm} e^{jkx_n u} \tag{3.63}$$

be the mth radiation pattern formed from elements at x_n weighted by a_{nm}. The coherent combination of M complex patterns is

$$f_M(u) = \frac{1}{M}\sum_m^M f_m(u) = \frac{1}{M}\sum_m^M \sum_n^N a_{nm} e^{jkx_n u} = \sum_n^N b_n e^{jkx_n u} \tag{3.64}$$

where

$$b_n = \frac{1}{M}\sum_m^M a_{nm} \tag{3.65}$$

and

$$\lim_{M \to \infty} b_n = \bar{a} = \text{constant} \tag{3.66}$$

As observed in (3.66), the addition of many sets of weights from a common population drives the resulting weights toward a constant value; that is, a uniformly excited array. Thus high-order (large M) coherent amplitude weight diversity has sidelobe properties equivalent to that of a uniformly excited random array. It was shown earlier (2.21) that a uniform excitation results in an average sidelobe level about 1 to $1\frac{1}{2}$ dB smaller than an array with the nonuniform current excitation necessary for amplitude weight diversity. The conclusion is that coherent amplitude weight diversity can at best recover the rise in the sidelobe level initially induced by perturbing the amplitude weights.

Noncoherent amplitude weight diversity similarly introduces an initial degradation in the average sidelobe level of 1 to $1\frac{1}{2}$ dB. Simulation experiments (Figure 3.11) indicate, however, that this loss is not recoverable. The overall conclusion is that amplitude weight diversity offers no advantage in sidelobe suppression.

CHAPTER FOUR

Methods of Self-Cohering

4.1. FOUR TECHNIQUES

In Section 1.1 the expected nonrigidity or general distortion of a large imaging aperture was given as the basic reason for the need for self-cohering. A microwave array may be distorted for many reasons: it may be too large to be surveyed properly, it may suffer from windloading, its installation may be faulty, or its distortion may be electrical rather than geometric. Medium turbulence can cause the integral of the dielectric constant over a path from a source or a target to the array to vary randomly with position in the array. Multipath and scattering of the echoes can alter the amplitudes and phases of the signals received across the array. In addition, electromagnetic coupling from the antenna elements to the local environment may vary randomly with position in the array, thus causing random errors in the driving point impedances of the antenna elements. The tolerance on the random variation is about 30° rms, or somewhat less than one-tenth wavelength (Section 2.4).

Many different techniques have been devised for cohering a distorted aperture. Three procedures and a small group of ideas still in the research stage are discussed in this chapter. The rudiments of each system are an antenna array, a bank of phase shifters, a summer or adder, a high resolution display, and a control network. These components are shown in Figure 1.1; RF, IF, and video circuits, as well as timing circuits and power supplies, are implicit.

The primary physical hardware of the various types of system is identical. The major differences are in the measured quantities and signal processing. It is possible to organize a radio camera based on *more than one* method of self-cohering, provided that the signal processor has the necessary capability and versatility.

The methods operate on different cues, correct for partially different distortion-inducing phenomena, and have different drawbacks. Thus a signal processor with multiple phase-synchronization capability leads to an attractive radio camera design.

Time-of-Arrival Measurements

The first method described in this chapter surveys the locations of the antenna elements with beacons that radiate to the array. (Alternately, the sources can be large, passive targets that reflect strong echoes from a radar transmitter to the array.) This technique consists entirely of open-loop processes. Propagation time-delay measurements of the beacon signals to the array elements are made. The signal processor then calculates element positions. On the basis of the element position estimates it then calculates phase shifts for beamforming and scanning. Afterward it forms beams and scans as in a conventional phased array. This procedure corrects for geometric distortion in the array. Hence it permits imaging in any direction. It does not, however, correct for medium distortion that may be caused by turbulence or any other source of spatial or temporal variation in refractive index of the propagation medium between the target scene and the array. Its primary limitation is the coarseness of its estimation of element position influenced by receiver noise on the time-of-arrival measurements of the signals received from the beacons.

Adaptive Beamforming

Although the position estimates are crude, they are often adequate when a retrodirective, self-cohering, adaptive beamforming process follows. Adaptive beamforming, the second method, is the technique used to obtain the radio camera angle scans in Chapter 2 and in later chapters. A single beamforming source in the general direction of the target area to be imaged provides the phase information needed by the signal processor to cophase the signals across the array, thereby synchronizing the system and focusing a beam on the source. It does not calculate element positions; it cannot do so with phase measurements from a single source. This procedure corrects for geometric distortions in the array *in the direction of* the synchronizing source; it also corrects for refractive index perturbations across the array in the direction of the source. Because it self-corrects only in the direction of the beamformer, its capability is limited to a narrow field of view. Its primary utility is for imaging targets or scenes in the general direction of the beamforming source and may be considered analogous to a telephoto optical system.

The adaptive beamforming system is partly closed-loop and partly open-loop. The closed-loop portion of the process is the retrodirective beamforming procedure, which is feedback-controlled. Scanning for image formation, on the other hand, is open-loop, as in a conventional phased array.

Multiple Phase Measurements

The third method, like the first, uses several beacons, but phase measurements are made instead of time-delay measurements. By making phase measurements from several sources located in different directions the system is able to calculate element positions and scan over a wide field of view. Its primary limitation is a limited tolerance on element-position uncertainty caused by the multiple-of-2π ambiguity in the phase measurements of the signals from the beacons.

Image Feedback Control

Although the first three methods are reasonably well understood, the fourth approach is still in the early research stage. Techniques in this category are entirely closed-loop. In one technique being studied an initial image is formed in a conventional fashion or by some self-cohering procedure such as adaptive beamforming, following which an assessment is made of the quality of the image. Vernier control signals are transmitted to the phase shifters to alter their values to maximize the image quality. This approach is called image feedback control. Its drawback is the requirement that the array be reasonably well organized to begin with; otherwise the feedback-control process may converge incorrectly. This limitation, however, pertains only to the radiation from fixed, spatially coherent sources, such as reflections from man-made generators. It has been shown by Muller and Buffington that correct convergence always can be achieved in optical telescopy with stellar sources [63].

A second technique being studied is the application of the Muller-Buffington theorem to microwave radar in which partial spatial incoherence is introduced by natural motions of targets. Because this work is in the early research stage, only a brief introduction, including a derivation of the Muller-Buffington theorem which shows dependence on spatially incoherent sources, is given (Section 4.5).

A third technique being investigated uses known properties of the class of expected signals, their autocorrelations or their Fourier transforms [32], [67]. It also has been successfully applied to the optical distortion problem that results from atmospheric turbulence [26], [27]. It introduces considerable heuristics into an iterative Fourier transformation process. After each successive transformation portions of the image or its Fourier transform are retained and others are deleted. Retention or deletion is based on *a priori* scene information and the physics of the process; for example, nonnegativeness of image intensity.

Techniques Are Compatible

The four procedures are compatible and together lead to strong algorithms for organizing and scanning a radio camera. If well surveyed beacons are available, it is folly not to capitalize on them. The beacons need not be active; they can be strong echoers that respond to the radar transmission. Their use would permit the signal processor to acquire reasonably good estimates of antenna element positions. As discussed in Chapter 7, this is an important consideration in scanning the narrow beam after adaptive beamforming has been accomplished. Thus the element survey technique is a useful adjunct to retrodirective adaptive beamforming.

Because retrodirective adaptive beamforming accommodates medium distortion, as well as variations in the electromagnetic coupling between the antenna elements and the local environment, it constitutes the central process of the radio camera. It permits imaging of scenes in the general direction of the beamformer. The quality, however, may be degraded by multipath, by errors in estimation of antenna element locations and in measurement of phase, by system noise, and by other troublesome factors. Although these difficulties in no way degrade the phase-synchronization process, which is a closed-loop operation, they do affect the scanning, which is open-loop; that is, when these errors exist the gain in the mainlobe decreases as the beam is scanned away from the direction of the aiming point. This loss may be corrected in future designs if means are developed for following adaptive beamforming by a closed-loop process such as image feedback control. Use of image feedback control, on the other hand, is likely to require that the initial system organization be reasonably sound; hence these two methods are natural companions in a radio camera.

The four techniques are described separately in the following sections. Only an introductory portion is given to adaptive beamforming with a single beacon because Chapters 5 and 6 are devoted to the details of phase synchronization and scanning by this method.

4.2. ELEMENT POSITION SURVEY BY TIME-OF-ARRIVAL MEASUREMENTS

Figure 4.1 shows several beacons that radiate RF pulses to the array. The travel times of the pulses are noted from each beacon at each of the array elements, **1**. Because travel times and not RF phases are measured, the exact frequencies of the beacons are not important, nor is it necessary that all beacons radiate at the same frequency or the same time. Assuming free space propagation with constant velocity, travel times from a given beacon convey distance information from the array elements to that beacon. Specifically, differential travel times with respect to an arbitrary reference element in the array provide differential distance measurements from beacon to array elements. This information is sufficient to permit calculation of the geometry of the entire system (Section 4.4) **2**. Once the coordinates of the element locations are known the phase shifts required to form a beam focused at an arbitrary point are easily calculated, **3**. The phase-shift instructions are then delivered to the phase shifters to scan the beam and form an image, **4**.

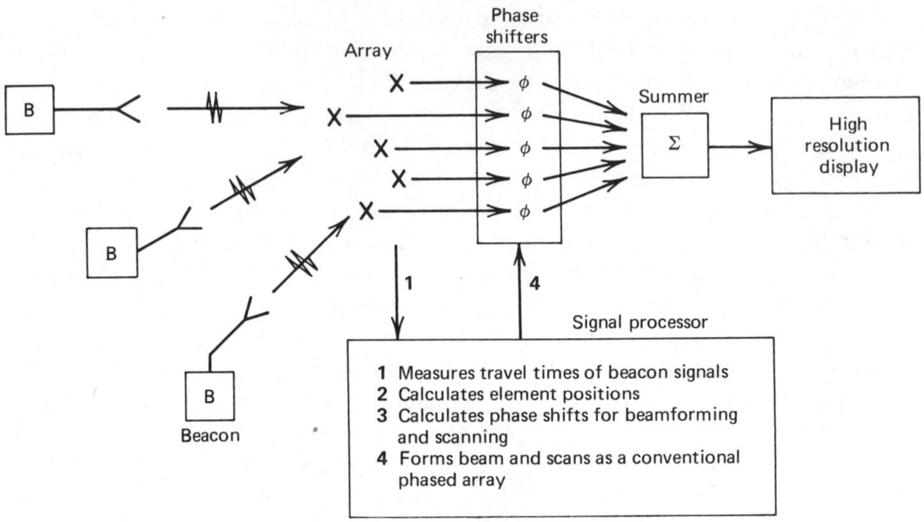

FIGURE 4.1. Element survey with beacons.

The procedure for focusing a beam after the element positions are surveyed is as follows: let the origin in Figure 4.2 be the location of an arbitrary element called the reference element. Let $\bar{r}_i$ be the known location of the ith element in the array. Let $\bar{R}_0$ and $\bar{R}_i$ be the vector distances from these elements to a field point P with coordinates R_0, θ, ϕ. The phase delay of a signal received at the origin from a source at P is $-kR_0$; similarly, the phase delay at the ith element is $-kR_i$. Hence the phase shift that must be added at the ith element to form a beam focused at P is

$$\Delta_i = k(R_i - R_0) \tag{4.1}$$

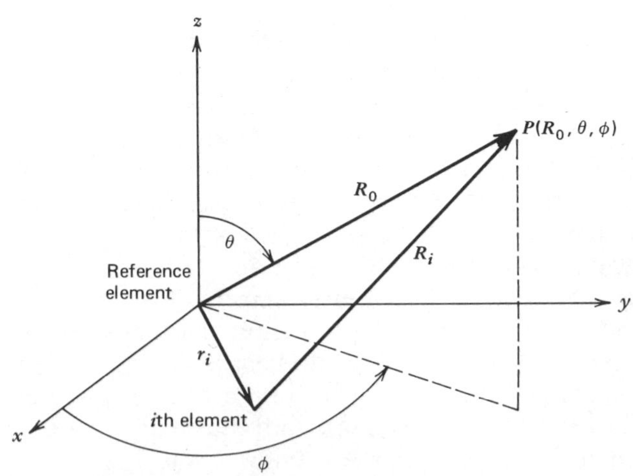

FIGURE 4.2. Array geometry and field point.

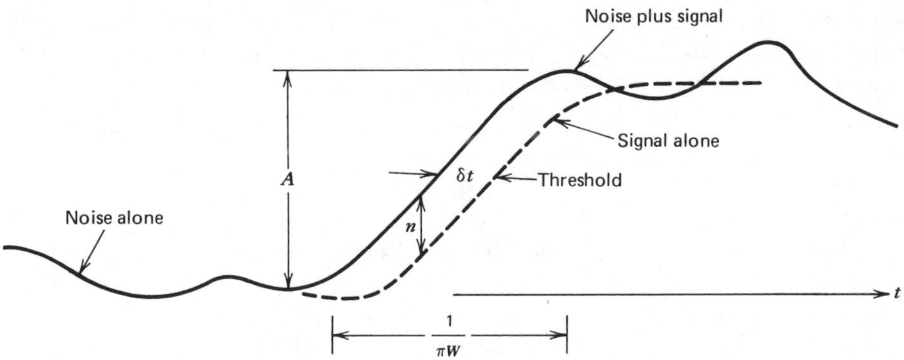

FIGURE 4.3. The accuracy of arrival time measurement is diminished by noise.

R_i may be calculated from the geometry of Figure 4.2:

$$R_i = \left(R_0^2 + r_i^2 - 2\bar{r}_i \cdot \bar{R}_0\right)^{1/2} \tag{4.2}$$

where

$$\frac{\bar{r}_i \cdot \bar{R}_0}{R_0} = x_i \sin\theta\cos\phi + y_i \sin\theta\sin\phi + z_i \cos\theta \tag{4.3}$$

This method of beamforming may be called an element survey technique. First the geometry of the system is calculated and then phase shifts are computed. It solves the entire problem of beamforming and scanning, provided that the measurement accuracy of the differential time delays is satisfactory. Unfortunately this is the case only at long wavelengths.

Consider a radio camera array operating as part of a pulsed radar with wavelength λ and IF bandwidth W Hz. The rise time of the receiver is about $1/\pi W$ s. The limiting accuracy of measurement of the time of arrival of a pulse is smaller than the rise time by about the voltage signal-to-noise ratio $(\sqrt{\mathrm{SNR}})$ of the signal. Shown in Figure 4.3 is the envelope of a bandlimited signal of amplitude A displaced upward by noise of amplitude n. The threshold crossing time is advanced by δt, given by the relation

$$\frac{\delta t}{n} = \frac{1/\pi W}{A} \tag{4.4}$$

The rms timing error is

$$\sigma_t = \frac{\sigma_n}{\pi W A} = \frac{1}{\pi W \sqrt{\mathrm{SNR}}} \tag{4.5}$$

where σ_n is the rms noise and σ_n/A is the voltage signal-to-noise ratio.

Time-difference measurements involve two independent timing errors, thereby increasing (4.5) by $\sqrt{2}$, and multiplication by the velocity of light c

converts it to differential distance units. The estimated coordinates of $\bar{r}_i$, calculated from the differential distances, are in error by no less (and usually somewhat more) than the errors in the differential distances. Thus the minimum rms position error is

$$\sigma_r \approx \frac{c}{W\sqrt{\text{SNR}}} \tag{4.6}$$

As an example, let the radar be an X-band system ($\lambda = 3$ cm) with a bandwidth of 10 MHz (for a range resolution of 15 m). Let the SNR of the beacon signal received in the array be 30 dB. Then $\sigma_r \approx 30\lambda$, which corresponds to an rms phase error of almost 200 rad. The low accuracy measurements are then used for the calculation of position locations, the accuracies of which can be no better. Yet it was shown earlier (Figure 2.5) that the rms phase error may not exceed $\frac{1}{2}$ rad without intolerable loss in array gain. Only an unrealistically high SNR would solve the problem. A 40-dB SNR on the beacon signal would reduce σ_r to about 10λ, which is still more than 100 times beyond tolerance. Greater than 80 dB would be required to achieve the desired result!

Thus envelope time-of-arrival measurements are not satisfactory for self-organization of the system. However, when an adaptive beamforming process is a companion technique within the system the relatively crude position estimates may be adequate for it to perform the remainder of the task. This indeed proves to be the case.

4.3. ADAPTIVE BEAMFORMING

Only One Beacon Required

Adaptive beamforming, as described in this section and in Chapters 5 and 6, makes use of less information about the array than a survey procedure. In a survey the system takes sufficient measurements to calculate all element positions. Once the element coordinates are known phase shifts can be calculated to form a beam in an arbitrary direction. Open-loop beamforming follows the survey. Adaptive beamforming, on the other hand, takes only enough measurements to compensate for array distortion in the direction of the phase-synchronizing source. A survey requires a minimum of three, and preferably four, beacons or prominent reflectors suitably distributed throughout the sector to be scanned; adaptive beamforming uses only one source and it must be in the general direction of the target to be imaged. Adaptive beamforming *does not* require knowledge of element locations to form a beam; it does, however, require some knowledge of the element positions to scan the beam. The envelope time-of-arrival survey method of Section 4.2 can be helpful in this regard. The tolerance theory of adaptive beamforming is described in Chapter 7.

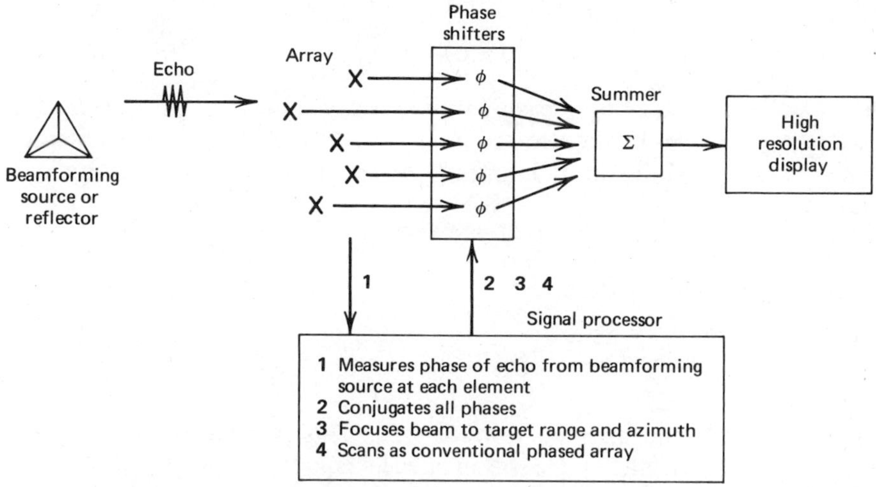

FIGURE 4.4. Adaptive beamforming with single source or reflector.

Figure 4.4 shows the functions performed by the signal processor in an adaptive beamforming system. Adaptive beamforming is a closed-loop process that serves to phase-synchronize the distorted and, possibly, time-varying array. An external signal is needed to provide a phase reference to synchronize the adaptive circuits. A signal pulse radiated from a single active or passive source reaches the array. A passive source is illustrated. Implied, therefore, is a microwave transmitter that illuminates the reflector. Phase measurements are made at each element **1**, in relation to a stable oscillator in the system or to the source signal received from one of the antenna elements in the array. All signals are rotated in phase by the negative of the measured phases, thereby cophasing them all, **2**. This process is called phase conjugation. The characteristics of good synchronizing sources are discussed in Chapter 6.

Phase Conjugation

Phase conjugation of the beacon signals at the array results in the formation of a beam pointing in the direction of the phase-synchronizing source. An array that forms and points a beam at a source is called retrodirective. Any method of retrodirective beam formation that the equipment can perform is acceptable [39], [47], [94].

After forming a beam focused on the synchronizing source the processor refocuses the beam to the target range and azimuth, **3**. It then scans the beam across the target to obtain its image, **4**.

It is easily shown that phase conjugation results in retrodirectivity. Let the array be in the x-axis. A source at a great distance located at angle $\theta_0 = \sin^{-1} u_0$ with respect to the normal to the x-axis produces a complex field proportional to

$$e(x) = e^{jkxu_0} \tag{4.7}$$

The phase of the radiation measured in an array coincident with the x-axis is $\phi_R(x;\, u_0) = kxu_0$ in relation to the phase at the origin. If a linear antenna is excited by this field distribution and its current distribution is given by

$$i_R(x) = e^{j\phi_R} \equiv e^{jkxu_0} \tag{4.8}$$

its far-field radiation pattern

$$f(u) = \mathcal{F}\{i_R(x)\} = \int e^{jkx(u+u_0)}\, dx \tag{4.9}$$

has its primary lobe directed toward $-u_0$. This is Snell's law of specular reflection in which the angle of incidence equals the angle of reflection. Similarly, a receiving array phase-weighted by kxu_0 forms a beam in the direction $-u_0$.

It is evident that if a receiving antenna coincident with the x-axis is to form a beam toward the source in the direction $+u_0$ its aperture excitation must be

$$i_{R^*}(x) = i_0(x) e^{j\phi_{R^*}(x;\, u_0)} \tag{4.10}$$

where $i_0(x)$ is its unphase-shifted aperture weighting function and $\phi_{R^*}(x;\, u_0) = -kxu_0$, for then

$$f(u) = \mathcal{F}\{i_{R^*}(x)\} = \int i_0(x) e^{jkx(u-u_0)} = f_0(u - u_0) \tag{4.11}$$

has its primary lobe directed toward $+u_0$. This is the condition for retrodirectivity. The required relation between the phase variation ϕ_R of the incident radiation and the phase variation ϕ_{R^*} of the antenna excitation is

$$\phi_{R^*}(x;\, u_0) = -\phi_R(x;\, u_0) \tag{4.12}$$

Phase reversal or complementing is called phase conjugation.†

Bulk Delays

Not shown in Figure 4.4 is a set of coarse time delays, called bulk delays, in series with the phase detectors. These delays sometimes are needed when the differential arrival times of the echoes from the reflector exceed a small fraction of the pulse duration of the beamformer signal. The function of the time delays is to cause the echoes to overlap in time in the receiver and are needed when

† The term phase conjugation is somewhat of a misnomer for it is not the *phase* that is conjugated but the *aperture excitation*. Thus $i_{R^*}(x) \propto e^*(x)$.

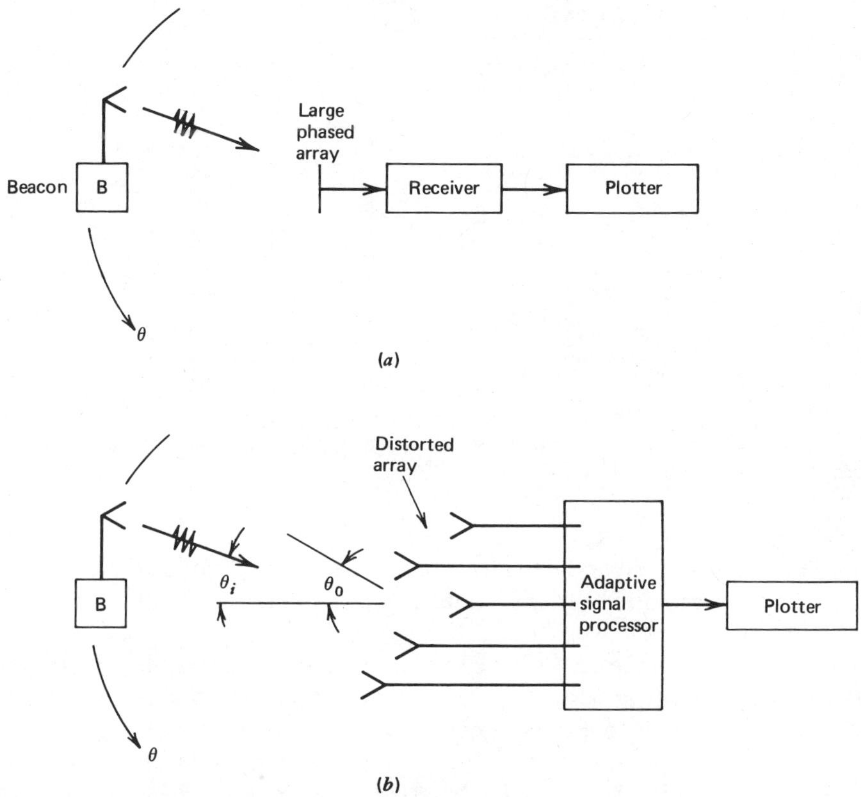

FIGURE 4.5. Measuring the radiation pattern of a large array: (a) conventional; (b) self-cohering.

phase measurements are made in phase detectors that beat two echoes together to extract the phase difference, in which case the two RF or IF pulses must coexist almost simultaneously. If, however, the phase measurements are made with respect to a continuous-wave reference oscillation, no bulk delays are required. Because RF and IF time delays are difficult components to design, systems that avoid their use are generally preferable.

Phase conjugation can be performed by hardware circuits like phaselock loops or by numerical operations in the digital signal processor. Both means are discussed in Chapter 5. The digital processor is used in the discussion that follows.

Pattern Measurement

Figure 4.5 shows a common procedure for measuring the antenna pattern of a large receiving array (a). The array takes a signal from a source, shown as a beacon, that radiates with constant amplitude as its moves around the horizon.

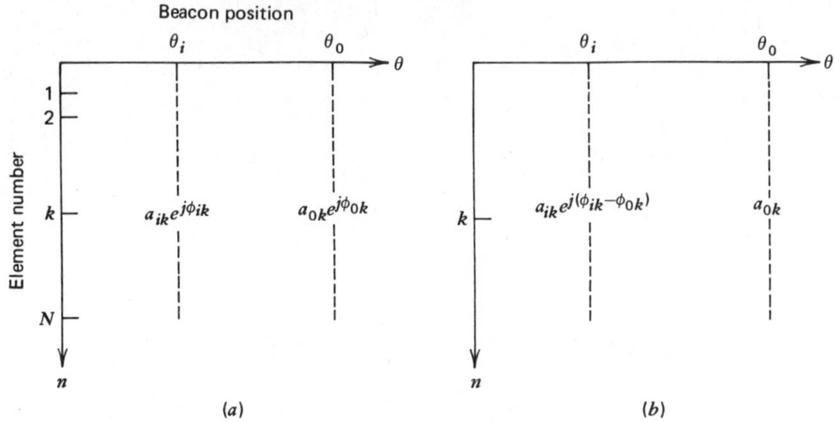

FIGURE 4.6. Beamforming procedure: (*a*) initial data organization in signal processor; (*b*) after phase rotation for beamforming at $\theta = \theta_0$.

The signal is amplified and detected in the receiver. The output drives a plotter to display the radiation pattern. Both *x-y* (rectilinear) and polar plotters are used.

The process is more complicated when the receiving array is distorted and an adaptive signal processor operates on the incoming waves (*b*). At each beacon position θ_i the received signal $a_{ik}\exp(j\phi_{ik})$ received by the kth antenna element is measured and stored. Figure 4.6 shows the data organization and the beamforming procedure. The abscissa is beacon angle θ in relation to the normal to the array. The ordinate is the element number, which is monotonic with position in the array.

The Beamforming Direction

Unlike pattern measurement practice for the conventional array, a beamforming direction θ_0 is first identified: θ_0 is the direction in which a retrodirective beam is formed by the distorted adaptive system. The radiation pattern has the properties shown in Figure 1.11. After beamforming in this direction the beam may be scanned in other directions.

To form a beam in the direction θ_0 the signal processor must cophase all signals received from the source when it is located at that direction. The cophased signals then may be summed and the output is the response of the cohered array to the beacon. The complex numbers in the $\theta = \theta_0$ column of Figure 4.6*a* are $a_{0k}\exp(j\phi_{0k})$. Beam formation at $\theta = \theta_0$ requires phase shifting the signals received at the kth element by $-\phi_{0k}$. When this phase rotation is applied to the data matrix of Figure 4.6*a* Figure 4.6*b* results. All signals from the direction $\theta = \theta_0$ become cophased and may be added. This is the same function performed by the phase shifters in a conventional phased array.

The logic behind this phase rotation is that phase differences between signals received by different elements reflect the distortions in the array. (They also reflect the other sources of distortion mentioned in the opening paragraph of this chapter.) Ideally, the phase adjustments for beamforming at θ_0 would be correct for other directions. This assumption is valid only in and near the θ_0 direction. With a single beacon perfect correction is possible in one direction only.

It is easily demonstrated that the correction is proper at $\theta = \theta_0$, even when measurement errors are present. Let ϕ_{ik} be the "geometric" phase; that is,

$$\phi_{ik} = \frac{2\pi}{\lambda} d_{ik} \qquad \text{mod } 2\pi \tag{4.13}$$

where d_{ik} is the distance between the beacon at angle θ_i and the kth element. Let the zero mean random variable $\delta\phi_{ik}$ be the error in the measured phase $\hat{\phi}_{ik}$, where the error is due to noise, element displacement, multipath, and other phenomena. The measured phase is $\hat{\phi}_{ik} = \phi_{ik} + \delta\phi_{ik}$ and the measured amplitude is assumed to be in error by the zero mean random variable δa_{ik}: $\hat{a}_{ik} = a_{ik} + \delta a_{ik}$. Phase rotation for beamforming at θ_0 is accomplished, as before, by removing the measured phase in the θ_0 column from each element. Thus the received signal $\hat{a}_{ik}\exp(j\hat{\phi}_{ik})$ is phase corrected to $\hat{a}_{ik}\exp[j(\hat{\phi}_{ik} - \hat{\phi}_{0k})]$. The sum

$$f(\theta_i) = \sum \hat{a}_{ik}\exp[j(\hat{\phi}_{ik} - \hat{\phi}_{0k})] \tag{4.14}$$

is the complex gain of the array in the direction θ_i. Of course, at $\theta = \theta_0$ the phase-rotated signals are all cophased and (4.14) becomes

$$\begin{aligned} f(\theta_0) &= \sum \hat{a}_{ik}\exp[j(\hat{\phi}_{0k} - \hat{\phi}_{0k})] \\ &= \sum \hat{a}_{ik} = \sum a_{ik} + \sum \delta a_{ik} \end{aligned} \tag{4.15}$$

the expected value of which

$$Ef(\theta_0) = \sum a_{ik} \tag{4.16}$$

If there had been no errors in the received signals the process would have resulted in the design pattern (subscript 0)

$$f_0(\theta_i) = \sum a_{ik}\exp[j(\phi_{ik} - \phi_{0k})] \tag{4.17}$$

At the beamforming angle

$$f_0(\theta_0) = \sum a_{ik} \tag{4.18}$$

which is the same as (4.16). Thus there is no loss in average gain at the beamforming angle, even in the presence of phase errors in the beamforming signals. The beamforming process is error-free because the phase which drives the adaptive process includes the initial phase error. The correction can be made with analog or digital phaselock loops operating in each of the antenna element channels or it can be made in a digital signal processor in which the data are contained in the format shown in Figure 4.6. Both approaches are discussed later.

Unlike the adaptive beamforming process, the scanning process is not adaptive and not error-free. Losses do develop when the radio camera is scanned away from θ_0. One cause of loss in gain is poor estimation of element position. The element survey technique in Section 4.2, even though crude, is useful; for example, it is shown later that the 10-λ rms coordinate error (for the 40-dB SNR example given at the end of Section 4.2) would be acceptable for imaging with adaptive beamforming over a field of view of about 1°. Although seemingly a small value, 1° is huge compared with the angle subtended by most radar targets. A large aircraft as close as 10 km from an airport (see Figure 1.2) subtends only about $\frac{1}{3}$ this angle. Thus the two systems together form a useful instrument for the airport application.

In addition, losses develop from many other causes. These matters are explored in later chapters.

Although 1° is large compared with target size, it is small compared with the normal field of view or angular sector scanned by most radars. Adaptive beamforming is excellent when only a limited field of view is needed, similar in function to the high-magnification, long-focal-length telephoto lens used in photography. Wider sector scanning requires other techniques like the one that follows.

4.4. MULTIPLE BEACON PHASE-MEASUREMENT SYSTEM

The limited field of view of the adaptive beamforming system (Section 4.3) can be broadened if more beacons are added. When this is done the system format looks like the one in Figure 4.1. In fact, with a single exception, the functions performed by the element-survey processor shown in that figure are identical to those required for a multibeacon phase-measurement system. The difference is in the function labeled **1**: instead of measuring arrival times of the beacon signals, the processor measures their phases. It is evident that the same system can be designed to measure both time of arrival and phase and thereby obtain data simultaneously for both processes.

Phase Is More Accurate

The advantage offered by phase measurement is that it can be measured much more accurately than arrival time. The phase-measurement error of a signal of amplitude A in the presence of noise with complex amplitude $n\exp(j\alpha)$ is

$$\delta\phi = \tan^{-1}\frac{n\sin\alpha}{A + n\cos\alpha} \simeq \frac{n\sin\alpha}{A} \qquad n \ll A \tag{4.19}$$

The rms value of (4.19), assuming that α is uniformly distributed in $[0, 2\pi]$ and is independent of n, is

$$\sigma_\phi \simeq \frac{\sigma_n}{\sqrt{2}\,A} = \frac{1}{\sqrt{2}\sqrt{\mathrm{SNR}}} \tag{4.20}$$

The equivalent timing error is (4.20) times the period of the wave T_0 divided by 2π:

$$\sigma_t = \frac{1}{2\pi\sqrt{2}\,f_0\sqrt{\mathrm{SNR}}} \tag{4.21}$$

where $f_0 = T_0^{-1}$ is the frequency of the signal. By comparing (4.21) with (4.5) it can be seen that the phase measurement is more accurate than the time of arrival measurement by the order of f_0/W, which is the Q of the signal waveform. This ratio in radar is typically the order of 10^3. Thus the phase measurement is a few orders of magnitude more accurate than the timing measurement.

The penalty paid for this large gain in accuracy is uncertainty or ambiguity about which RF cycle the measurement pertains to. Phase is measured modulo 2π, which implies that the corresponding distance inferred from the phase is measured modulo λ.

Wide Field of View

Although the format of the phase measurement system is the same as the differential time-delay processor, conceptually the operation is similar to adaptive beamforming. Moreover, because there are beacons in several directions, the phase-measurement system can calculate element positions and therefore scan over a wider field of view. On the other hand, because of the limitations imposed by the modulo 2π nature of the phase measurements, it requires more accurate estimates of element coordinates than adaptive beamforming.

The algorithm assumes straight-line propagation at constant speed. It further assumes that propagation is free of multipath and scattering, that there is no noise in the phase measurements, nor is any other phenomenon present that could distort the phasefront across the array or cause phase-measurement error. Errors in these assumptions lead to errors in the calculated element positions. The beacons may be active sources or passive reflectors; the physical source of the beacon radiation is unimportant. A two-dimensional geometry is

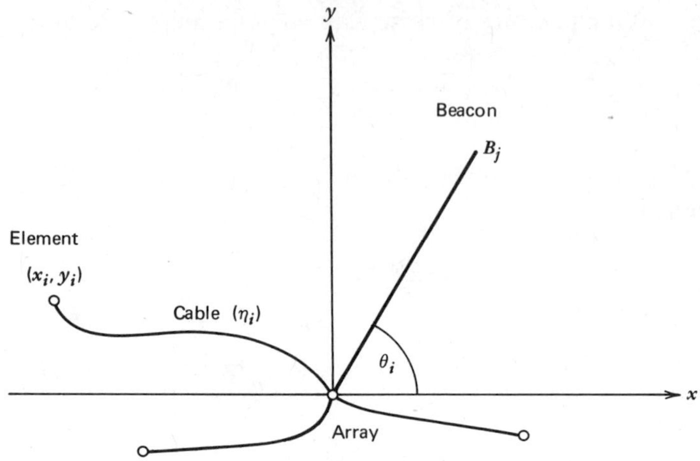

FIGURE 4.7. Array and beacon geometry.

shown in Figure 4.7, although the self-survey concept can also be applied to three-dimensional geometry. The beacons are located in the far field of the array at azimuth angles θ_j; the subscript designates the beacon. (The azimuth angle for this discussion is measured from the array axis.) The elements are located at x_i, y_i. The beacons must radiate at the same frequency and the radiation frequency must be the same as that of the radar transmitter of the radio camera. Frequency synchronization of the entire system permits meaningful phase measurements to be made at the array.

If the beacons are passive reflectors of the radar transmission, frequency synchronization is automatically accomplished. Separation of the reflected signals in the receivers is achieved on the basis of time of arrival of the echoes following the transmission of the radar pulse, provided that the distances of the reflectors to the array differ by more than the pulse duration of the radar.

If the beacons are active, their signals may be transmitted successively and their phases measured at each of the array elements. In some cases they may be operated simultaneously, in which case it is necessary to code their waveforms to distinguish their signals in the receiver.

The Phase Measurement

Let the measured phases be ψ_{ij}; the first subscript designates an antenna element, the second, a beacon. It is not necessary that the beacon coordinates be known accurately. Each phase measurement is highly ambiguous, for it is proportional to a distance modulo 2π; that is,

$$\psi_{ij} = \frac{2\pi d_{ij}}{\lambda} (\mathrm{mod}\, 2\pi) \tag{4.22}$$

Except for this ambiguity, a set of such measurements permits the solution of a sufficiently large number of simultaneous equations to deduce the coordinates of each antenna element. The problem is simplified somewhat because the beacon locations can always be known to conventional radar accuracies, for the radar system associated with the radio camera can measure range and bearing. Furthermore, only phase *difference* measurements are made (in relation to the phase of the signal received at some reference element in the array) and the wavelength (or multiples of 2π) uncertainty is usually small. The radar measurements, however, are insufficiently accurate to perform the survey of the system to the required tolerance of approximately $\lambda/10$. In addition, the difficulty is compounded because in each RF section of each array module some unknown phase shift biases the phase measurement in the array module.

The analysis that follows is taken from the work of Dorny [23]. Figure 4.8 illustrates a large array that, because of its large size, is not rigid and may be time-varying. One element in the array acts as a phase reference for the signals received from the beacons. A signal received at the reference element is transmitted by cable to each of the other elements of the array so that a phase comparison can be made. The phase difference between the signals received at the ith and reference elements, due to the signal radiated from the jth beacon, is ψ_{ij}. Except for the ambiguity caused by the mod 2π nature of the phase measurement process, the ψ_{ij} measurements are indicators of the differential beacon-to-element time delays. These delays are used to compute estimates of the element positions x_i, y_i, in relation to the reference element, and the phase delay η_i in the cable connecting the ith and reference elements; η_i also includes all differential phase shifts in the antennas and receiver circuits of the two elements.

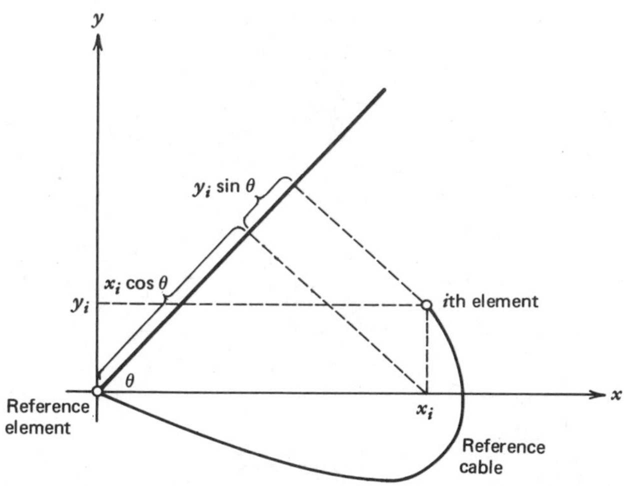

FIGURE 4.8. Differential phase delay. (From C. Nelson Dorny, *IEEE Trans. Antennas and Propagation*, AP-26, November 1978.)

These computed values of x_i, y_i, η_i are then used to cohere the array; that is, phase shifts $\phi_i(\theta)$ are calculated for the array elements and when applied as element weights result in a beam formed in the direction θ and focused at infinity. The self-survey procedure is repeated as often as necessary to track the motions of the flexing array.

Let the origin of an x, y coordinate system be placed at the reference element in Figure 4.8. The phase ψ_{ij} of the beacon signal received at element i in relation to the corresponding signal received at the reference element and delivered by cable to element i is

$$\psi_{ij} = k(x_i \cos\theta_j + y_i \sin\theta_j) + \eta_i, \qquad k = \frac{2\pi}{\lambda} \tag{4.23}$$

where θ_j is the angle to the jth beacon measured from the x-axis. Because of errors, however, the phase of the measured signal differs from (4.23). Call the measured phase $\hat{\psi}_{ij}$.

The geometry is shown in Figure 4.8. The beacons are assumed to be located in the far field of the array. The phase difference equations (4.23) are solved in the least-square-error sense to obtain the element coordinates and the phase delays in the cables and RF circuits [21].

Following Dorny [23], let the coordinates and phase delay for the ith element be represented by the matrix

$$\mathbf{X}_i = \begin{pmatrix} x_i \\ y_i \\ \eta_i/k \end{pmatrix} \tag{4.24}$$

Define the beacon-angle matrix

$$\mathbf{B} = \begin{pmatrix} \cos\theta_1 & \sin\theta_1 & 1 \\ \vdots & & \\ \cos\theta_J & \sin\theta_J & 1 \end{pmatrix} \tag{4.25}$$

and the phase-measurement matrix

$$\mathbf{M}_i = \begin{pmatrix} \hat{\psi}_{i1} \\ \vdots \\ \hat{\psi}_{iJ} \end{pmatrix} \tag{4.26}$$

The matrix equation

$$k\mathbf{B}\mathbf{X}_i = \mathbf{M}_i \tag{4.27}$$

expresses the relations in (4.23). When $J = 3$ the solution is given by $k\mathbf{X}_i =$

$\mathbf{B}^{-1}\mathbf{M}_\mathrm{i}$. In general, J will be larger than 3. Among the various possible solutions the one wanted minimizes

$$\epsilon_i = \Sigma_j\left(\psi_{ij} - \hat{\psi}_{ij}\right)^2 \tag{4.28}$$

The solution is

$$\mathbf{X}_\mathrm{i} = \frac{1}{k}(\mathbf{B}^T\mathbf{B})^{-1}\mathbf{B}^T\mathbf{M}_\mathrm{i} \tag{4.29}$$

The solution $\mathbf{X}_\mathrm{i}$ thus obtained presents a choice of x_i, y_i, and η_i that permits the best focusing of the array in the directions of the beacons. The same linear solution (4.29) can be extended to the three-dimensional array when the beacons are in the far field. The process can also be extended to apply to near-field beacons, in which case, however, (4.29) is replaced by a set of nonlinear equations that can be solved iteratively by Newton's method [21].

Minimum Error Criterion

Because the measured quantities represent differential distances modulo wavelength, errors in initial estimation of coordinates can result in false solutions to (4.28); that is, (4.29) yields a unique solution for each choice of numbers of wavelengths. The solution to the modulo wavelength (or mod 2π) problem is obtained by computing (4.29) for each combination of the likely values of the integers that represent integral multiples of wavelength displacements and then selecting as the correct solution the one that produces the smallest squared error (4.28).

Figure 4.9 illustrates the process of selecting the possible values of these integers. The reference element is shown at the origin of the coordinate system and the ith element has uncertain coordinates represented by a circle of diameter equal to wavelength. Given the beacon direction shown, the differential distance estimate based on phase measurement is ambiguous; it is either 2λ or 3λ greater than $\hat{\psi}_{ij}/k$. In addition, the phase delay in the ith cable and RF circuits can increase the measured differential phase by any value between zero and 2π, thereby possibly causing as much as a unit increase in the values of the integer. Thus the possible values shown in Figure 4.9 are 2λ, 3λ, and 4λ. Solutions must be calculated for all three possible values and in all combinations of the possible values for the other elements. Then the best solution is chosen as the one with the lowest mean-squared error.

Clearly, the amount of computation rises rapidly with the initial uncertainty in element position. The coarse survey method in Section 4.2 has questionable value for this algorithm. In the example given in that section the position uncertainty was 10λ when the beacon SNR was 40 dB; it would require more than 65-dB SNR to reduce the uncertainty in the element coordinates to that assumed for Figure 4.9. This is an unrealistic value on which to rely. Therefore

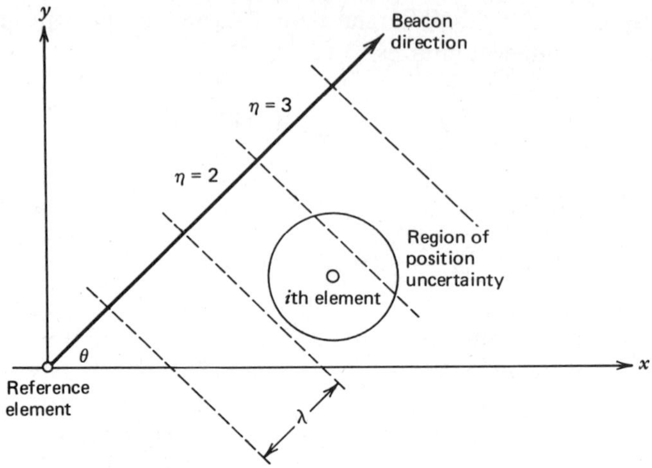

FIGURE 4.9. Determination of ambiguity number. (From C. Nelson Dorny, *IEEE Trans. Antennas and Propagation*, AP-26, November 1978.)

other methods are needed to reduce the initial multiple-of-wavelength ambiguity. Direct measurement of element placement is one obvious choice. Wideband signals or multiple frequencies are helpful.

Beamforming

Calculation of the element coordinates is not the overall objective of the procedure, however; the overall objective is to form a beam in an arbitrary direction notwithstanding the *a priori* lack of knowledge of element positions. Thus phase shifts

$$\phi_i(\theta) = -k\mathbf{R}\mathbf{X}_i \tag{4.30}$$

are calculated, where

$$\mathbf{R} = (\cos\theta \sin\theta) \tag{4.31}$$

Two Types of Errors

Now let two kinds of error be introduced. The first is an error in the estimation of the beacon directions. The second is an error in the measured phases due to multipath and scattering of the beacon signals as they propagate to the array and to noise in the receivers. Call the first error $\delta\theta_j$ and the second, $\delta\psi_{ij}$. Let their variances be σ_θ^2 and σ_ψ^2, respectively. Then the measured value ψ_{ij} changes from (4.23) to

$$\hat{\psi}_{ij} = k\left[x_i\cos(\theta_j + \delta\theta_j) + y_i\sin(\theta_j + \delta\theta_j)\right] + \eta_i + \delta\psi_{ij} \tag{4.32}$$

The measurement errors lead to others in the calculated element positions (4.29) and result in errors in the calculated phase shifts (4.30). These errors introduce two important effects: loss in mainlobe gain and mispointing of the beam. The former results from errors in the measured phases that normally are random and independent from element to element: the phase-measurement error $\delta\psi$ tends toward this type. Beam-pointing errors result from correlated phase errors across the array: beacon-angle errors $\delta\theta$ produce phase errors with this tendency.

It is shown in [23] that the error in (4.30), due to independent phase-measurement errors and correlated beacon-angle errors, is zero mean. Let σ_ψ^2 be the variance in the phase-measurement error. The error variance in the computed phase shifts (4.30) due to σ_ψ^2 is

$$\sigma_\phi^2 = \sigma_\psi^2 \mathbf{R}(\mathbf{B}^T\mathbf{B})^{-1}\mathbf{R}^T \tag{4.33}$$

Its effect is a loss of mainlobe gain given by (2.15) or (2.16).

The portion of the phase error due to beacon-angle errors turns out to be proportional to the distance of an element from the origin. If the array is linear, beacon-angle errors will result in a pure tilt of phase across the array and consequently induce a pure pointing error on beam formation. A flexing array that is approximately linear should also exhibit essentially pure pointing error due to beacon-angle errors. The mean beam-pointing error is zero and the standard deviation is

$$\sigma_u \approx \sigma_\theta \left[\mathbf{R}(\mathbf{B}^T\mathbf{B})^{-1}\mathbf{R}^T\right]^{1/2} \tag{4.34}$$

The multiplying factor in (4.33) and (4.34) depends on the beacon geometry (via **B**) and the beam-pointing angle (via **R**). It is evident from the definition of **B** in (4.25) that $\mathbf{B}^T\mathbf{B}$ is proportional to the number of beacons J; therefore the multiplier term $\mathbf{R}(\mathbf{B}^T\mathbf{B})^{-1}\mathbf{R}$ is inversely proportional to that number. The proportionality constant can be estimated by noting that the **R** and **B** matrices consist of the same kind of terms and the same range of values. Hence the proportionality constant is the order of unity and the multiplier is expected to be the order of J^{-1}. A computer simulation experiment reported in [23] verified this supposition [22]. The results of the simulation are reproduced in Figure 4.10 for $J = 4$. The worst case resulted from grouping the beacons near the edges of a wide field of view[see Figure 4.10(c)]. The remaining evidence suggests that with less clustering of the beacons the multiplier can be held to about $4/J$. Thus the expected loss in mainlobe gain [from (2.16)] and the rms beam-pointing error reduce to

$$\Delta G(\text{dB}) \approx \frac{18\sigma_\psi^2}{J} \tag{4.35}$$

$$\sigma_u \approx \frac{2\sigma_\theta}{J^{1/2}} \tag{4.36}$$

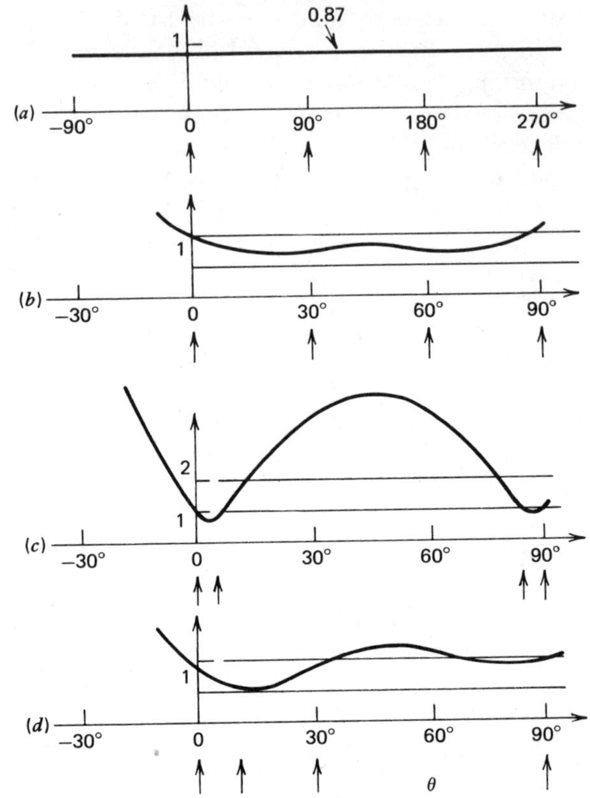

FIGURE 4.10. Error magnification factor $(\mathbf{R}(\mathbf{B}^T\mathbf{B})^{-1}\mathbf{R}^T)^{1/2}$ versus beam-pointing angle: $J = 4$. Vertical arrows show far-field beacon positions. (From C. Nelson Dorny, "Self-Cohering by Self-Surveying," QPR No. 22, Valley Forge Research Center, Moore School of Electrical Engineering, University of Pennsylvania, August 1977.)

4.5. IMAGE FEEDBACK CONTROL

The element survey technique is an entirely open-loop process. Adaptive beamforming is partly closed-loop (beamforming) and partly open-loop (scanning). The question is whether it is possible in principle to develop an entirely closed-loop approach in which the control signals that drive the processor would be developed entirely from the image. This process is called image feedback control. It may be possible to dispense entirely with beacons or passive beamforming sources.

The concept of image feedback control, placed in the context of earlier figures in this chapter, is shown in Figure 4.11. The array is driven by echoes from the target or from the scene to be imaged. As in the other systems, the received signals pass through a bank of phase shifters, after which the signals are summed and delivered to the high resolution display **1**. The phase shifters

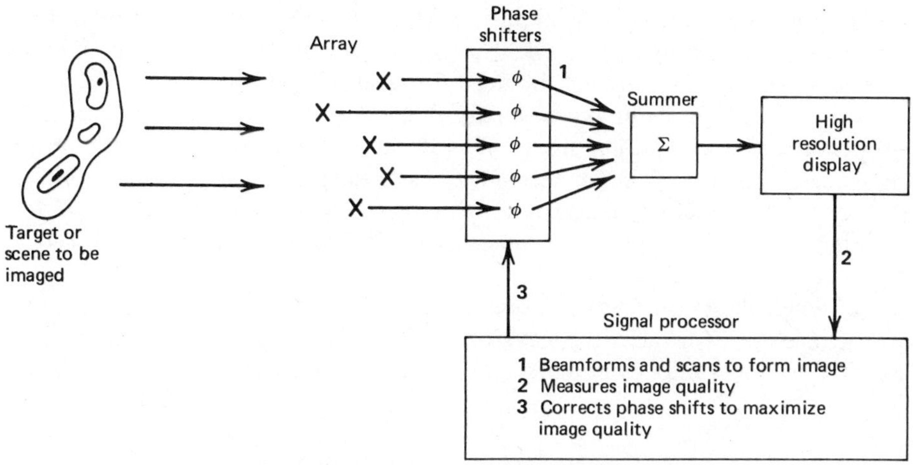

FIGURE 4.11. Image feedback control.

are adjusted to form a beam and to scan it across the target. This first step can be performed by conventional phased-array procedures or by adaptive beamforming and scanning. The signal processor measures the quality of the image in the high resolution display **2** and corrects the phase shifts in the bank of phase shifters to maximize image quality **3**. Two main design problems are (1) the metric to measure image quality without *a priori* knowledge of the scene being imaged and (2) an efficient search procedure for correcting the phase shifters.

Notwithstanding the success already achieved with adaptive beamforming in obtaining diffraction-limited images from distorted arrays, there are two reasons for interest in the development of image feedback control. Although adaptive beamforming is error-free, the scanning process that follows it is not. As shown in Chapter 7, the errors in element position locations reflect themselves in phase errors across the array as the beam scans away from the direction in which it was formed. Phase errors develop for several other reasons. Multipath and scattering contribute to the radiation field at the array during adaptive beamforming, thereby distorting it from a simple plane or spherical wave; because the self-cohering algorithm (Chapter 5) is based on the assumption of sphericity of the wavefront, phase errors inevitably result when propagation is not in free space. Phase errors also derive from receiver noise during the measurement process, from quantization in A/D converters, and from noise in phaselock loops. Therefore the image is degraded by phase errors across the array, notwithstanding the robustness of the adaptive beamforming process.

By beginning with a reasonably well formed picture of the scene, image feedback control offers the possibility of correcting these errors. Furthermore, it has been shown in the field of incoherent optics that a unique quality measure can be abstracted from an image, even in the absence of any *a priori* information regarding the scene. The possibility therefore exists that image feedback control may obviate the need for adaptive beamforming.

The second reason for image feedback control relates to the poor sidelobe properties of the random array. Chapter 3 was devoted to techniques of reducing the effects of the highest sidelobes in the image. Another possibility is that image feedback control can contribute to their reduction by capitalizing on the known or expected properties of microwave images discussed in Chapter 2.

Although there are several potential techniques, none has yet been satisfactorily developed for radarlike waveforms. The study of image feedback control applied to the radio camera is still in its infancy. This section is merely introductory.

Phase Error Correction

Image feedback control has been successfully demonstrated for controlling the positions of multiple mirrors in an optical telescope to create a large, high resolution image from a collection of smaller mirrors. A remarkable theorem due to Muller and Buffington shows that, given a spatially incoherent source (as in passive optics), certain integrals of the image intensity are monotonic with the mean-square difference between the image and the scene [63].

Let the scene be represented (in one dimension) by a complex source function $s(u)$. The complex image

$$\hat{s}(u) = s(u) * f(u)$$

$$= \int s(v) f(u - v)\, dv \tag{4.37}$$

where the $(\hat{\ })$ indicates that the image is the estimate of the source and $f(u)$ is the complex radiation pattern of the imaging aperture represented by a weighting function $w(x)$. Hence

$$f(u) = \int w(x) e^{jkux}\, dx \tag{4.38}$$

and

$$\hat{s}(u) = \int s(v)\, dv \int w(x) e^{jk(u-v)x}\, dx \tag{4.39}$$

For incoherent light, however, the intensity $I = |s|^2$ becomes the quantity of interest. The intensity in the image plane $|\hat{s}|^2$ is related to the intensity in object plane $|s|^2$ by the convolution [10]:

$$|\hat{s}(u)|^2 = |s(u)|^2 * |f(u)|^2. \tag{4.40}$$

Model the phase errors encountered during the receiving process by introducing an error distribution $\delta(x)$ over the aperture, that is, $w(x) = w_0(x)e^{j\delta(x)}$, where $w_0(x)$ represents the unperturbed aperture. Thus (4.39) becomes

$$\hat{s}(u) = \int s(v)\, dv \int w_0(x) \exp\{jk[\delta(x) + (u - v)x]\}\, dx \qquad (4.41)$$

and the image intensity evaluates to

$$|\hat{s}(u)|^2 = \int dv\, dv'\, s(v)s^*(v') \int dx\, dx'\, w_0(x)w_0^*(x') \times \exp\{jk[\delta(x) - \delta(x') + (x - x')u]\} \exp[-jk(vx - v'x')] \qquad (4.42)$$

Note that $|\hat{s}(u)|^2$, as given by (4.42), is not equal to $|s(u)|^2 * |f(u)|^2$; the reason is that the image is formed coherently. Consequently (4.42) does not lead to the general result of Muller and Buffington.

Condition (4.40), which implies spatial incoherence, must be invoked to obtain the theorem in [63]. The proof of the theorem follows the derivation in [37]. Equation 4.40 can be rewritten as

$$|\hat{s}(u)|^2 = |s(u)|^2 * P(u) \qquad (4.43)$$

where $P(u) = |f(u)|^2$ is the power pattern of the array. The fundamental sharpness function of Muller and Buffington is

$$S_1 = \int du \left(|\hat{s}(u)|^2\right)^2 \qquad (4.44)$$

From the Fourier transform relationship

$$|\hat{s}(u)|^2 \leftrightarrow \mathcal{F}\{|s(u)|^2\}\mathcal{F}\{P(u)\} = R_s(X)R_w(X)$$

where

$$R_w(X) = \int_{-\infty}^{\infty} w(x)w^*(x + X)\, dx \qquad (4.45)$$

and

$$R_s(X) = \int_{-\infty}^{\infty} e(x)e^*(x + X)\, dx$$

where $e(x) = \mathcal{F}\{s(u)\}$ is the electric field in the aperture. By using Parseval's theorem we can rewrite (4.44) as

$$S_1 = \int_{-\infty}^{\infty} \left(|\hat{s}(u)|^2\right)^2 du = \int_{-\infty}^{\infty} |R_s(X)|^2 |R_w(X)|^2 \, dX \tag{4.46}$$

$R_s(X)$ is determined by the source distribution alone. Maximizing S_1 is therefore equivalent to maximizing $|R_w(X)|$ for all X. By substituting $w(x) = w_0(x)e^{j\delta(x)}$ in (4.45) we obtain

$$R_w(X) = \int_{-\infty}^{\infty} w_0(x) w_0^*(x + X) e^{j[\delta(x) - \delta(x+X)]} \, dx \tag{4.47}$$

Now assume that the aperture weighting $w_0(x)$ is real. The absolute value of R_w is maximized when the argument of the exponential is constant for all values of x:

$$\delta(x) - \delta(x + X) = \phi(X) \qquad \text{independent of } x \tag{4.48}$$

Expanding $\delta(x)$ in a Taylor series

$$\delta(x) = a + bx + cx^2 + dx^3 + \cdots$$

and

$$\delta(x) - \delta(x + X) = -bX - c(X^2 + 2xX) - \cdots = \phi(X)$$

Thus

$$c = d = \cdots = 0$$

and no terms beyond the linear term can exist, which leaves

$$\delta(x) = a + bx \tag{4.49}$$

A linear variation of $\delta(x)$ causes a shift in image position but no image distortion. This proves that maximizing S_1 restores all but the original position of the undistorted image.

The sharpness function S_1, which is the simplest integral in the class of integrals considered useful by Muller and Buffington, is essentially a contrast measure. The processor adjusts the wideband delays at each element to maximize the contrast in the image. The image that results appears to be unique, provided that the sources are incoherent and all cross products average down into the noise. This is not the case in radar. The sidelobes that appear in range and azimuth (see Figure 2.10) can become centers of local contrast maximization, thereby preventing the signal processor from reaching the global optimum.

The radar situation is solved by returning to (4.42). By substituting (4.42) into (4.44), we obtain

$$S_1 = \int du \int dv_1\, dv_1'\, dv_2\, dv_2'\, s(v_1) s^*(v_1') s^*(v_2) s(v_2') \int dx_1\, dx_1'\, dx_2\, dx_2'$$

$$\times w_0(x_1) w_0^*(x_1') w_0^*(x_2) w_0(x_2') \exp[jk(x_1 - x_1')u] \exp[-jk(x_2 - x_2')u]$$

$$\times \exp\{-jk[v_1 x_1 - v_1' x_1' - v_2 x_2 + v_2' x_2']\}$$

$$\times \exp(jk\{\delta(x_1) - \delta(x_1') - \delta(x_2) + \delta(x_2')\}) \tag{4.50}$$

The integral over u can now be performed to yield a Dirac delta function, $\delta_D(k[(x_1 - x_1') - (x_2 - x_2')])$ [3]. The integral over this delta function can be taken immediately by eliminating one of the x variables. Defining a new variable $z = x_1 - x_1'$, we obtain (substitute $x_1 - x_1' = x_2 - x_2' = z$ in the integrand and divide by k):

$$S_1 = k^{-1} \int dx_1\, dx_2\, dz w_0(x_1) w_0^*(x_1 - z) w_0^*(x_2) w_0(x_2 - z)$$

$$\times \exp[jk\{\delta(x_1) - \delta(x_1 - z) - \delta(x_2) + \delta(x_2 - z)\}]$$

$$\times \int dv_1\, s(v_1) \exp(-jkx_1 v_1) \int dv_1'\, s^*(v_1') \exp[jk(x_1 - z)v_1']$$

$$\times \int dv_2\, s^*(v_2) \exp(jkx_2 v_2) \int dv_2'\, s(v_2') \exp[-jk(x_2 - z)v_2'] \tag{4.51}$$

Recognize the following Fourier transforms:

$$\tilde{s}(kx_1) = \int s(v_1) e^{-jkx_1 v_1}\, dv_1$$

$$\tilde{s}^*(k(x_1 - z)) = \int s^*(v_1') e^{jk(x_1 - z)v_1'}\, dv_1'$$

$$\tilde{s}^*(kx_2) = \int s^*(v_2) e^{jkx_2 v_2}\, dv_2$$

$$\tilde{s}(k(x_2 - z)) = \int s(v_2') e^{-jk(x_2 - z)v_2'}\, dv_2' \tag{4.52}$$

S_1 reduces to

$$S_1 = k^{-1} \int dz \left| \int dx\, w_0(x) \tilde{s}(kx) w_0^*(x - z) \tilde{s}^*(k(x - z)) e^{jk\{\delta(x) - \delta(x - z)\}} \right|^2 \tag{4.53}$$

Defining

$$G(x) = w_0(x)\tilde{s}(kx)e^{jk\delta(x)} \tag{4.54}$$

allows (4.53) to be rewritten as

$$S_1 = k^{-1}\int dz \left| \int G(x)G^*(x-z)\,dx \right|^2 \tag{4.55}$$

Note that $G(x)$ represents the total received weighted field at a point x on the linear aperture. This quantity depends largely on the particular scene being imaged. It is shown below, subject to condition (4.49), that S_1 is maximized only if the scene consists of a single point source and the weighting is uniform. To demonstrate place two point sources at angles u_1 and u_2. Thus

$$\tilde{s}(kx) = e^{jku_1x} + e^{jku_2x} \tag{4.56}$$

$$G(x) = w_0(x)(\exp\{jk[u_1x + \delta(x)]\} + \exp\{jk[u_2x + \delta(x)]\}) \tag{4.57}$$

and

$$\begin{aligned} G(x)G^*(x-z) = {} & w_0(x)w_0^*(x-z) \\ & \times (\exp\{jk[u_1z + \delta(x) - \delta(x-z)]\} \\ & + \exp\{jk[u_2z + \delta(x) - \delta(x-z)]\} \\ & + \exp\{jk[(u_2 - u_1)x + u_1z + \delta(x) - \delta(x-z)]\} \\ & + \exp\{jk[(u_1 - u_2)x + u_2z + \delta(x) - \delta(x-z)]\}) \end{aligned} \tag{4.58}$$

To maximize S_1 the inner integral of (4.55) should be maximized. This is achieved when $w_0(x)$ is real and the four phasors of (4.58) are cophased. This occurs only if $u_1 - u_2 = 0$; that is, if the scene is reduced to a single point source, for then condition (4.49) reduces to

$$\delta(x) - \delta(x-z) = C \tag{4.59}$$

where C is a constant. It is clear that (4.49) satisfies (4.59) because z is kept constant as the inner integration of (4.55) is performed.

The conclusion is that the Muller-Buffington sharpness function (4.44) is irrelevant to the general microwave imaging problem. It is extremely scene-dependent and works only for the special case of an isolated point source. The

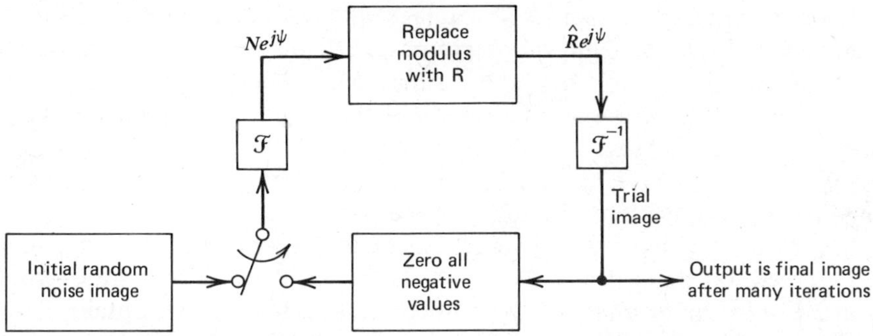

FIGURE 4.12. Gerchberg-Papoulis loop applied to speckle interferometry.

reason that the technique generally fails is the spatially coherent nature of the reradiation field from a microwave source. This fact made it impossible to separate the scene-dependent terms from the error-dependent, as shown in (4.55). As suggested earlier, however, the natural motions of radar targets may reduce the coherence enough to permit some benefit to be achieved by a modified form of the theorem.

Maximization of Spatial Bandwidth

By analogy with the focusing properties of cameras, microscopes, and telescopes maximization of spatial bandwidth would appear to be a candidate for image feedback control. An optical image is sharpest when the instrument is in focus, which implies that, and occurs when, the spatial bandwidth in the image is maximized. Defocusing is a two-dimensional low-pass filtering process. Maximizing the spatial bandwidth is equivalent to focusing the instrument.

This conclusion is correct when the uncertainty or error is in the quadratic or curvature term in the phase variation across the aperture [see (2.44) and Section 2.6)]. It is not correct when the phase errors are random, for then the spatial bandwidth is the reciprocal of the beamwidth of the array, which is determined by its size-to-wavelength ratio and is not related to the random phase errors (see Section 2.4). The phase errors reduce the gain of the main-lobe and therefore the image contrast. Hence the Muller and Buffington approach is sound for the treatment of random phase errors, whereas spatial bandwidth maximization is not.

Introduction of Heuristic Information

Figure 4.12 illustrates another approach to image feedback control. This method, initiated by Gerchberg and Saxton [32], and Papoulis [67], has proven successful in the reconstruction of images in speckle interferometry [26], [27]. In this procedure heuristic knowledge is applied as constraints on both sides of the Fourier signal processor as the image and its correlation function are iterated around a loop. Time-varying atmospheric distortion causes the details of a telescopic image to vary many times per second. The moduli of the two-dimensional autocorrelation function of these images are relatively stable, however. Averaging several of them results in a usable estimate $\hat{R} = (1/N)\Sigma_{i=1}^{N} |\mathcal{F}\{\mathcal{I}_i\}|$ of the modulus of the autocorrelation function of the scene being imaged, where $\mathcal{I}_i$ is the intensity of the ith sample image.

The process begins with a two-dimensional "random noise" image, computer created, introduced into the loop. The phase of its Fourier transform ψ_n is grafted onto the estimate of the autocorrelation modulus $\hat{R}$ obtained from experiment. The inverse transform is the first trial image. Because of the arbitrariness of the phase function attached to $\hat{R}$, the trial image will exhibit negative intensity values. All negative values are zeroed. Following Fourier transformation the modulus is again replaced by $\hat{R}$. After a large number of iterations an acceptable image results. This work also has been reported in a radar context [61].

CHAPTER FIVE

The Radio Camera Algorithm, Experiments, and Techniques

5.1. INTRODUCTION

This chapter deals with imaging based on adaptive beamforming, a subject introduced in Chapter 4 for the purpose of overcoming the effects of aperture distortion.† It is evident from an examination of (5.1) that a microwave array designed for high angular resolution imaging is likely to suffer from one or more of the difficulties mentioned in the opening paragraph of Chapter 4:

$$\hat{s}(u) = \int_L i(x)\left[\int s(u')e^{jkxu'}\,du'\right]e^{-jkxu}\,dx \tag{5.1}$$

Equation 5.1 is the simplest form of the integral equation which relates a source function or scene s and its image or estimate $\hat{s}$ when the radiation field due to the source is measured by a line aperture of extent L with aperture-weighting i. The source is assumed to be in the far field of the aperture; this permits its description in terms of the one-dimensional reduced angular variable $u = \sin\theta$; θ is measured from the normal to the aperture. The aperture is also assumed to be one-dimensional. The exponential kernels are Fourier kernels and both integrals are Fourier integrals. The inner integral is the

†Sections 5.1 to 5.5 and 5.8 are based on [87].

radiation field S at the aperture due to the source. Equation 5.1 can be written in terms of S:

$$\hat{s}(u) = \int_L i(x)S(x)e^{-jkxu}\,dx$$

$$= \mathscr{F}\{i(x)S(x)\} \tag{5.2}$$

where $\mathscr{F}\{\cdot\}$ means Fourier transform. By the multiplication-convolution property of the Fourier transform

$$\hat{s}(u) = \mathscr{F}\{i(x)\} * \mathscr{F}\{S(x)\}$$

$$= f(u) * s(u) \tag{5.3}$$

where the asterisk refers to convolution and $f(u)$ is the radiation or diffraction pattern of the aperture. Because convolution broadens or spreads a function, the closer that $f(u)$ approximates a delta function, the better the estimate $\hat{s}(u)$ of $s(u)$. Because the width of $f(u)$ is always the order of λ/L, it is evident that the larger the aperture size L, the more closely s is represented by $\hat{s}$. The same conclusion pertains to a target or scene in the near field.

To achieve a resolving power of 10^{-4} to 10^{-5} rad typical of cameras and small telescopes a microwave antenna must be hundreds of meters to tens of kilometers in size. As discussed in Chapter 1, these antennas are too large to be constructed as single structures like the parabolic dish. Instead, they must be phased arrays, highly thinned (mean interelement spacing $\gg \lambda/2$) to limit cost and with aperiodic element distribution to eliminate grating lobes [85]. It is most likely that this large size will preclude accurate knowledge of element locations; hence the array properties will tend to degenerate to those of the random array and the system design must accommodate the poor sidelobe performance expected (Section 2.4). As a consequence, array distortion is a highly likely property of a huge array designed for microwave imaging. Its effect, however, can be neutralized to a considerable extent by introducing the adaptively controlled phase corrections discussed in Section 4.3. The corrections are based on measurements at the array of the radiation field from a point source. Adaptive, retrodirective beamforming techniques then focus the distorted array on the source [47], [39], [68]. The instrument that embodies this procedure is called the radio camera.

This chapter describes an algorithm for phase synchronizing or self-cohering a distorted array on such a source, then moving the focused beam in range and angle to a target area, and scanning it across the target to image it. Experimental evidence of the validity of the technique is given. The experiments described in this chapter were conducted with an X-band ($\lambda = 3$ cm) radar by using distorted, quasi-linear, random antenna arrays, 27 and 39 m long.

5.2. THE RADIO CAMERA

Figure 5.1, which is similiar to but more detailed than Figure 2.8, shows a badly distorted receiving array and a point source of radiation in its near field. It also shows a pulsed transmitter. The near-field source may be active, like a beacon or passive like a large cross section reflector which echoes the radiation from the transmitter. This source, or reflector, is called the adaptive beamformer; it is also called the phase synchronizer. Notwithstanding the distortion in the array, the signal processor organizes the signals received at each element from the phase synchronizer and forms a radiation pattern with a beam focused on the synchronizing source.

Error Correction

To accomplish this task the signal processor assumes free space radiation. The signals received by the antenna elements differ in phase according to the differential distances of the elements, $r_n - r_0$, to the source, where r_n = the distance from the source to the nth element and r_0 = the distance to some reference element in the array. The signal processor measures the phase differences and phase rotates the signals to cophase them.

The array in Figure 5.1 is shown measuring the phase of the signal received at each element in relation to the phase at some reference element within the

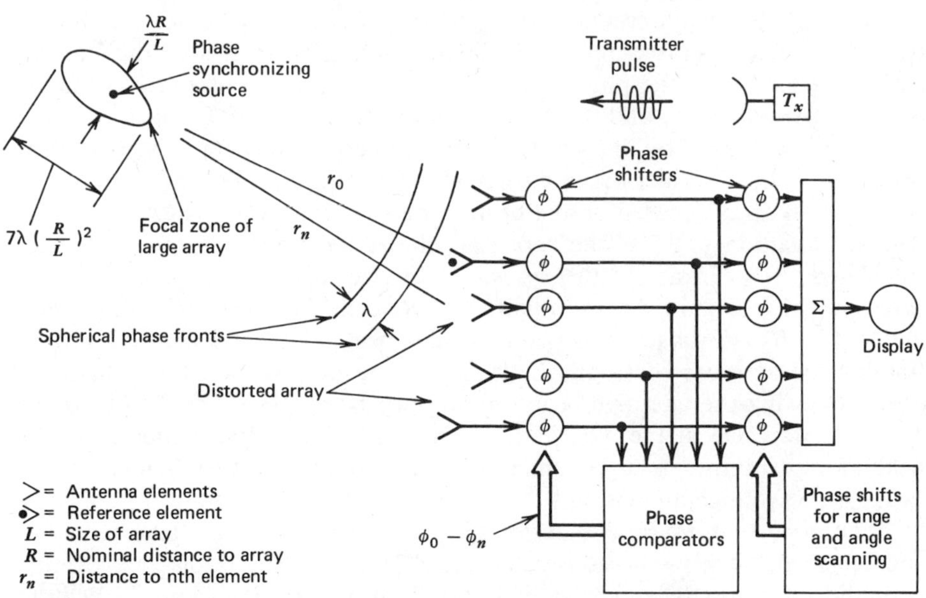

FIGURE 5.1. Distorted array self-coheres on a phase-synchronizing source: > = antenna elements; •> = reference element; L = size of array; R = nominal distance to array; r_n = distance to nth element.

array. In the absence of multipath the phase ϕ_n of the received signal at the nth element is $(\omega r_n/c_n) + \xi_n$, where ω is the radiation frequency, r_n is the path length, c_n is the average speed of propagation over the path, and ξ_n is the phase shift due to electromagnetic coupling of the element to its environment as well as in the receiving element and associated circuits; ϕ_n may be rewritten as

$$\phi_n = \frac{\omega(r_{0n} + \delta r_n)}{c_0 + \delta c_n} + \xi_n \tag{5.4}$$

to show explicitly the displacement δr_n of the element in the direction toward the adaptive beamformer from its correct distance r_{0n} and the deviation δc_n of the mean propagation speed over the nth path from the average speed c_0 in the medium. Because $\delta c_n \ll c_0$,

$$\phi_n \simeq \frac{\omega(r_{0n} + \delta r_n)(1 - \delta c_n/c_0)}{c_0} + \xi_n$$

$$\simeq kr_{0n} + k\delta r_n - \frac{kr_{0n}\delta c_n}{c_0} + \xi_n \qquad k = \frac{\omega}{c_0} \tag{5.5}$$

The phase differences from element to element have four components. The primary, kr_{0n}, is due to source location in relation to the undistorted array and is the same as in any phased array. The residuals are due to the distorted geometry of the array (second term), to variations in propagation conditions from the source to different parts of the array due to its large size (third term), and to variations in the phase shifts in the elements and in electromagnetic coupling between them and their surroundings (fourth term). By phase shifting each receiver channel by the negative of these phase differences all signals from the adaptive beamformer become cophased. The output, following summation, is that of an array focused on the beamforming source. The system has a well defined radiation pattern: the spot size of the focal zone in the near field, calculated from diffraction theory, was discussed in Section 2.6. The dimensions shown in Figure 5.1 are nominally $\lambda R/L$ in the cross-range dimension and $7\lambda(R/L)^2$ in range; λ/L is the angular beamwidth of a focused diffraction-limited aperture of length L. It is also the beamwidth of the distorted array following the adaptive beamforming procedure. Then $7\lambda(R/L)^2$ is the approximate value of the 3-dB range beamwidth due to the limited depth of field of an aperture focused on a near-field target. Further, $\lambda R/L$ is the far-field cross-range dimension.

It is important to note that all the distortion-inducing phenomena are compensated by this process (see Section 4.3). Array distortion is associated with the error δr_n. The second distortion-inducing phenomenon is medium turbulence. The measured phase at each array element is proportional to the integral of the refractive index from the source to that element. Thus the "distance" correction made by the processor includes the geometric distortion in the array and the "electrical distance" error due to medium fluctuations.

The third distorting phenomenon which is automatically corrected by the cophasing process is random impedance variations from element to element in the array. Sources of these errors are differences in electromagnetic coupling from the elements to the terrain and differences in antenna and RF circuit impedances due to aging and mistuning.

Although not explicitly written in (5.4), the effect of multipath and scattering of the wave from the phase synchronzing source to the array is also automatically corrected. A multipath signal or one that arrives by a scatter path would add vectorially to the direct wave and shift its phase. The effect is an additional additive random term to (5.4), which can be viewed as combined with the ξ_n used there.

Scanning

Also shown schematically in Figure 5.1 is a second set of phase shifters located ahead of the summer. Unlike the first phase shifting operation, the second operation is open-loop and nonadaptive. After the array is phase synchronized the focused beam may be used to form an image, that is, to take a picture of a target. To do so the beam must be refocused to the target coordinates (range and bearing) and then scanned in angle across the target. Scanning is accomplished by phase shifting at each array element much as in an ordinary array. Scanning in range and angle is accomplished by open-loop corrections calculated from the geometry. The calculations are made exactly as in a conventional phased array.

Thus the first process, retrodirective self-cohering or adaptive beamforming, requires no knowledge of element position. The second process, refocusing and scanning the beam, does. It is shown in Chapter 7 that once the beam has been formed adaptively the tolerance on element position accuracy is greatly eased. This is the basis for high-resolution radio camera imaging with nonrigid or crudely surveyed apertures. The theory [83], [101], [85], supported by experiment (Section 7.3), indicates that the position accuracy tolerance relaxes by two or more orders of magnitude in typical imaging problems. It is this extraordinary fact that makes the radio camera an attractive imaging or picture-taking instrument.

During near-field scanning the spot size and shape remain as indicated in Figure 5.1. In the very near field the resolution of the two-dimensional image that results from the scanning operation is about equal to the spot size. When the target is in the far field of the large array the depth of field becomes semiinfinite and the resolution in range is determined by the pulse duration of the transmitter. At intermediate distances the range resolution is the smaller of the depth of field and radar pulse length.

The Beamforming Source

The beamforming source must produce a radiation field at the array of predictable character. The only practical field to establish for this purpose is one with a plane or spherical wavefront. Two conditions are required of this synchronzing source: it must approximate a point source and propagation to the array must approximate free space propagation. Although neither is necessary in the self-cohering step, both are needed for high-quality scanning. These are necessary conditions because the phase correction $\phi_n - \phi_0$ is the negative of the measured phase difference, which is assumed to be proportional to the free space differential distance. If either condition fails, the phasefront of the radiation across the array is not spherical and self-phasing will be exact only at the array focus. The detailed requirements are developed in Chapter 6.

The first requirement, approximation to a point source, is due to the necessity that a single lobe of the radiation pattern of the source encompass the entire array. The second requirement, that propagation closely approximates free space conditions, limits the permissible level of multipath and scattering. Both limitations are described later, when it is shown that these phenomena can generally be ignored if the array is smaller than $\lambda R/2T$, where T is the size of the beamforming source, and the scattered field at the array is 6 dB or more below the field of the direct ray.

The Operations

Both phase shift operations shown in Figure 5.1 can be analog or digital. The phaselock loop is the natural analog circuit for adaptive beamforming. The experimental equipment described later uses digital phase shifting. Figure 5.2 describes the procedure explicitly. A ground-to-ground imaging situation is portrayed. On the left are a radar transmitter and receiver. The target region to be imaged is on the right. A transmitted pulse illuminates the target area and a passive phase-synchronizing target, which is sketched as a corner reflector. The corner reflector is shown closer to the radar than the target region. Each antenna element receives an echo trace that contains clutter, the beamforming corner reflector, and the target. The echo waveform from each element is coherently demodulated, sampled in range, and stored in the format shown for image formation. To produce the image the processor must perform several functions: it must find the range element that contains the phase synchronizing source, self-focus a beam on it, refocus the beam to the target range, and scan the beam in azimuth across the target. When these operations are completed, the output of the system is a one-dimensional scan or image of the target. A two-dimensional (range, angle) image is built up from a succession of angle scans.

The range trace delivered by each antenna element is stored as a row of complex numbers. The data format is the same as discussed in Section 4.3.

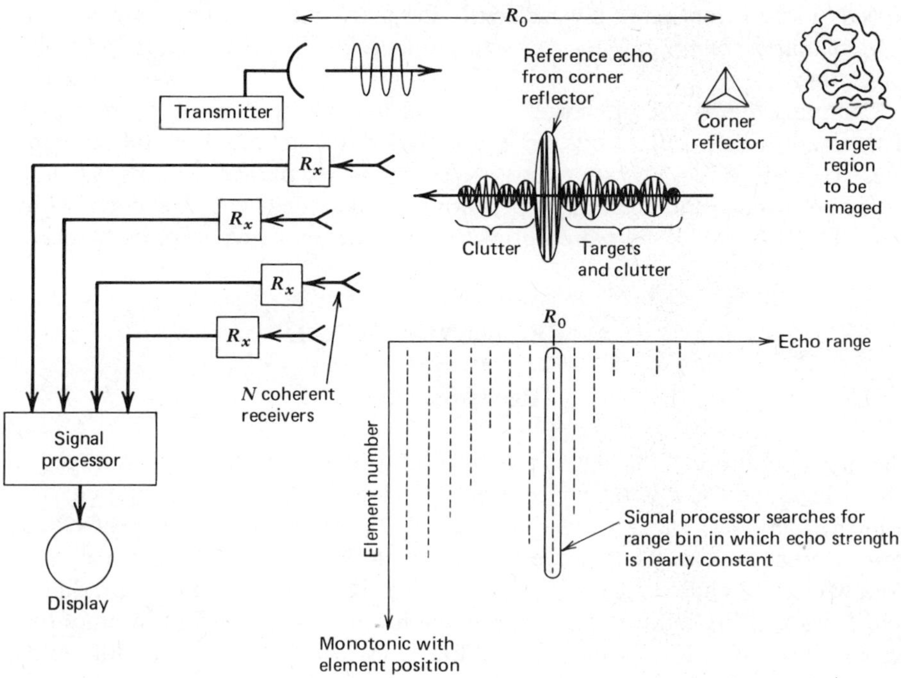

FIGURE 5.2. Radio camera data format and procedures. (From B. D. Steinberg, *IEEE Trans. Antennas and Propagation*, Ap-29, No. 5, September 1981.)

Successive rows correspond to successive elements in the array. The position of a sample in a row is proportional to range and designates the range bin. The position of a sample in a column designates the array element number and is monotonic with but not necessarily proportional to element position because the array may be distorted.

Range-Bin Correction

For simplicity it is assumed that $|r_n - r_m| \ll \Delta R$, all n, m, where ΔR is the length of a range cell. Thus all echoes from a common target will appear in a single range bin in the format shown in Figure 5.2. Actually, target echoes can appear in different range bins in different parts of the array for two reasons. The first is due to the large size of the array because the differential arrival time from a source at angle θ exceeds the pulse length when $L \sin\theta > \Delta R$. This effect is easily calculated and may be corrected in the signal processor. The second cause is array distortion in the direction toward the source, although most arrays designed to be linear or planar systems will not suffer enough distortion to warrant range-bin correction. Nevertheless, all distorted arrays require adaptive phase correction if the unknown *a priori* geometric distortion exceeds a small fraction of a wavelength. Large irregular arrays with deviations from planarity comparable to or exceeding ΔR will also require range-bin correction.

Assuming that range-bin correction, if needed, has been accomplished, three distinct additional operations on the stored data are required for imaging. These constitute the algorithm and are covered in separate sections. Two are the beamforming and scanning operations discussed earlier. The third, which precedes these two, is the search procedure for the phase synchronizing source.

5.3. REFERENCE SIGNAL

Range Search

The first operation of the signal processor after data sampling and storage is a search in range for a good phase synchronizing source; that is, for that echo sequence across the array that most closely approximates the expected field from a point source. That range is designated R_0 and is the reference range from which the phase synchronizing reference signal is obtained. The radiation field from a point source in free space would be nearly constant in amplitude across the array, whereas the phase differences would disclose the differential distances (modulo wavelength) to the source as well as the differential phase shifts through the receiving elements. A simple test to find the target with a radiation field that most closely approximates that of a point source is to measure the normalized echo amplitude variance at each range; R_0 is that range for which the test value is minimum.

A large target may or may not have the desired properties. Unless the target is physically small ($< \lambda R/L$), its radiation field across the array will not be nearly constant. Thus a strong target echo is insufficient to ensure that the target is an adequate synchronizing source. This fact is illustrated in Figure 5.3, which shows the difference in the variation of the amplitude of the echo across an array when the source is a compact target, compared with a more distributed target. The ordinate is echo amplitude in arbitrary units; the abscissa is element location in a 100-element, approximately linear random array. The data were taken with the X-band coherent radar and rooftop array described in Section 5.7. The upper curve shows the echo strength variation from a range bin in which there was a large corner reflector in clutter. The rms variation, normalized to local mean echo strength, is about 5%. This target makes an excellent phase-synchronizing source. The lower curve is the amplitude variation from a range bin in which the target was a pickup truck with a camper top that produced the radio camera image in Figure 5.12. The complicated spatial structure of the echo variation is evident. This target is a poor synchronizing source, notwithstanding the fact that its peak echo strength is large. The corner reflector is the ideal adaptive beamformer, yet even its echo amplitude pattern is not constant with element position.

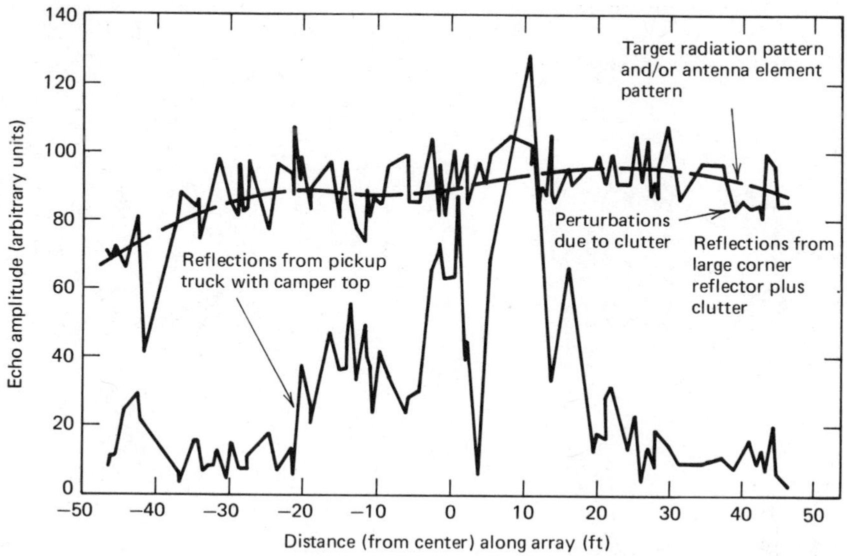

FIGURE 5.3. Echo strength from two range bins versus element position in array. The corner reflector approximates a point source (From B. D. Steinberg, *IEEE Trans. Antennas and Propagation*, Ap-29, No. 5, September 1981.)

Sources of Echo-Amplitude Variation

There are four contributing factors to the general variation of echo amplitude with element position. The first is due to the size of the reflector or source, the second, to clutter, and the third, to multipath. The fourth is due to the radiation of the antenna element used in the array.

A reflector of size T reradiates a lobular pattern in which the nominal lobe spacing is λ/T rad (Figure 5.4, *left*). The nominal lobe spacing at the array, at a distance R, is $\lambda R/T$, and the echo sequence across the array is expected to have a spatial period of this value. If $\lambda R/T \lesssim L$, a low spatial frequency or long period amplitude modulation would be evident in the measured data. Such a long period spatial modulation appears in the corner reflector echo sequence in Figure 5.3 (although the cause is not due to the size of the reflector but to the radiation pattern of the antenna element, which is discussed later).

The high spatial frequencies or short period fluctuations seen in the corner reflector echo sequence in Figure 5.3 are due to clutter (Figure 5.4, *center*). The transmitter beamwidth is broad compared to the high resolution receiver beamwidth; the beamwidth $\Delta\theta = \lambda/a$, where a is the size of the radiating antenna. The clutter patch illuminated by the transmitter at distance R is $R\Delta\theta = R\lambda/a$. The lobes of the back-scattered radiation have a nominal spacing $\lambda/(R\lambda/a) = a/R$ and the nominal cross section at the array is equal to $R(a/R) = a$. This last expression means that the correlation distance of the spatial amplitude modulation of the scattered field at the array, due to the

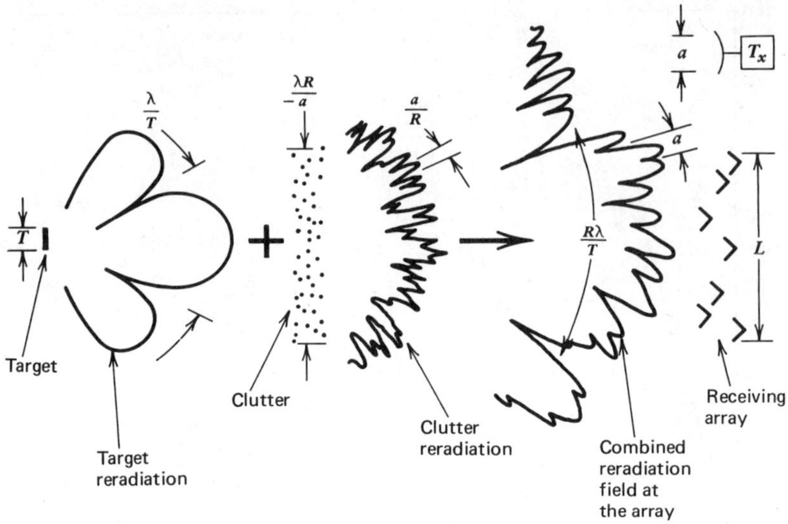

FIGURE 5.4. Target size and clutter contribute to echo amplitude variation across array. (From B. D. Steinberg, *IEEE Trans. Antennas and Propagation*, Ap-29, No. 5, September 1981.)

clutter, equals the size of the transmitting antenna. In the experiment reported in Figure 5.3 this size was less than the average interelement distance; hence the perturbations due to the clutter appear to be decorrelated from element to element. Because the transmitting antenna is always small compared with the large receiving array, the clutter modulation will always appear, as in Figure 5.3, as a high spatial frequency or short period fluctuation.

The effect of multipath is similar. Let β be the angular separation at the array between the direct arrival and the multipath signal. The complex amplitude of the received signal along the array (x-coordinate) has the form $1 + \alpha \exp(jkx\beta)$, where α is the reflection coefficient of the multipath scatterer and $k = 2\pi/\lambda$ is the wavenumber. The period of the amplitude modulation of the sum of the two signals is λ/β. Rarely is $\beta > \Delta\theta$; therefore multipath in general will introduce a lower spatial frequency into the echo sequence across the array.

The fourth contributing factor to the variability of the reference echo sequence is the radiation pattern of the receiving element used in the array. Figure 5.5 illustrates four array elements, each of length d, in an array of length L. The beamwidth of an element pattern is λ/d and its cross section at the beamforming target is $\lambda R/d$. It is shown that the element gain to the target varies with element position in the array when $L > \lambda r/d$.

Echo Variance Test

The echo variance calculation should be based on the first, second, and third factors, which measure the quality of the radiation field for adaptive

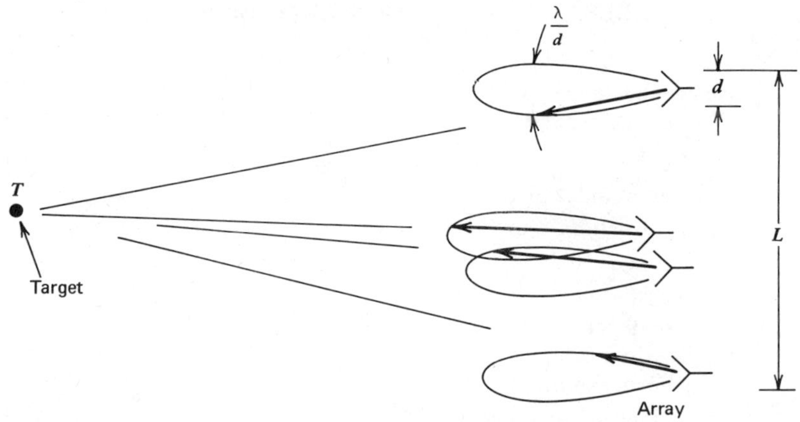

FIGURE 5.5. The element pattern also contributes to amplitude variation. (From B. D. Steinberg, *IEEE Trans. Antennas and Propagation*, Ap-29, No. 5, September 1981.)

beamforming, but not the fourth factor, in which the measuring instrument induces a variation. The effect of the fourth factor is minimized by dividing V_{in}, the echo amplitude from the ith range bin received by the nth element, by the estimated element pattern gain $\hat{f}_n$ from the nth element to the target. (A single subscript suffices for the element pattern gain unless the target is in the near field of the element, which is unlikely.) The corrected amplitude is $A_{in} = V_{in}/\hat{f}_n$. Its mean $\overline{A_i} = (1/N)\Sigma_{n=1}^{N} A_{in}$ and mean square $\overline{A_i^2} = (1/N)\Sigma_{n=1}^{N} A_{in}^2$ are calculated, where N is the number of antenna elements. Its variance $\sigma_i^2 = (1/N)\Sigma_{n=1}^{N}(A_{in} - \overline{A_i})^2 = \overline{A_i^2} - \overline{A_i}^2$; σ_i^2 is normalized to $\overline{A_i^2}$ to remove the effect of target strength and range from the calculations. The normalized variance is $1 - \overline{A_i}^2 / \overline{A_i^2}$, which is a minimum at that range element for which $\overline{A_i}^2 / \overline{A_i^2}$ is the largest value. The rule for finding R_0, then, is to calculate $\overline{A_i}^2 / \overline{A_i^2}$ for all range elements and search for the largest value. The signal sequence from that range bin becomes the reference signal. An experiment conducted with airborne radar clutter data indicated that a value in excess of 0.98 resulted in satisfactory adaptive beamforming (Chapter 9).

5.4. ADAPTIVE BEAMFORMING

Table 5.1 shows all the steps in the procedure. Step 1 initiates the process: $V_{in}\exp(j\psi_{in})$ is the complex envelope of the echo from the ith range bin received by the nth element. Steps 2 and 3 are the search for R_0. Step 4 is the adaptive beamforming step. The measured phases ψ_{in} can be broken into the sum of two terms $\psi'_{in} + \psi''_{in}$. The first term contains the conventional target and array geometry and is all that would be expected in the absence of array distortion, medium turbulence, multipath, and scattering. The second term

TABLE 5.1. Steps in Radio Camera Imaging [87]

Step		
1	Measure and store complex envelopes of echo samples	$V_{in}e^{j\psi_{in}}$ range bin—↗ ↖—element number
2	Correct amplitudes by dividing by element pattern estimate $\hat{f}_n$	$A_{in}e^{j\psi_{in}}$
3	Find R_0 such that $A_{0n} \approx A$, all n	$Ae^{j\psi_{0n}}$
4	Phase rotate at R_0 by phase conjugate in relation to reference element, $\exp j(\psi_{00} - \psi_{0n})$	$Ae^{j\psi_{00}}$
5	Phase rotate at all range elements	$A_{in}e^{j(\psi_{in}-\psi_{0n}+\psi_{00})}$
6	Focus at each range R_i	$A_{in}e^{j[\psi_{in}-\psi_{0n}+\psi_{00}+(kx_n^2/2)(1/R_i-1/R_0)]} \triangleq B_{in}$
7	Phase shift linearly with angle	$B_{in}e^{-jkx_nu}$
8	Sum at each range element	$\hat{s}_i(u) = \sum_{n=1}^{N} B_{in}e^{-jkx_nu}$

represents the errors that the phase synchronization process must overcome. In the absence of error the image of the target or clutter from the ith range element would be obtained from the integral of the phase-weighted signals received across the array from the ith range bin. As in any phased array, the phase weighting is the conjugate of the kernel of the diffraction integral. For ease of discussion the diffraction integral can be approximated by the Fresnel integral, which reduces to the Fourier integral when the target is in the far field. The Fresnel integral, in turn, is best represented by a sum, as in (5.6), because the array is discrete. The sum

$$\hat{s}_i(u) = \sum_{n=1}^{N} A_{in}e^{j\psi'_{in}}e^{-jk(x_nu - x_n^2/2R_i)} \tag{5.6}$$

is the image $\hat{s}_i$ of the scattering sources in the ith range bin, provided that $\psi'' = 0$. As in (5.1), $u = \sin\theta$ and θ = scan angle from the normal to the array. The discrete variables x_n and R_i are, respectively, the x-coordinate in the array (transverse to the array normal) of the nth element and the distance to the ith range bin. Although $\hat{s}_i(u)$ is written as a one-dimensional image (in the reduced angular variable u) of the echoes from the ith bin, it is, in reality, a two-dimensional image in u and R when the targets are in the near field of the array. Properties of the two-dimensional image are described in Section 2.7.

The function of the adaptive processor is to compensate for ψ''_{in} to allow the operation described by (5.6) to be completed. In the earlier steps of the process the data were searched to find the range bin in which the echo amplitude across the array had the smallest variation. That range bin became the reference range and its signal became the synchronizing signal. Its complex envelope following normalization to $\hat{f}$ is $A_{0n}\exp(j\psi_{0n})$, where the amplitudes are all nearly the same. The phases of the echoes, however, are random because of the perturbed geometry of the array, the spatial variations of the refractive index, the impedance variations from element to element, or any combination of these factors. If the reference source were ideal, the phase differences would be due entirely to them. In addition, the phases are perturbed by multipath and scattering and by the finite lobe width of the reradiation from the synchronizing source. But because it is generally impossible for the system designer to obtain *a priori* information about these conditions the signal processor must necessarily ignore them.

The fourth step in the process is to compensate for the phase variations, which are assumed to be due to the first set of factors (i.e., geometry, refractive index, and impedance). Correction is made by phase-rotating the complex envelopes of the signals from R_0 received by the different array elements to cophase them. Let some element in the array be designated the reference element and the phase of its signal from the reference range be designated ψ_{00}. The proper phase shift for the nth element is the negative of the phase difference $\Delta\psi_{0n} = \psi_{0n} - \psi_{00}$. The complex signal envelope at the nth element becomes $A_{0n}\exp(j\psi_{0n})\exp[-j(\psi_{0n} - \psi_{00})] \simeq A\exp(j\psi_{00})$. This correction, or phase conjugation, is exactly what a phased array or lens would do when focusing on the reference source. The output signal from the array when it is focused is the sum of these phase-corrected echoes from R_0.

$\Delta\psi_{0n}$, or the differential distance $\Delta\psi_{0n}/k$ that it represents, is the sum of two quantities. The first is due to the geometric distortion and includes the

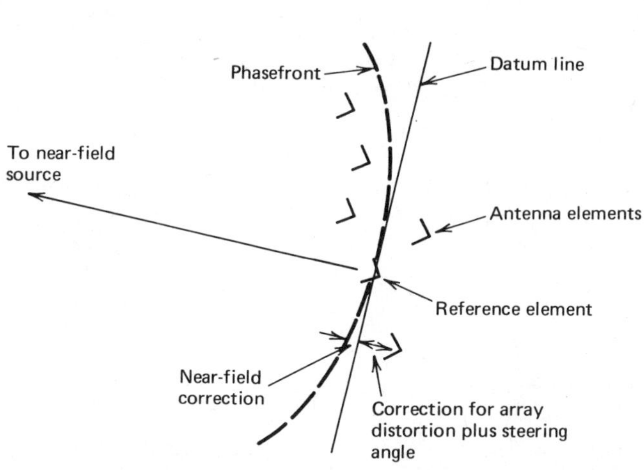

FIGURE 5.6. Components of the phase shift in adaptive beamforming.

steering direction to the source. It is the distance of each element from a datum line or plane in the array that contains the reference element and that is oriented normal to the direction to the synchronizing source, which is the desired correction. When the source is in the near field of the array $\Delta\psi_{0n}/k$ also has a second component. It is the added differential distance $x_n^2/2R_0$ between the spherical phasefront and the datum line. These components of $\Delta\psi_{0n}/k$ are illustrated in Figure 5.6.

5.5. SCANNING

The fifth step (performed simultaneously with Step 4) is to phase-rotate all the signal samples from each antenna element by $\psi_{0n} - \psi_{00}$. The samples of the complex envelope from the ith range bin now become $A_{in}\exp[j(\psi_{in} - \psi_{0n} + \psi_{00})]$. The sum of a set of complex samples represents the output of a misfocused phased array or lens because $\psi_{in} \neq \psi_{0n}$, except at the range R_0 of the reference reflector.

The sixth step is to focus the array at all ranges simultaneously. This task is accomplished for an arbitrary range R_i by refocusing the array from the reference range R_0 to R_i. The phase correction is approximately quadratic, given by (5.6). Assuming that the earlier steps were performed properly, the self-cohering process forced the quadratic component of the phase shift of the signal in the nth channel to become $-kx_n^2/2R_i + kx_n^2/2R_0$. To focus the array to range R_i this term must be set to zero, which requires a further phase addition of $(kx_n^2/2)(1/R_i - 1/R_0)$. This step requires a knowledge of the range R_0 of the reference reflector. Fortunately the value of R_0 can be measured (as in conventional radar) by the round-trip travel time of the pulse to the phase synchronizing source and it is read directly into the signal processor from the radar receiver. The accuracy of measurement is determined by the range resolution of the system, which is the order of the reciprocal of the signal bandwidth [in distance units, see (4.5) and (4.6)] or the near-field range beamwidth, whichever is smaller.

Step 7 imparts a linear phase rotation to the range-focused complex envelope (designated B_{in} in Table 5.1) for each scan angle u. The phase shift is $-kx_nu$.

The last step forms the sum of the linearly phase-weighted, range-focused samples to obtain the image $\hat{s}_i(u)$ for the ith range bin:

$$\hat{s}_i(u) = \sum_{n=1}^{N} B_{in}e^{-jkx_nu} \tag{5.7}$$

5.6. ELEMENT POSITION TOLERANCE

Steps 3, 4, and 5 require no knowlege whatsoever of element position. The phase synchronization process is purely retrodirective. Steps 6 (refocusing in range) and 7 (scanning in angle) do require coordinate information. The tolerances on element position error have been worked out [83], [101], [85, Chapter 13], and are reviewed in Chapter 7. The most stringent tolerance is invoked by Step 7. Briefly, the theory consists of the following points:

First, the loss in mainlobe gain, in dB, due to all the random phase errors across the array was shown in (2.16) to be $\Delta G \simeq 4.3\sigma_\phi^2$, where σ_ϕ^2 is the variance of the phase errors in square radians.

Second, the phase variance due to random position errors is approximately $\sigma_\phi^2 \simeq k^2\sigma_x^2\theta^2$, where σ_x^2 is the position error variance in the array in the direction perpendicular to the beamforming direction and θ (rad) is the scan angle measured from the direction of beamforming. Combining these expressions and allowing 1-dB loss in array gain as an acceptable gain–loss tolerance, the rms element tolerance becomes $\sigma_x \simeq (\lambda/4\pi)\theta_{max}^{-1}$.

The factor $\lambda/4\pi$ is the conventional tolerance in phased array, mirror, or lens design; θ_{max} is the maximum scan angle from the direction of the reference source and is half the field of view. Because an individual radar target always subtends a small angle at the radar, $\theta_{max} \ll 1$ and the allowed element position error is exceedingly large; for example, if the system were designed to image a target as large (in angle) as the moon, θ_{max} would be approximately 10^{-2} rad. In this case the position accuracy tolerance increases by two orders of magnitude to about 10 λ. It is precisely the extraordinary liberty in position tolerance, *following adaptive phase synchronization*, that permits radio camera imaging with a nonrigid or ill-surveyed array.

5.7. EXPERIMENTS

The first radio camera system experiments were made with the setup shown in Figure 5.7 [95]. A 27 × 1.2 m section of the laboratory roof of the Valley Forge Research Center, University of Pennsylvania, was marked with 100 randomly selected locations. These 100 sample points represented the element locations of the random array of the radio camera. The x- and y-coordinates were uniformly distributed. A corner reflector was placed at a distance that varied from 150 to 250 m from the array. This corner reflector was used to provide the synchronization echo for adaptive beamforming. The line of sight to the random array was nearly 30° from broadside. The corner reflector was 1.2 m on each side.

The target to be imaged was placed at a greater or lesser distance than the phase-synchronizing reflector. The target in the figure is shown as two corner reflectors separated by 1.4 m and placed 33 m farther from the array than the beamforming reflector.

Targets are placed on wheeled carts that roll on a two-rail wooden track. The three corner reflectors, carts, and track are visible in the photograph in Figure 5.8, which was taken from the center of the array. The beamforming reflector, being closest, appears larger than the target reflectors.

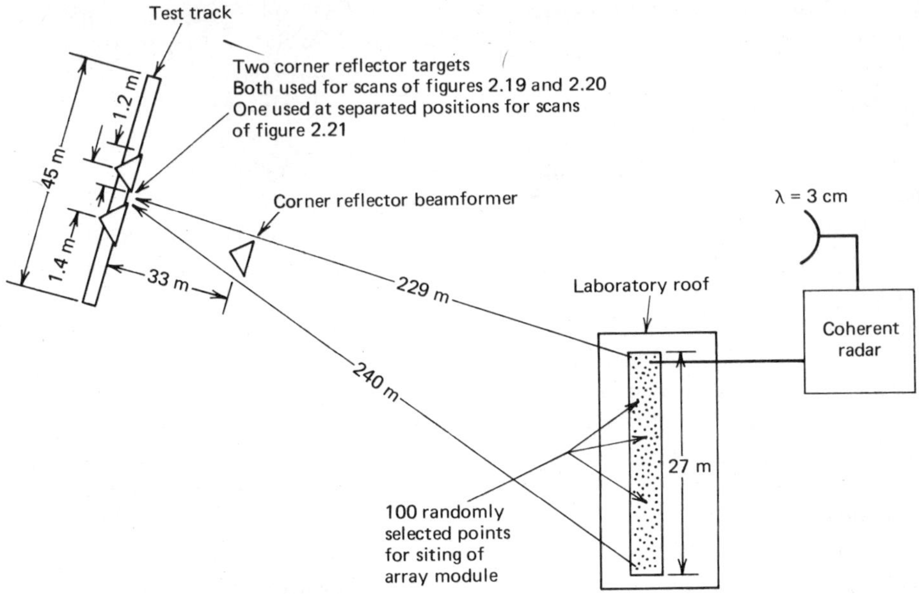

FIGURE 5.7. Geometry of experiment with time-shared radio camera.

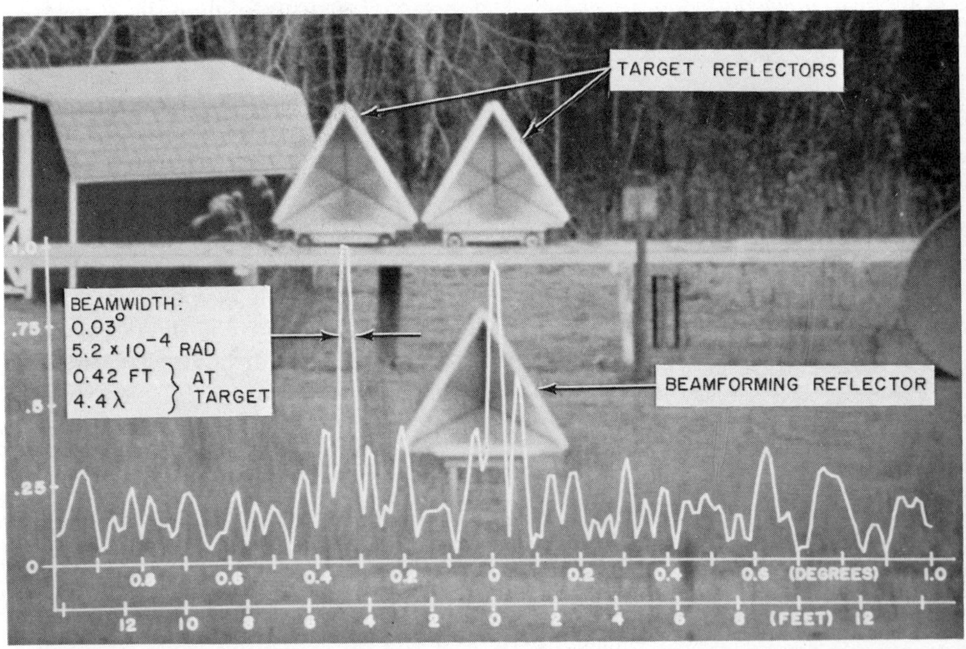

FIGURE 5.8. High-resolution radio camera image (at 3-cm wavelength) of two corner reflectors. Deflection modulated display. Resolving power of 5.2×10^{-4} rad observed. Background is optical photograph from center of array. (From B. D. Steinberg et al., *Proc. IEEE Letters*, Vol. 67, No. 9, September 1979.)

The experimental setup was a time-shared system with the transmitter and receiver colocated. A single array module was moved from position to position on the roof. The equipment was a low-power (10 mW), coherent pulse radar operating at 3 cm wavelength. The antenna was a 12° *X*-band horn. Polarization was vertical. The pulse duration was 100 ns. At each of the 100 randomly selected positions the radar transmitter was pulsed; the receiver then recorded the echoes from a group of range bins that included the reference reflector and the target. Detection was coherent, thereby permitting phase as well as amplitude of each echo to be measured. Quadrature coherent demodulators were used and the quadrature components of the echoes were stored in a microprocessor.

The equipment performed the functions described in the earlier sections. The phase of the reference echo was measured at each point in the array. This phase was subtracted from the phase of the target or clutter echoes in each range bin to correct for the distortion in the aperture. At this stage the simple addition of all the phase-corrected target echoes from the range element that contained the reference corner reflector would produce an output corresponding to a beam focused at the reference reflector. Further phase corrections were made to refocus the beam to the target range. The 100 target-echo samples, phase-corrected for aperture distortion and range displacement, were then transformed [according to the diffraction integral (2.38)] to form a one-dimensional, deflection-modulated image.

Because a single receiver/transmitter was time-shared across the aperture, the radio camera that was formed by the signal processor was a synthetic aperture system which may be called a synthetic aperture radio camera or SARC for short, as discussed earlier in Chapter 1.

The one-dimensional microwave image of the two corner–reflector target is the trace in the lower portion of Figure 5.8. The horizontal scales of the optical photograph and the radar image are the same. The radar output corresponds to an angular scan of the radio camera from 0.9° to the left of the beamformer to 0.9° to the right. The data from which the output scan was derived did not contain the echo from the beamforming reflector; only the range-gated echoes from the range of the target reflectors contributed to the microwave image of Figure 5.8. The echoes from the beamforming reflector served to phase-synchronize the radio camera array so that the target could be imaged.

The target image consisted of two narrow angular responses coincident with the spines of the target reflectors. The width of the main response is 0.03° or 5.2×10^{-4} rad. The theoretical expected value of the beamwidth for the system operating without errors, that is, diffraction-limited, is 5.1×10^{-4} rad. The beamwidth is half that of a conventional 27-m array at *X*-band because the experiment was conducted as a synthetic aperture (Section 2.4). The linear width of the main response at the target is 12.6 cm or 4.4λ. To the left and right of the main responses are sidelobe responses of the array to the two targets. The measured average sidelobe level is 17 dB below the target responses in accordance with the theory given by (2.34).

Because the beam cross section of the radio camera was almost 10 times smaller than each of the target reflectors, it was possible to "see" what a corner reflector looked like at microwaves. It is evident that it is black or nonreflecting everywhere except at the spine, as expected from the theory of the corner reflector [71].

Figure 5.9 compares the two-target image with the calculated optimum response of the distorted array *had it self-organized perfectly*. It is evident that the system was able to image the targets properly, notwithstanding the severe distortion in the aperture. The corner reflectors are drawn to scale below the radio camera image.

In Figure 5.10 a single corner reflector was moved to two different locations on the track. The horizontal axes of the output images were displaced to align the images. The origin (0) on each of the two abscissas shows the direction to

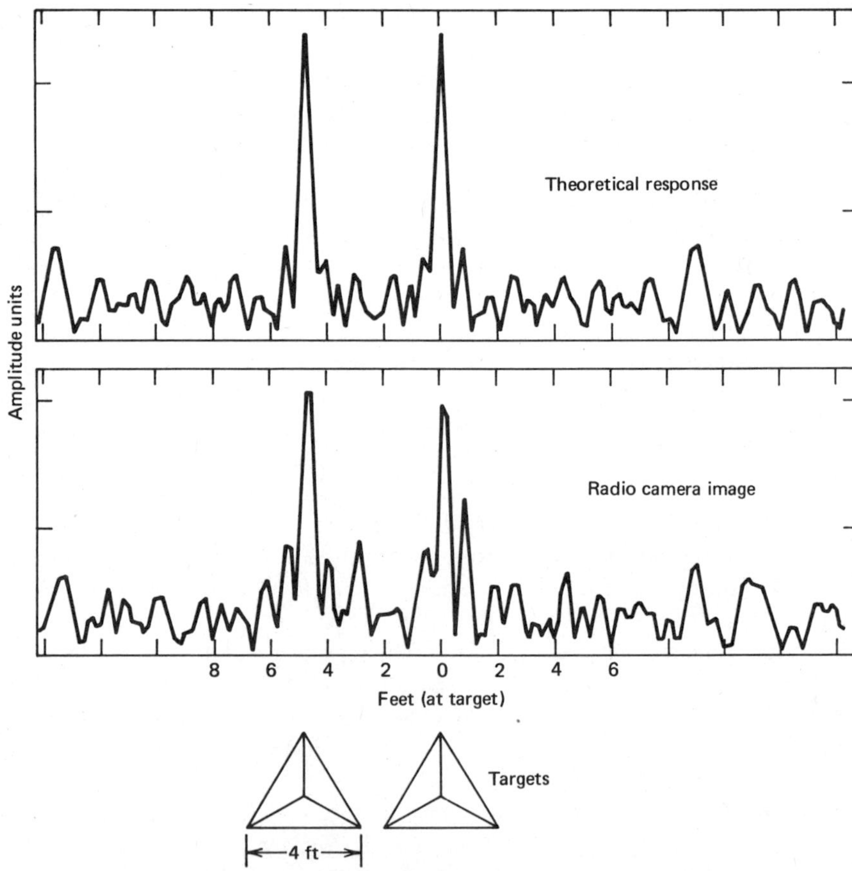

FIGURE 5.9. One-dimensional image of two corner-reflector targets with target-of-opportunity reference for adaptive beamforming. The target range is 240 m. Theoretical free-space response is shown for comparison.

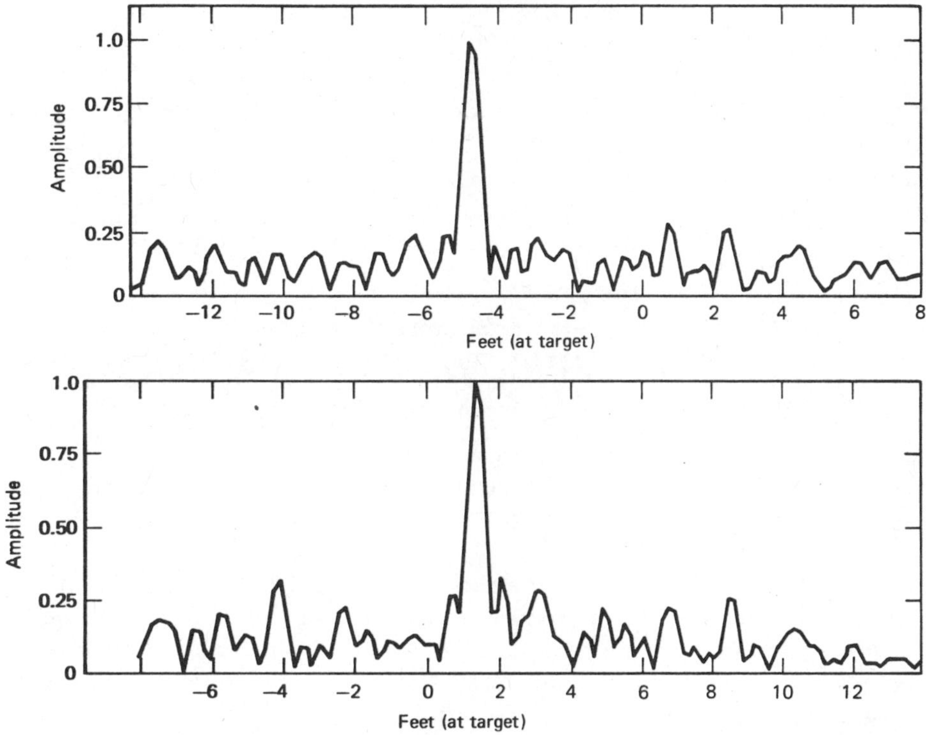

FIGURE 5.10. Microwave images of single corner reflector at two different angular positions. The target range is 240 m.

the beamformer in each case. Thus the upper image was made when the target was to the left of the beamformer; the opposite was the case for the lower image. The correlation between the target images is high, thus indicating that the target-reflecting characteristics did not change with position on the track. The correlation is also high with the two-corner target images in Figures 5.8 and 5.9. The degree of correlation between the two images in Figure 5.10 to the left and right of the targets implies that the remainder of the display is due in part to the radiation pattern of the random array and in part to clutter, receiver noise, or multipath.

Figure 5.11 is a one-dimensional image of a 0.6-m corner reflector and a rectangular, 0.4 × 0.5 m flat-plate reflector, oriented perpendicular to the line of sight to the array.

Figure 5.12 is a one-dimensional image of a pick-up truck with camper top. Also in the figure is a sketch of the vehicle which indicates its location and size. The high scattering regions are evident in the image. In Figure 5.13 the target is a motorcycle; 200 sample points were used in this experiment, which accounts for the lowered sidelobe level.

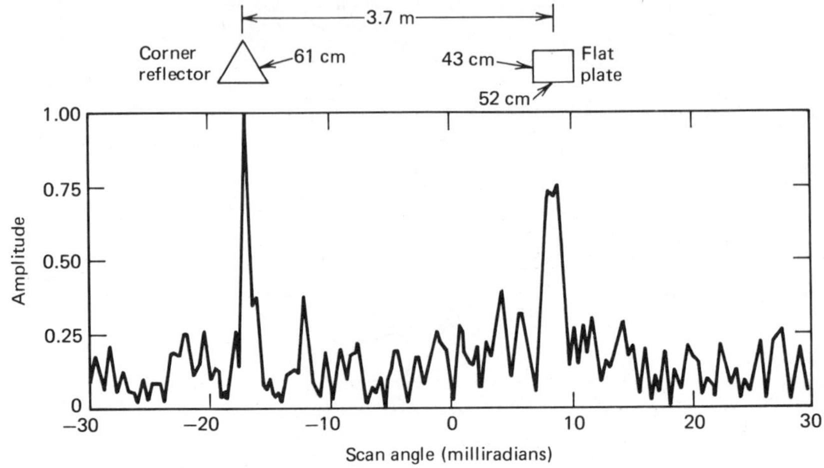

FIGURE 5.11. Measured system response to two dissimilar targets located in the same range cell. The target range is 145 m.

Because the random array sketched in Figure 5.7 is large only in the horizontal dimension, it can produce high angular resolution only in azimuth; its lack of vertical aperture precludes high angular resolution in elevation. The pulsed nature of the radar permits resolution in range, however. In this way a two-dimensional display in range and azimuth can be obtained from the radio camera. Furthermore, if the target is close enough to the radar, range resolution may be obtained from the small depth of field of the system, even if the transmission were continuous. The depth of field is given by (2.42).

In Figure 2.10 a two-dimensional contour map is shown which demonstrates range-angle imaging. To produce that figure the radio camera was scanned in

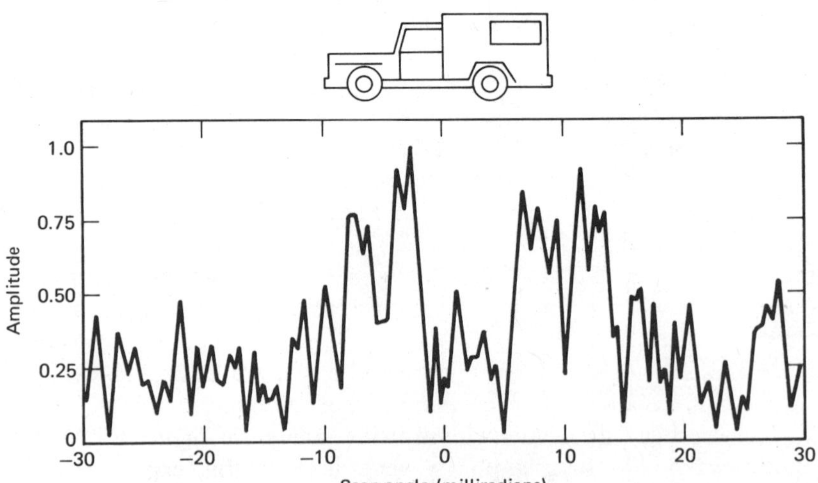

FIGURE 5.12. One-dimensional image of a large, complex target. The target range is 145 m.

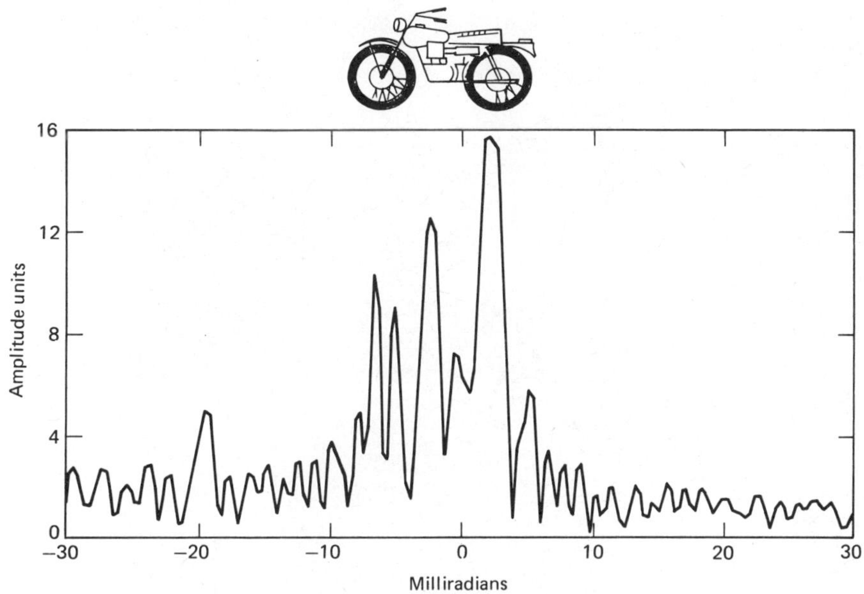

FIGURE 5.13. A one-dimensional image of a motorcycle as target. The target range is 145 m.

both range and angle. A single corner reflector was used for the target. As discussed in that chapter, the many artifacts are due to the range and angle sidelobes of the 100-sample-point array. Evident is the effect of the small depth of field of the system. Based on (2.42), the depth of field should be 8.5 m or 28 ft. (Remember that the effective wavenumber is not $2\pi/\lambda$ but twice that value because the experiment was conducted as a synthetic aperture.) This is also the measured value. Note that this interval is shorter than that defined by the pulse duration of the radar, which was approximately 15 m. Thus it is theoretically possible to achieve fine range resolution without a short pulse or wide bandwidth transmission. This theoretical potential of the radio camera is not likely to be realized, however, except in unusual circumstances.

The result of a more conventional two-dimensional imaging technique is shown in Figure 5.14*b*. In this image the range dimension of the resolution cell is determined by the pulse duration of the radar, whereas the cell's angular or cross-range dimension is determined by the size of the aperture. A bistatic experiment was conducted to produce this image. The transmitter was a modified unit from an APQ-102 Air Force radar. Its wavelength was 3 cm and its pulse duration, 60 ns.

The experiment is shown schematically in Figure 5.15. A single receiver was moved along a cable 10 m above the ground at the Valley Forge Research Center of the University of Pennsylvania. The cable was 39 m long and free to sway with the wind, which it did. The radar transmitter was pointed toward the town of Phoenixville, Pennsylvania, a distance of 6 to 7 km. Figure 5.16 is a map of the region. The reference radiation field was obtained from the echo from a 1.2-m corner reflector placed in a park in Phoenixville. The target area

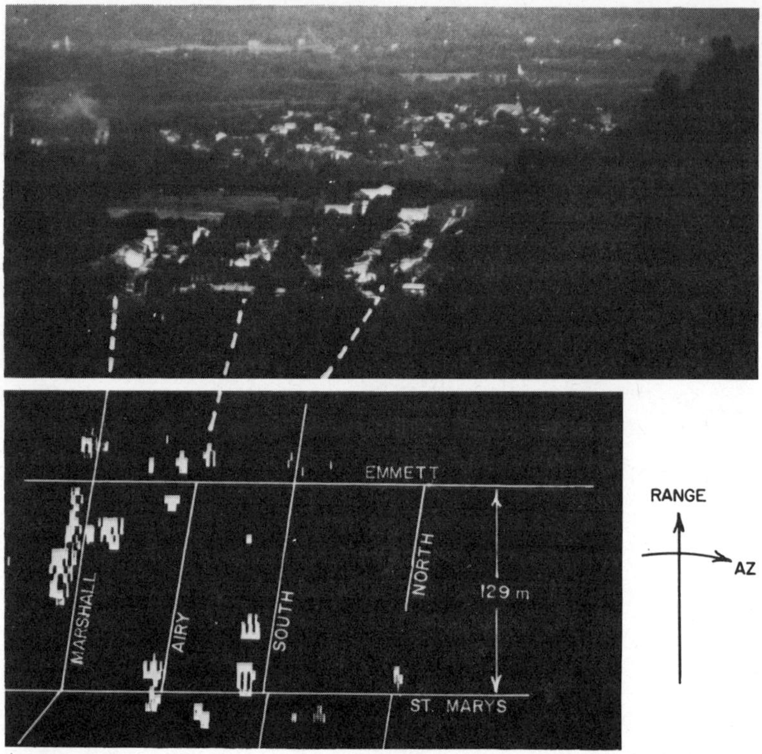

(a)

(b)

FIGURE 5.14. Images of a few square blocks in Phoenixville, Pa. (*a*) Infrared telephoto from radio camera array. (*b*) Radio camera image (streets superimposed manually). (The streets shown are in the upper central portion of encircled "target area" in Figure 5.16. The distance is 6.5 km.)

was a few square blocks located several hundred meters closer to the radar than the corner reflector. The angle of the target area from the array normal was 25° and the transmitter was pulsed 1000 times per second. The receiver was turned on 200 times at random intervals as it was moved over the 39-m course. The movement of the system took approximately 2 min. The estimated uncertainty in antenna location at each of the 200 samples points was about one wavelength rms.

The time-shared system should not be confused with classical synthetic aperture radars. SAR, described in Section 1.3, shares a common receiver–transmitter which is moved at constant velocity. The physical antenna is small in comparison to the synthesized aperture. In the experiment shown in Figure 5.15 the transmitter was separate from the receiver; only the receiver was time-shared. This time-shared bistatic imaging radar falls in the category of systems illustrated in Figures 1.1 through 1.5.

The two-dimensional radar map in Figure 5.14*b* is a small portion of Phoenixville (part of the target area circled on Figure 5.16). It is a black-and-white photo of a color display, hence carries no gray level. Clearly visible

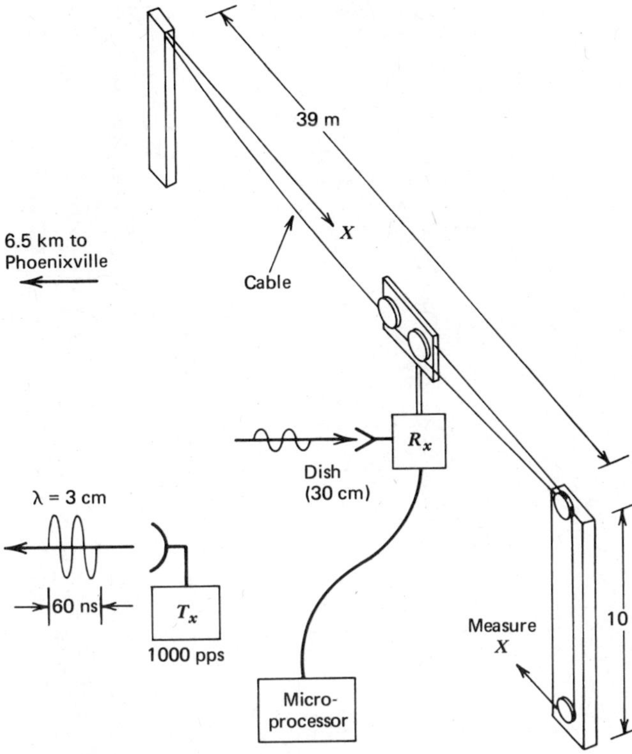

FIGURE 5.15. Cable-supported array for mapping Phoenixville.

are individual houses and street patterns (street locations have been drawn in). Part (*a*) of Figure 5.14 is an infrared photograph of the mapped area; the camera was located at the array center—white dotted lines were added to the photo to assist in the location of named streets. The resolution in part (*b*) is consistent with the radar and geometric parameters: it is 0.85 mrad in azimuth, which is less than 6 m at the distance to the target area and 9 m in range, corresponding to the 60-ns pulse radiated.

Although the microwave image in Figure 5.14 was made with a corner reflector for a beamforming source, experimentation has shown repeatedly that ample targets of opportunity exist to serve this function. Figure 5.17*a* is a radio camera map of approximately 1 km^2 of Phoenixville made with the same equipment but without using a corner reflector for the beamformer. Instead, the echoes from a large smokestack (some 2 km further in range) proved to be suitable. Figure 5.17*b* is an aerial photograph of the mapped area. The radio camera image was made from radar echoes at nearly horizon level. Because of the low radar grazing angle, most of the city is obscured by trees or buildings. An exception is the row of houses in the lower left in the area marked *E*. The individual houses can be seen clearly. The buildings' back surfaces face the radar and therefore are visible, but the facades are shadowed by the houses.

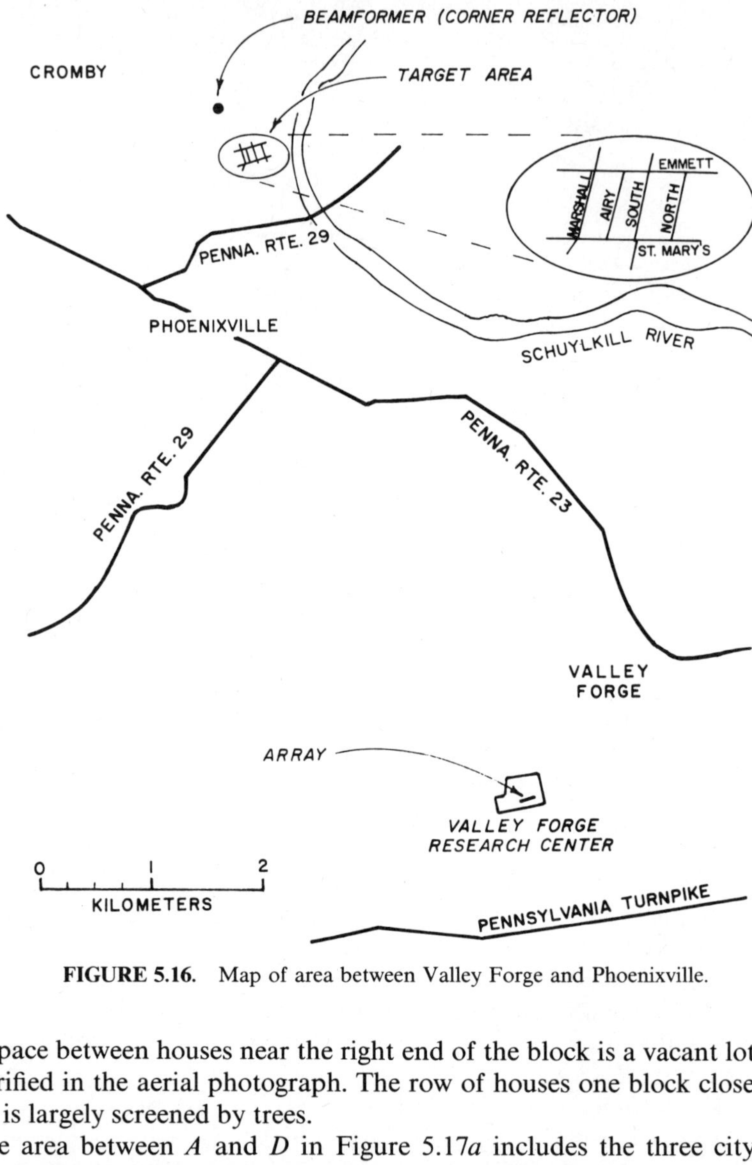

FIGURE 5.16. Map of area between Valley Forge and Phoenixville.

The space between houses near the right end of the block is a vacant lot, as can be verified in the aerial photograph. The row of houses one block closer to the radar is largely screened by trees.

The area between *A* and *D* in Figure 5.17*a* includes the three city blocks shown in Figure 5.14*b*. A church at the corner of North and St. Mary's streets, at *A* in Figure 5.17*a*, is radar-visible, whereas the field above it and to the right is nonechogenic; this region, a cemetary, is tree- and grass-covered, as shown in the aerial photograph (Figure 5.17*b*). The target at *B* in Figure 5.17*a* is a commercial water tower, seen in Figure 5.17*b* at the right edge of the square building roof just above the center of the aerial photograph. The target at *C* in Figure 5.17*a* is significant only because it provides a triangulation point for locating the street grid.

Figure 5.18 shows images of another target, the Cromby power plant, some 8.2 km distant from the array (i.e., about 2 km beyond the target area in

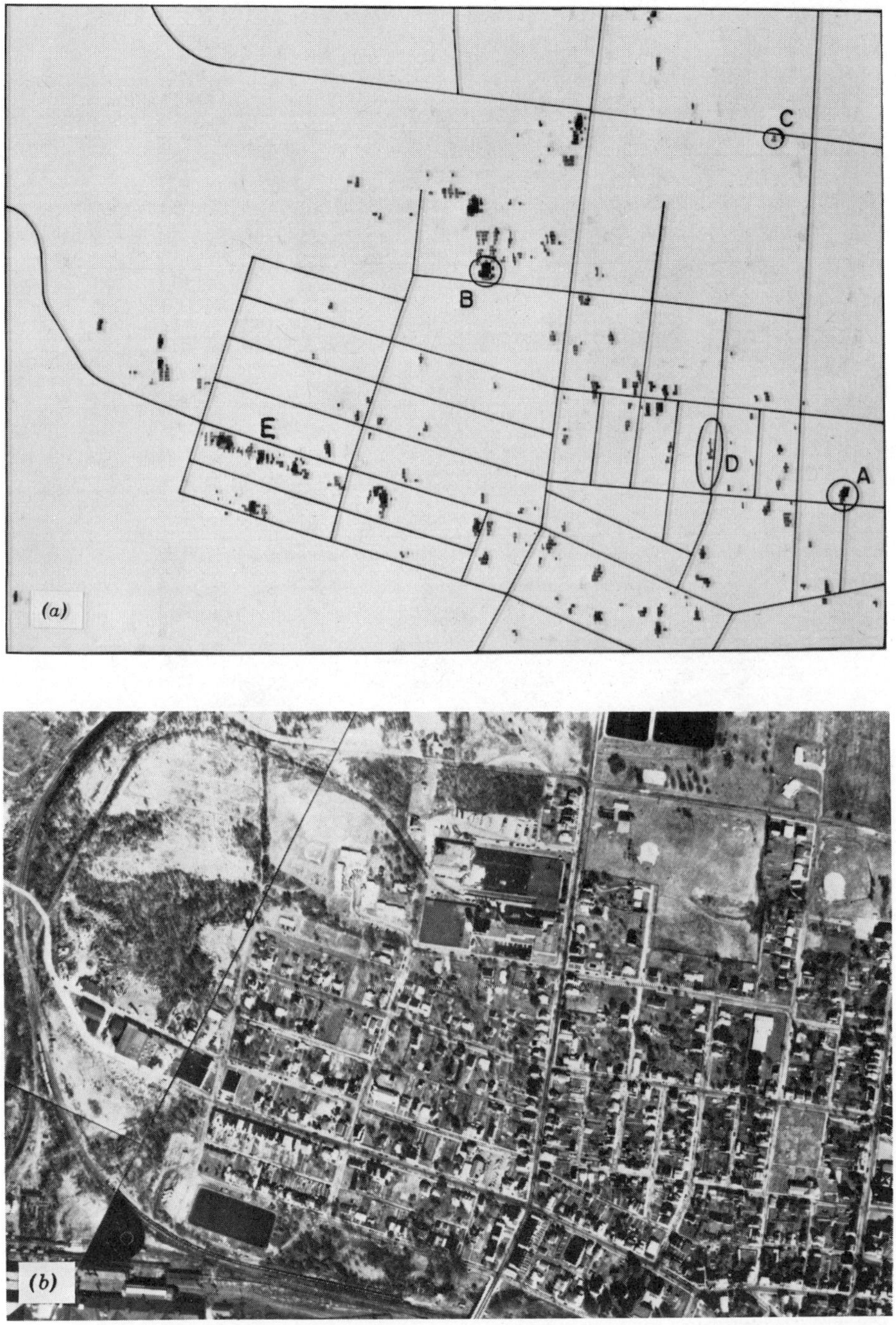

FIGURE 5.17. Images of a square kilometer area in Phoenixville. (*a*) Radio camera image (streets superimposed manually). (*b*) Aerial photograph. (The streets shown are to the left of the encircled "target area" in Figure 5.16. The distance is 6.5 km.)

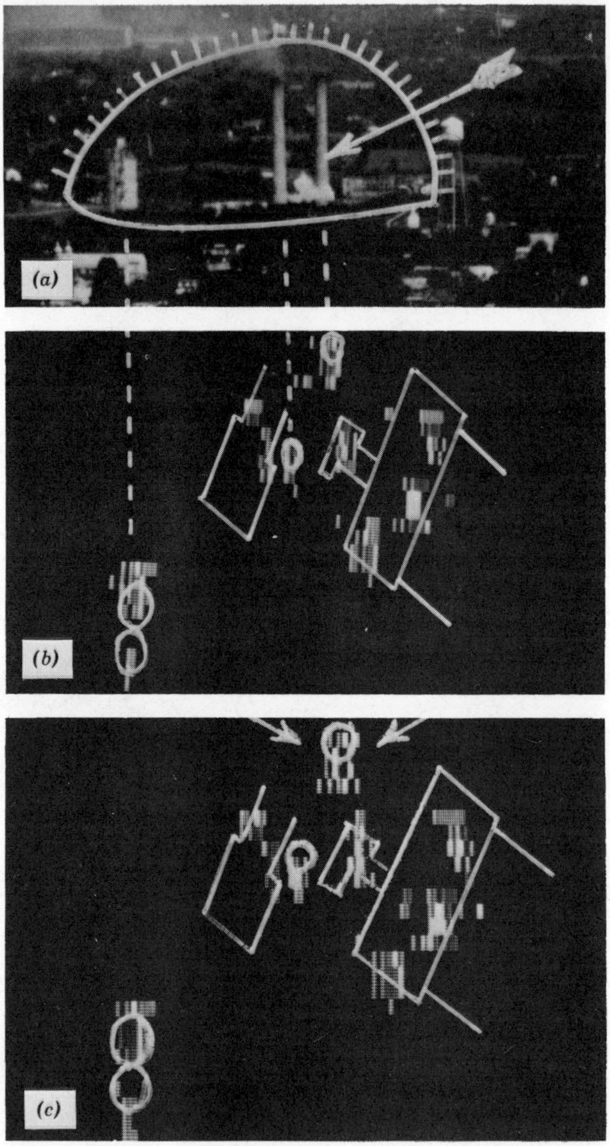

FIGURE 5.18. Optical and radar images of portions of a power plant. The distance is 8.5 km. Part (a) is an optical telephoto from the radio camera array. Part (b) is a radar image. The beamformer was a 1.2-m corner reflector. Part (c) is a radar image. The beamformer was a target of opportunity. The arrows here and in part (a) indicate the image and photo of the smokestack used as a beamformer.

Figures 5.14, 5.16, and 5.17). Figure 5.18*a* is an optical telephoto (from the antenna site) of the radar-visible features of the plant (inside the delineated area); parts (*b*) and (*c*) in Figure 5.18 are radar images of these features. (The water tower at *B* in Figure 5.17*a* is also visible in the photograph in Figure 5.18*a* just to the right of, and touching, the white-line–delineated area. This water tower is some 2 km closer to the radio camera than the power plant; therefore it does not appear on the radar images.) A 1.2-m corner reflector was used for part (*b*), whereas the radar target of opportunity used for Figure 17*a* was also used for adaptive beamforming in Figure 5.18*c*. That target of opportunity was the northernmost smokestack of the power plant and is marked by arrows in Figures 5.18*a* and 5.18*c*.

Figures 5.18*b* and 5.18*c* show the same targets, indicating that man-made structures may be adequate for phase-synchronizing the array. Outlines of the power plant's major buildings, stacks, and structures have been added manually to the figures to indicate the related positions of the plant's radar-visible features (i.e., radar-visible only above the masking trees and ground elevations near the plant).

In Figure 5.18 the left-hand target in parts (*b*) and (*c*), shown photographically in the delineated area and above the left-hand dotted line in part (*a*), is of particular interest. The target itself consists of two huge, cylindrical, silo-type structures, touching each other along a common cylinder element, and oriented one behind the other along the line of sight to the radio camera. Normally, only the nearest element of the front silo would be radar-visible. The two silos, however, seem to appear as two targets separated in range (vertical dimension in figures). The center of density of the radar image of the nearer (*lower*) silo is correctly located at the front element of the nearer cylinder (note the relation of the lower part of the target image to the superimposed lower circle that indicates the true location of the silo). The center of density of the apparent radar image of the more distant silo (*upper*) appears to be centred just in front of the rear element of the rear silo. This apparent anomaly was clarified by a physical inspection of the power plant premises; the top of the near silo is flat and clear of additional structure, whereas at the top of the rear silo, and extending above and slightly in front of the rearmost element, is a large openwork girder structure that encompasses the silo-loading facilities [visible at the top of the silo's photograph in part (*a*)]. Thus the radar image of the silo complex is dimensionally correct and nonanomalistic.

5.8. ISOLATED TARGET

Common Target for Beamforming and Imaging

The situation is somewhat different for an isolated object like an aircraft or ship, the reference source of which is on the target. One difference is that no range refocusing (Step 6 in Table 5.1) is required because the size of the

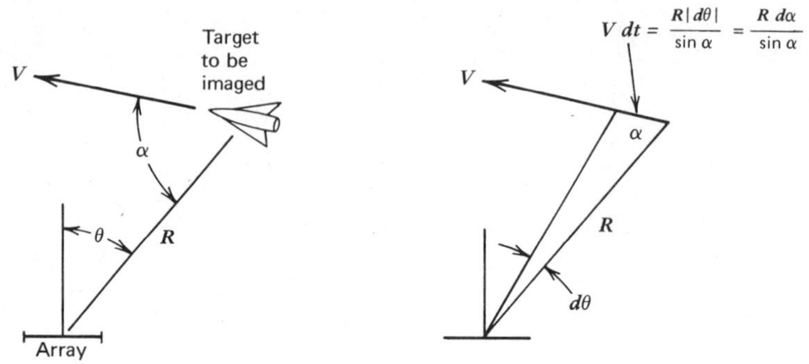

FIGURE 5.19. The aspect angle α changes with time. (From B. D. Steinberg, *IEEE Trans. Antennas and Propagation*, AP-29, No. 5, September 1981.)

aircraft will usually be small compared with the depth of field of the array. A more significant difference lies in the fact that target echoes are used for adaptive phase synchronization as well as for imaging. The requirements of the echo characteristics are opposed for these two processes. Phase synchronization requires a dominant point source, whereas the purpose of picture taking is to reproduce the angular backscatter profile of a complicated target for classification or characterization.

The conflict resolves itself when the target is moving in relation to the observer. Figure 5.19 shows an aircraft at distance R with its velocity vector V making an angle α with the direction to the array. The aspect angle α changes with time at the rate $d\alpha/dt = V\sin\alpha/R$. Because of this turning motion, the radar echo changes with time, sometimes having the desirable properties for phase synchronization; at other times the return signal is good for imaging purposes. From time to time a strong highlight appears, due, say, to a broadside specular return from the fuselage or a two-plane corner reflector formed by wing and fuselage or in the tail assembly. The highlights are severely aspect-angle-dependent, as demonstrated in Figure 5.20, in which a small flat-plate reflector of length T is shown momentarily oriented broadside to the direction of the array. As the line-of-sight angle θ changes as a result of platform motion, the reradiation pattern moves through twice that angle. Consequently the time that the mainlobe dwells on the array is determined by the turning rate $d\alpha/dt$ and the reradiation beamwidth ($\approx \lambda/T$). The duration or correlation time can be estimated as the time required for the differential phase of the echoes from two scatterers separated a distance equal to the target size T to change about $\pi/2$. Figure 5.20 (*right*) shows the geometry. The change in the differential distance δR from two scatterers is approximately $T\,d\alpha$. The rate of change is $d(\delta R)/dt \simeq T\,d\alpha/dt \simeq TV\sin\alpha/R$ and the phase difference between their echoes changes at a rate $d(\delta\phi)/dt = 2k\,d(\delta R)/dt \simeq 4\pi TV\sin\alpha/\lambda R$. Hence the correlation time is the order of

$$T_c \simeq \frac{\lambda R}{8TV\sin\alpha} \tag{5.8}$$

Modulation Bandwidth Test

Consider a Mach 1 aircraft with a reflecting region 1 m in size, normally oriented to the radar line of sight. When observed with 3-cm radiation from a distance of 20 km and at a 30° angle the correlation time is about $\frac{1}{2}$ s. During this time the large array on the ground may be synchronized to form a narrow beam focused on the specular reflector. Rather than attempting to determine *at what range* to phase-conjugate or focus, as in the earlier discussion (Step 3 in Table 5.1), the signal processor now decides *when* to perform this operation. Its criterion is exactly the same; that is, reasonable uniformity of echo strength across the array. Furthermore, it can apply an additional and highly sensitive measure: the correlation time, which is the reciprocal of the echo modulation bandwidth, varies inversely with target size. When the highlight disappears the "target" in (5.8) becomes the size of the aircraft. If the 1-m flat-plate reflector were on a 30-m aircraft, the modulation rate would jump by a factor of 30 following the $\frac{1}{2}$-s phase synchronizing period. Similarly, the echo modulation bandwidth would drop by this factor when a stable echo appears, thereby permitting the signal processor to determine when to phase-synchronize.

Observation of the drop in echo modulation bandwidth can be made in the low-frequency control branch of the phaselock loop, if such a circuit is used for adaptive beamforming, or in the data store of a real-time digital processor, in which case the data may be organized in the format shown in the lower part of Figure 5.2. The exception is that the horizontal coordinate is no longer range but time. All the data come from the same range bin and the columns are spaced by one interpulse period. Successive columns are successive range-gated samples of the echoes at each antenna element. The processor observes the data bandwidth and the amplitude stability to determine when to phase synchronize (Step 3). During the dwell time of a highlight the signal processor ceases the imaging process and phase-conjugates instead (Step 4). Next it observes when the modulation bandwidth returns to normal, at which time it begins the imaging process again (Steps 7 and 8). Following the high-resolution

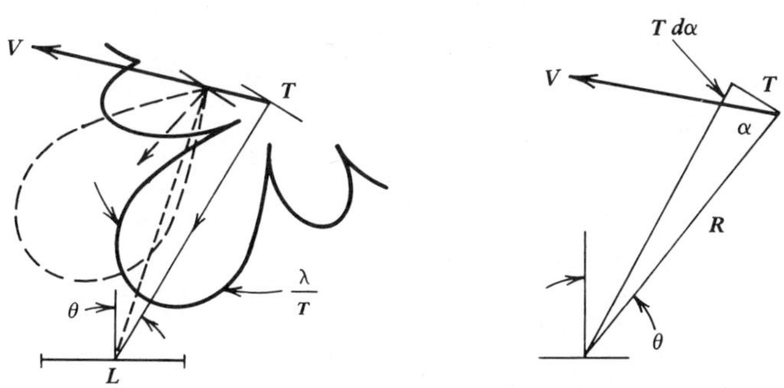

FIGURE 5.20. Reradiation pattern rotates with platform motion. (From B. D. Steinberg, *IEEE Trans. Antennas and Propagation*, AP-29, No. 5, September 1981.)

angle scan, it awaits the next synchronizing interval to repeat the process. The cycle time is the order of 1 s. During several seconds the aircraft has moved and rotated insignificantly insofar as the observer is concerned but has altered its orientation sufficiently to cause successive images to be statistically independent. Hence the logical next step is to form "multiple exposures"; that is, sums of intensities of several successive images. In the context of Chapter 3 this process might be called time or position diversity. Its application reduces the sidelobe peaks of the random array, builds up the information content in the image, and fills in the speckles or glint that result from the narrowband (near monochromatic) radiation.

Measured Correlation Intervals

An experiment has measured highlight dwell times and the percentages of time that aircraft targets are in the "synchronizing" and "imaging" modes [17]. The basis for the measurement is the expected change in the pdf of the echo amplitude when a highlight occurs. Without it the echo is the vector sum of a large number of small, randomly phased echoes. The sum is a two-dimensional, zero mean, random Gaussian variate, the amplitude of which is Rayleigh distributed; that is, if A is the echo amplitude and $2\sigma^2$ is its mean-square value, the pdf of A is $f(A) = (A/\sigma^2)\exp(-A^2/2\sigma^2)$. When a strong scatterer of strength A_0 is added, the pdf changes to the Ricean distribution $f(A) = (A/\sigma^2)\exp[-(A^2 + A_0^2)/2\sigma^2]I_0(AA_0/\sigma^2)$, where I_0 is the modified Bessel function of the first kind and zeroth order; $f(A)$ reduces to the Rayleigh pdf when $A_0 = 0$ [70].

The experiment was designed to gather simple statistics on the short-time pdfs of radar echoes from aircraft. The radar was an L-band AN/TPS-1D located at the Valley Forge Research Center of the Moore School of Electrical

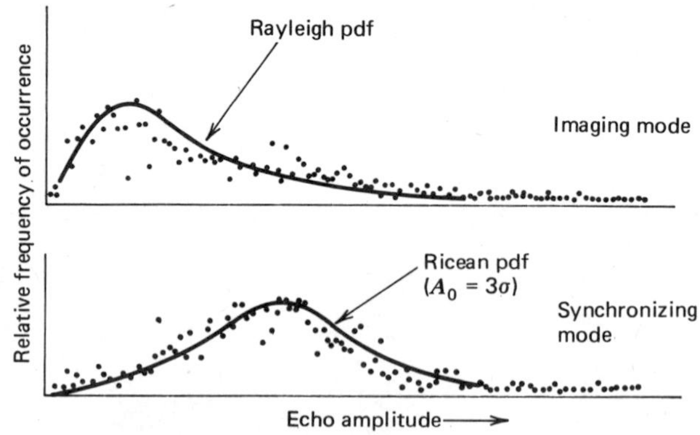

FIGURE 5.21. Histograms of aircraft echo sequences. (From D. L. Carlson, "The Phase Synchronization of Large Random Arrays Using Specular Aircraft Radar Return Signals," M.S. Thesis, University of Pennsylvania, 1975.)

TABLE 5.2. Relative Frequencies of Occurrence of the Imaging and Phase-Synchronizing Modes [17]

View	Number of Aircraft	Imaging Mode (Rayleigh pdf)	Synchronizing Mode (Ricean pdf)	Neither
Broadside	3	.594	.286	.120
Nose	2	.501	.363	.136
Tail	4	.668	.048	.284
All	9	.606	.193	.201

TABLE 5.3. Approximate Correlation Times for Aircraft Echoes [17]

View	Number of Aircraft Observed	Average Correlation Time (s)	Average Deviation (s)
Broadside	16	1.66	0.68
Nose	7	0.30	0.04
Tail	4	0.17	0.11

Engineering, University of Pennsylvania, at Valley Forge, 40 km west of Philadelphia. Commercial aircraft flying in the Philadelphia–New York–Scranton triangle were observed and tracked. Distances varied from 15 to 100 km. Echo sequences of varying numbers of pulses were sorted by amplitude in real time in a histogram generator and the nature of the pdf was judged to be Rayleigh, Ricean, or nondescript.

The results are shown in the tables above for broadside, nose, and tail aspects. Sample histograms are shown in Figure 5.21. As expected, Rayleigh and Ricean pdfs did occur, the former with about three times the frequency of the latter (Table 5.2). About 20% of the sample runs showed no preference for either one but exhibited instead, for example, tendencies toward uniformity or bimodality. The average correlation time in each mode was also measured (Table 5.3). Note that the $\frac{1}{2}$-s dwell time estimated earlier for the sync mode is within the range shown. The measurements confirmed the expectation that aircraft will alternately permit array synchronizing and imaging.

5.9. CIRCUITS AND SYSTEMS

The essential ingredient required for self-cohering a radio camera is the phase of the echo received at each receiver from the reference reflector. Measurement of phase requires that each array module has a local oscillation to which the received echo can be compared in phase, that the local oscillator (LO) is correctly frequency-locked to the transmitter frequency, and that its phase has some fixed and known relationship to the array geometry or to the remainder of the system. The phaselock loop (PLL) is a natural circuit for controlling an oscillator in the receiver chain; it ensures that the final detection waveform, that is, the LO for the final mixer, is both frequency- and phaselocked to the radar transmitter.

Among the many useful designs for self-synchronization (e.g., [47], [39], [18]) two that have been found valuable in the Valley Forge radio camera program at the University of Pennsylvania are described in this section. The first, called the broadcast reference mode, is broadly applicable to ground-based and airborne designs [68], [94]. The second is a time-shared processor which is most appropriate for the synthetic aperture mode of operation [95], [6]. Both provide digital signals to a central processor in a form suitable for applying the algorithm in Sections 5.3–5.5.

Broadcast Reference

Figure 5.22 illustrates the broadcast reference system in one of its many possible configurations. With each receiving antenna in the distributed array is a microwave receiver that heterodynes the RF echo to IF, amplifies it, and quadrature-demodulates the signal to video coherently, where the quadrature components are sampled and A/D-converted. The phase of the digital output contains not only the desired phase information due to radar-target-receiver geometry but also an arbitrary (and sometimes time-varying) phase due to the accumulated phase shifts in the sections of the receiver.

The broadcast reference system is designed to compensate for the latter random component of the echo phase perceived in the video section of the receiver. It does this by providing a reference signal that enters at the antenna and passes through the entire receiver. The reference thereby experiences (almost) exactly the receiver phase shift acquired by the target signal. The phase difference between the two therefore removes the effect of receiver phase shift.

The transmitter in Figure 5.22 radiates a gated and amplified stable RF oscillation, derived from a stable local oscillator (STALO). This high-power pulse is the main radar pulse. Accompanying it is a low level (30-dB down) controlled RF wave with a window at target range phase-synchronous with the high-power pulse. This wave is launched to the target by the mainlobe of the radar antenna and also illuminates the receiving array elements by leakage through the sidelobes. In this way a nearly continuous RF wave is broadcast across the array, being turned off only for a short interval (10 to 100 pulse durations) at the time that the echo of the target to be processed is expected.

This quasi-CW signal is a phase-stable, delayed replica of the stable oscillator waveform in the transmitter. Its function is to synchronize the receiver. During the synchronization interval the receiver is organized as shown in the left circuit in Figure 5.22. The wave train passes through the antenna and is mixed with a free-running local oscillator. The amplified IF wave drives

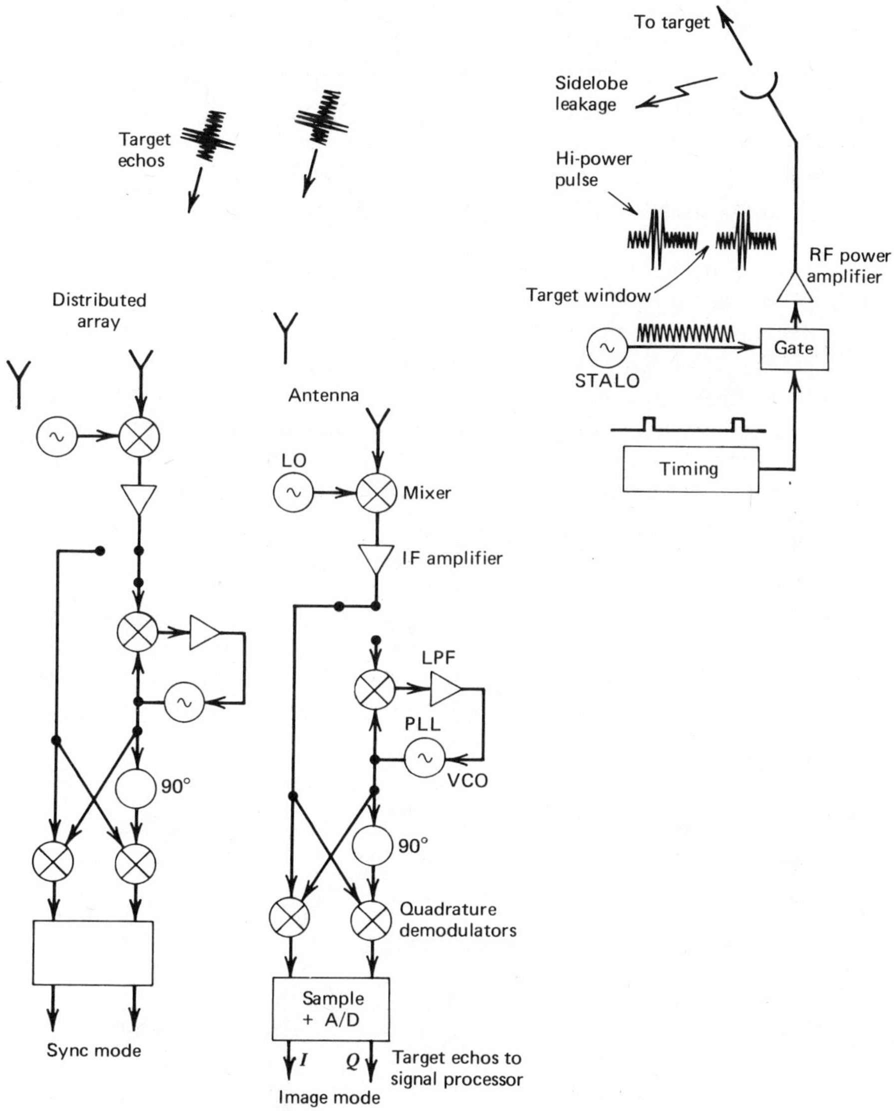

FIGURE 5.22. A broadcast reference system.

a PLL, the output of which becomes phaselocked to the transmitter oscillator via the free-running receiver local oscillator. When this occurs the phaselock-loop output becomes the proper detection waveform for the target echo.

The target echo arrives during the time that the reference wave is turned off. To detect the target a four-pole IF electronic switch converts the receiver from the synchronizing mode to the target detection mode. The left module in Figure 5.22 shows the synchronizing mode, the right module, the detection mode. During the target-window interval the switch connections are changed and the IF target echo is delivered to the quadrature demodulators, from which the inphase (I) and quadrature (Q) video components of the echo are derived.

The local oscillator frequency may vary from module to module. Its exact frequency in a module is unimportant because it operates on both reference wave and target echo. It must, however, satisfy two conditions. The first is short-term stability: it must be phase-stable during the interval between the arrival of the reference wave and the target echo. This interval is normally not greater than hundreds of microseconds, a condition easily satisfied in commercial local oscillators.

The second condition pertains to its frequency. Although its exact frequency is not critical, the beat frequency from the PLL mixer must be within the acquisition range of the loop. Thus some attention must be paid to its tuning.

Figure 5.23 shows an alternate transmitter configuration in which a separate low-gain antenna, excited by the low-level stable oscillator, broadcasts to the array. Several alternate synchronizing waveforms appear in Figure 5.24.

Synthetic Aperture Mode

The radio camera technique can be applied to synthetic aperture radar to obviate the need for accurate knowledge of platform trajectory. As mentioned in Chapter 1, the combination is called SARC, which stands for synthetic aperture radio camera; it has also been called TASAR, for thinned adaptive

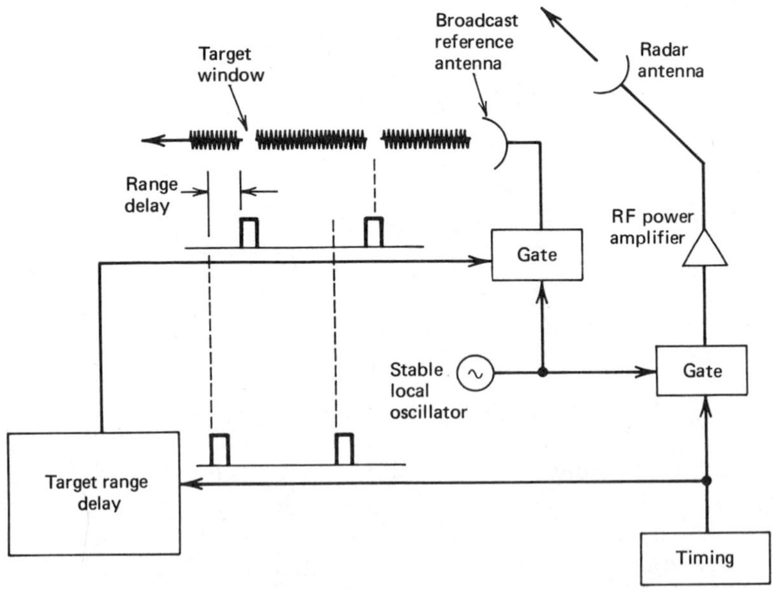

FIGURE 5.23. An alternative transmitter configuration.

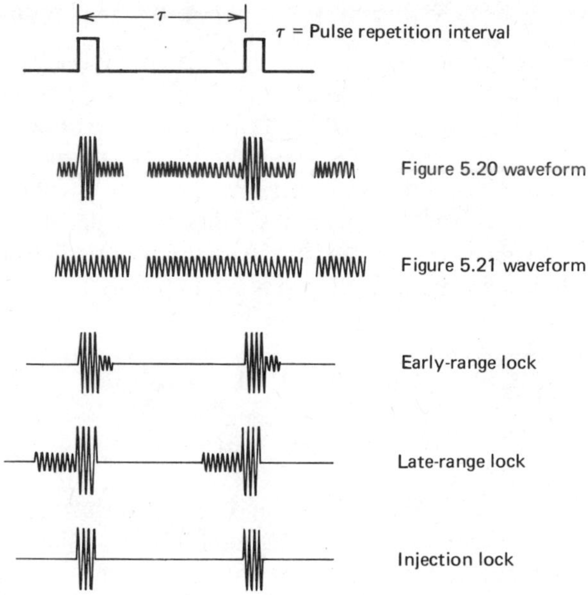

FIGURE 5.24. Various waveforms for the broadcast reference method.

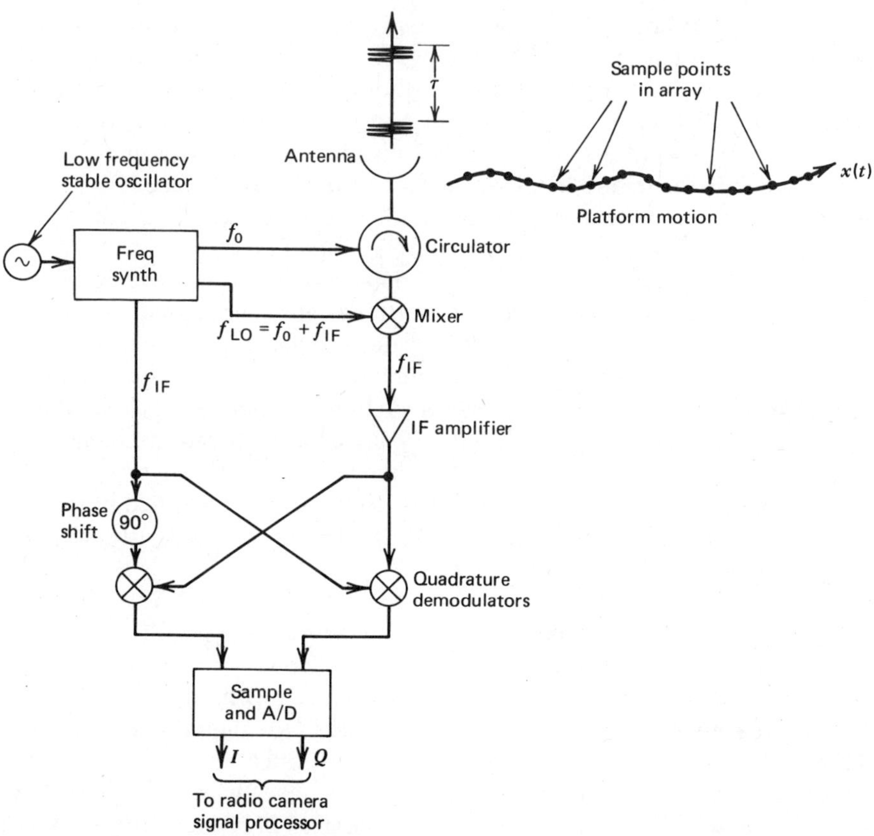

FIGURE 5.25. SARC: synthetic aperture radio camera.

synthetic aperture radar [6]. Figure 5.25 shows an airborne platform moving erratically. The locations of the radar when pulses are transmitted can be viewed as element locations in a distorted array. The radio camera signal processor can be driven directly by the coherent radar (as shown). In that embodiment a stable low-frequency (typically 10 to 100 MHz) oscillator drives a phase- and frequency-locked frequency synthesizer to generate phase-related waves for RF transmission, for the microwave local oscillator and for the coherent IF oscillator, a system described in Section 5.7.

5.10. ALTERNATE TECHNIQUES

Reference Search

The algorithm described in Sections 5.4 and 5.5 performs four distinct functions. It is not the only procedure for imaging, however. Other algorithms may be used, provided that the same functions are performed; for example, the first function is the search for and location of the reference source. The minimum amplitude-variance test is a sensitive and powerful test for the best beamforming source, but it is not the only one. A simpler test is maximum target strength (normalized by the fourth power of range to account for propagation loss). Although not so satisfactory as the minimum echo-variance test, which is sensitive to the physical size of the target as well as to propagation conditions, this test tends to identify strong, specular reflectors. If the range search for the strongest echo is made at a single location in the array and tests are made at several other locations to corroborate the fact that the echo is large at other points, a useful reference-search procedure results. Similarly, the modulation bandwidth test in Section 5.8 is useful for determining *when* a moving target is suitable for serving as an adaptive beamforming reference source.

The system in Figure 5.26 which illustrates reference-range selection based on detection of the strongest echo in a single receiver shows a main radar and an array module. The main radar has a high-gain antenna through which a high-power transmitter pulse is radiated. The same high-gain antenna drives the receiver. Because of the use of the high antenna gain on reception, the reference reflector has sufficiently high SNR to be readily identified in the echo trace, thereby permitting reference range R_0 selection by conventional range-gate techniques.

Phase Control at the Antenna Module

The second function of the algorithm is adaptive beamforming and focusing on the reference source. This function was performed in the digital signal processor described in the earlier section. It can be performed in hardware circuits instead; for example, adaptive beamforming can take place in an adaptively controlled phase shifter. The logical circuit for this purpose is the phaselock

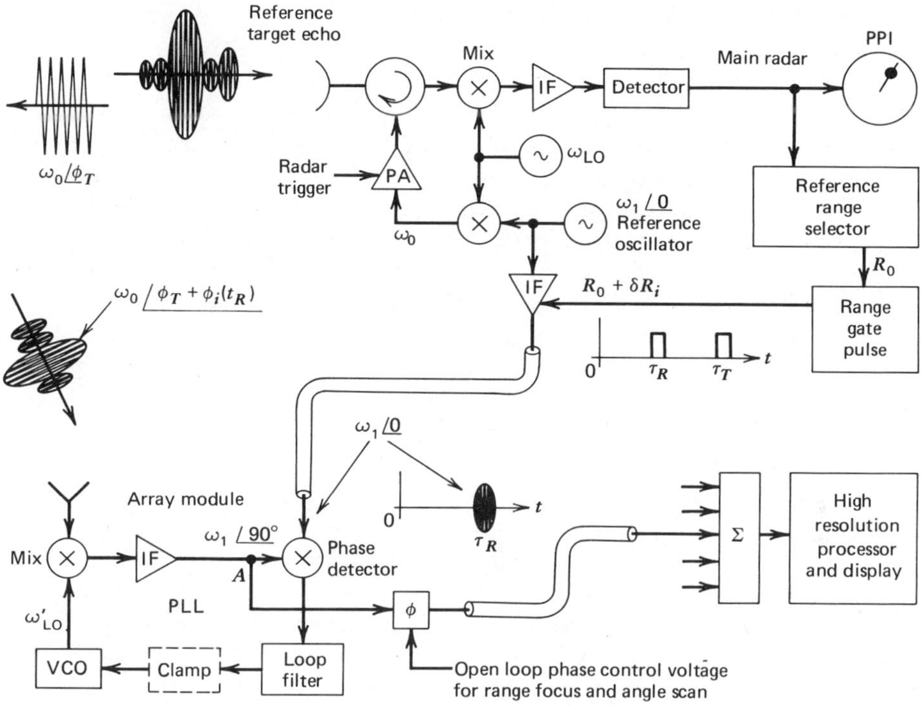

FIGURE 5.26. Adaptive phase control within each array module.

loop. Figure 5.26 shows one way in which the PLL can be used to perform adaptive phase control for the self-cohering operation. At the lower left is an array module which consists of an antenna, mixer, and local oscillator. The LO is a voltage-controlled oscillator (VCO). The IF signal is amplified and mixed in a phase detector with a reference wave of the same frequency and reference phase. The beat product is smoothed in the loop filter to control the VCO.

The reference wave is delivered from the main radar (in the upper right), which provides the microwave illumination and the reference oscillation for the entire system. The reference oscillator in the main radar at $\omega_1 \angle 0$ mixes with a local oscillator at ω_{LO} to produce the radiated frequency ω_0. The continuous wave at ω_0 is gated and amplified in the power amplifier (PA) and delivered, via the circulator, to the radar antenna. The radiated transmitted pulse has arbitrary phase ϕ_T. The same antenna receives the echo sequence shown, which includes a large reference-target echo. After detection and amplification the large target echo at time t_R is selected by the operator or by an automatic detector and delivered as a range-gate pulse to the line-driver amplifier of the reference oscillator. Thus the reference-range selection shown in Figure 5.26 is based on the alternate reference search test described at the beginning of this section.

The gate pulse permits the amplifier to pass the reference oscillation at the time when the reference target echo arrives at the array module. The reference oscillator wave therefore acts as a carrier of information regarding the expected arrival time of the reference echo. The gate pulse actuates the IF amplifier by a switch (not shown) which is closed only while the system is in the synchronizing mode. (When the system is performing the imaging operation the switch is open and reference oscillation is applied continuously to the array module.) In this way an IF pulse at reference frequency is delivered to the phase detector in the receiver of the array module. Not only must its pulse be at reference frequency it must also have reference phase. (Delivery of reference phase can be a complicated matter. It is treated separately later in Chapter 8.) By gating the reference wave in this fashion the circuit develops the control voltage from the reference-target echo, which arrives at the module at the same nominal time as at the main radar. Thus the loop responds to the reference target.

Time of Arrival Correction

Because the array of a radio camera may be large, the reference target echoes may arrive at the array modules at slightly different times; for example, let the array length be L, the direction to the synchronizing source, θ_0, and the pulse duration, δ. The differential arrival time across the array is $\Delta t = L \sin \theta_0 / c$. Unless $\Delta t \ll \delta$, the signal-to-noise ratio of the system will suffer. To avoid this loss the correct range $R_0 + \delta R_i$ is calculated from the geometry of the system and a gate pulse at the proper time $t_R + \delta R_i / c$ turns on the line-driver amplifier. Although the positions of the antenna modules are not known accurately enough for phasing the array (small fraction of a wavelength), the geometry is generally known with adequate accuracy for this task (small fraction of a pulse duration).

The reference target echo is shown arriving at the ith module with frequency ω_0 and phase $\phi_T + \phi_i(t_R)$. When heteodyned with the local oscillator its frequency is $\omega_1' = \omega_0 - \omega_{LO}'$ and its phase is some arbitrary value, say ϕ_1. After loop closure the frquency ω_1' is forced to equality with ω_1; hence ω_{LO}' becomes equal to ω_{LO}. Because the loop is a null-seeking servo, the two IF inputs to the phase detector are in phase quadrature. Thus the reference target echo arriving at the phase detector with arbitrary phase ϕ_1 is phase-rotated into a fixed phase relation with the system reference. This operation is the functional equivalent to the second task performed by the signal-processing algorithm in Sections 5.4 and 5.5.

Continuous Frequency Lock

Prior to the initiation of this process the PLL normally would be unlocked and the VCO would run at its free-running frequency ω_{LO}'. When the reference target is selected in the main radar the loop has to converge in frequency ($\omega_{LO}' \rightarrow \omega_{LO}$), after which it locks inphase. The phaselocking process, which is the self-cohering operation for the array, can be accelerated by keeping the PLL in frequency lock even when no reference target is present. This can be accomplished by removing the range-gate pulse in the main receiver to allow the phase reference wave to run continuously. In this way the loop locks on clutter, which has the correct frequency, albeit some meaningless phase. The VCO frequency is therefore correct at all times, and only its phase has to be altered when a target reaches the radar beam, thus shortening the phase acquisition time.

After phase acquisition on the reference target the job of the PLL is done. The input to the VCO is then clamped (shown in the dashed box in Figure 5.26) to the residue voltage out of the loop filter which drove the loop into phaselock. The loop now acts as a fixed phase shift between the antenna and point A in the circuit. Following this operation, the range selector in the main radar moves the range-gate position for the reference oscillator to the target range or time t_T, which is depicted as following the reference range t_R in the sketch in Figure 5.26.

Open-Loop Functions

The major remaining functions, range refocusing and angle scanning, involve open-loop phase shifting *after* the closed-loop phase shift has been completed. Assume that a PLL circuit, shown at the lower left in Figure 5.26, is part of each of the array modules. The IF signal at point A in each loop is then in quadrature with the reference wave. All waveforms are cophased and therefore ready for the subsequent open-loop phase-shift operations. Each voltage can be delivered to an open-loop phase shifter for range-focus and angle-scan voltage control. This phase shifter can be an IF component, as shown, in which case it would be part of the array module. Its output would be delivered to a voltage summer in the signal processor which, most likely, would be colocated with the main radar. The output would be demodulated, processed to enhance image quality, and displayed.

Doppler Correction

This design has an interesting added feature for a ground-based system so large that the Doppler-shifted echoes from an airborne target differ in frequency across the array. The reference target echoes are used to phase-synchronize the array in phaselock loops. These loops act as the bank of phase shifters for the adaptive beamforming process. When the target reference is aboard a high-speed moving object, the loops synchronize in phase, notwithstanding the fact that the geometry, including the transmitter, target, and array, is time-varying. Thus the instantaneous Doppler shift in each echo is corrected in each module.

There are two penalties for this attractive feature. One is the time required for the phaselock loops to close, which in some applications may be excessive, and the other is in the method of delivery of the reference signal from the main radar to each array module.

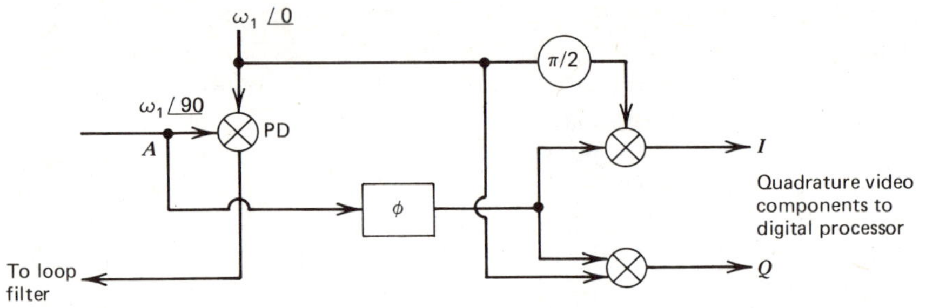

FIGURE 5.27. Transmitting the coherent video to the signal processor in the main radar.

Delivery of the phase-shifted IF signals to a voltage summer, shown in Figure 5.26, requires that all IF circuits between the open-loop phase control and the signal processor maintain identical phase shifts through them. In a large system it is easier to maintain the required phase integrity from element to element by demodulating each IF signal coherently at the module and transmitting the inphase and quadrature video components to the signal processor. This is shown in Figure 5.27. Figure 5.28 illustrates another modification to the circuit in which the RF VCO is replaced by a fixed-frequency LO running at ω'_{LO} and the voltage-controlled component is moved to the IF portion of the receiver. The advantage of this circuit over the preceding one is the availability of PLL-integrated circuits at IF.

Because maintaining the phase integrity of IF cables from open-loop phase shifters to summer is a problem, so too is the delivery of the phase reference of the reference oscillator from the main radar to the antenna elements. This difficulty is a central problem in systems that transmit as well as receive through the same distorted array (Chapter 8).

A second complication in this circuit is that the input to the loop filter that drives the PLL into phaselock is a low duty-cycle pulse train. Although the loop can be locked in frequency by the nearly continuous sequence of clutter echoes, phaselock occurs only after a reference target is identified. The loop filter therefore must operate with a sampled-data input. Both analog and digital circuits can provide the long-term memory needed to integrate the voltage control for the VCO and to hold it between input pulses. One circuit is

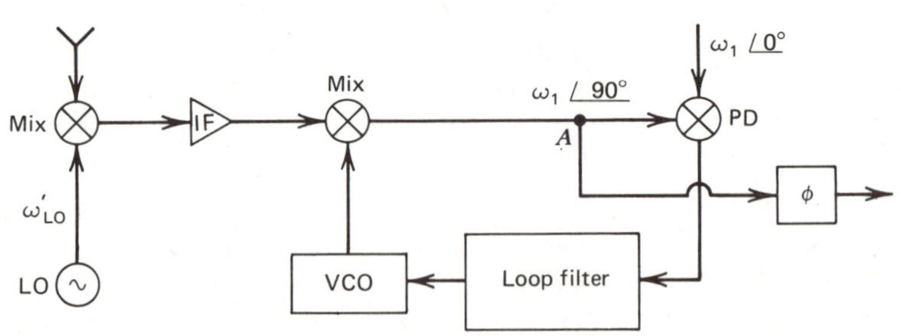

FIGURE 5.28. Voltage control in the IF section of the receiver.

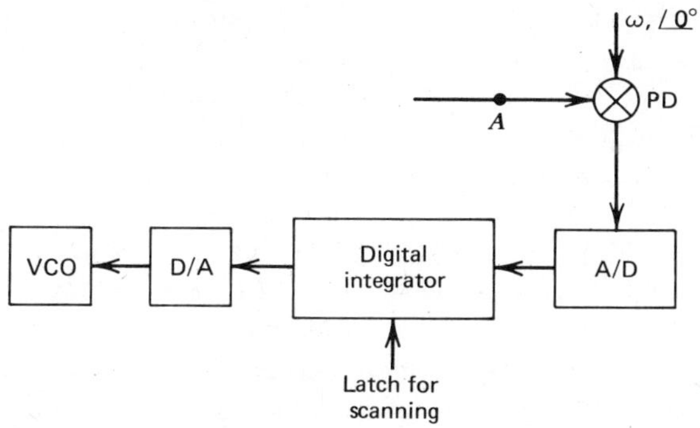

FIGURE 5.29. Digital loop filter for sampled-data voltage control.

the sweep integrator [80], which can be implemented in analog and digital circuits. Another is a simple digital integrator following an A/D converter driven by the pulses from the phase detector (Figure 5.29). This circuit has the added virtue of latching (holding constant) the VCO control voltage after adaptive beamforming during the period when the beam is scanned.

Data Rate Reduction

When the array is small enough for it to be practical to deliver the RF element signals to a central processor a digital implementation that eliminates both problems becomes available. Central phase referencing (a term introduced by

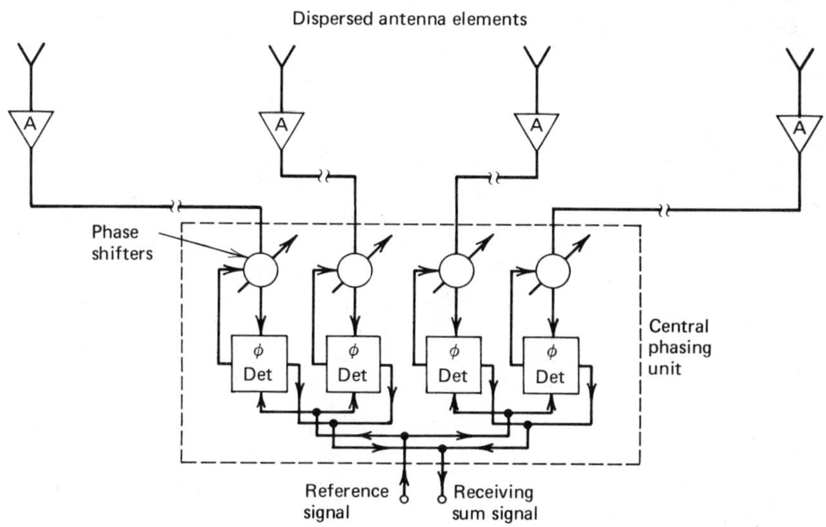

FIGURE 5.30. A dispersed-element receiving array with a central phasing unit. (From R. C. Hansen, Ed., *Microwave Scanning Antennas*, Academic, New York, 1964.)

Margerum in [39]) obviates the need for delivery of the phase reference to each module (Figure 5.30) and the low-duty-cycle voltage-control problem is avoided by the broadcast-frequency-reference technique. Another complexity enters instead; the required speed of operation is high. Let the array have N elements, let each echo sample be quantized into b bits, and let the radar pulse duration be δ s.

The bit rate within the processor will then be the order of Nb/δ bits/s. Taking $N = 1000$, $b = 10$, and $\delta = 10^{-7}$ s (a 15-m pulse), Nb/δ evaluates to 10^{11} bits/s, which is nearly three orders of magnitude too fast for digital circuits. Hence it is impractical to connect the I and Q outputs of each of the N modules in Figure 5.22 directly to a single input commutator to the central signal processor, as sketched in Figure 5.31. Instead the data rate must be reduced.

Three techniques are shown in Figures 5.32 to 5.34, among which only the first relieves the signal processing load of the central processor. All three, however, reduce the data transfer problem materially from the array modules to the central processor. In the first method of data reduction a restricted interval of the radar echo trace is selected for high-angular resolution process-

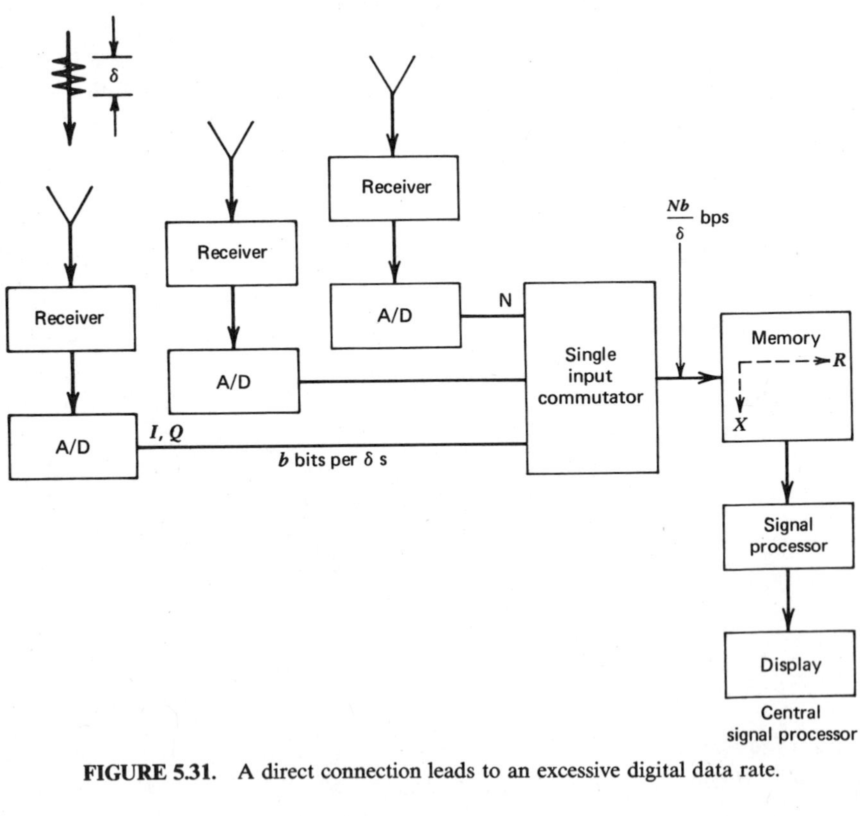

FIGURE 5.31. A direct connection leads to an excessive digital data rate.

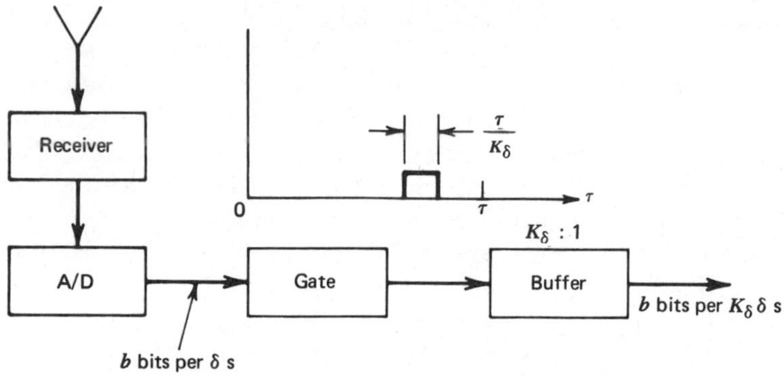

FIGURE 5.32. Data reduction by range-interval selection.

ing, a practical approach to data reduction because only in rare applications of the radio camera will it be useful to process the entire range sweep. The radar range sweep is typically 1000 or more times the size of the range cell. Consider the example $\delta = 10^{-7}$ s, which implies a range cell of 15 m. A radio camera user might have need for a magnified image of 10 to 100 range cells but rarely the entire range sweep. Therefore data reduction by a factor $10 \leq K_\delta \leq 100$ can be achieved by gating only a limited portion of the echo trace into a buffer store and reading the data out of the memory at a rate reduced by the factor K_δ. The gating waveform is shown in Figure 5.32 to have duration τ/K_δ, where τ is the interpulse period of the radar.

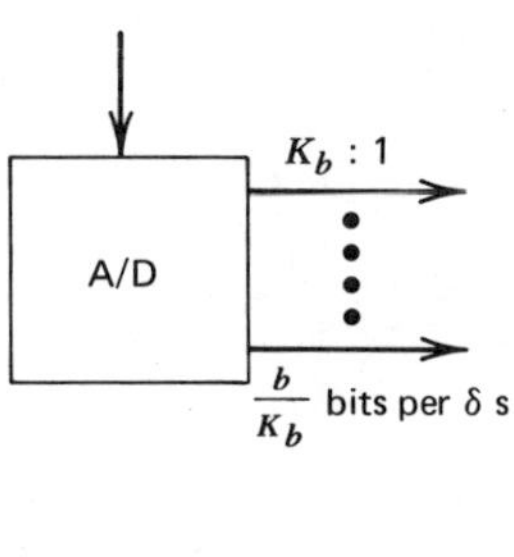

FIGURE 5.33. Multiple outputs from the A/D converter.

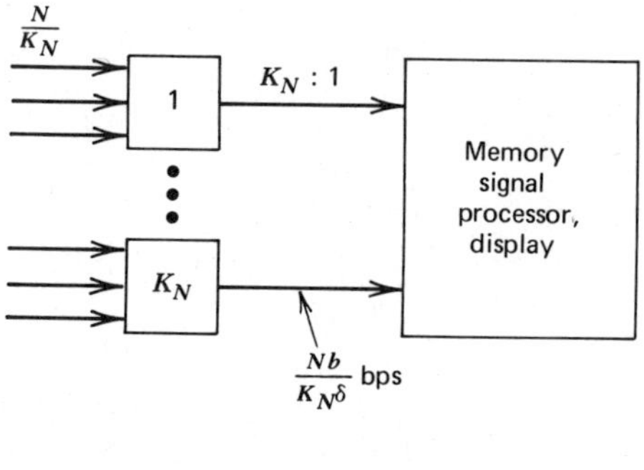

FIGURE 5.34. Multiple input commutators to the central signal processor.

The internal data rate can also be reduced by delivering the digital output of the A/D converter in several parallel lines (Figure 5.33), each carrying a single bit. The local data rate reduction is $K_b = b$. Last, the single input commutator to the central signal processor can be replaced by K_N parallel commutators, each handling N/K_N antenna modules (Figure 5.34). The data rate handled by each commutator is reduced by the factor K_N. By combining all three techniques the digital data rate per channel into the central processor is reduced by the factor $K = K_\delta K_b K_N$. In this way the data rate in the delivery subsystem can be reduced to manageable proportions.

CHAPTER SIX

Phase Synchronizing Sources

6.1. INTRODUCTION

Adaptive Beamforming is Retrodirective

This chapter, which follows [89] closely, examines the retrodirective beamforming source and determines the required properties for satisfactory operation of a radio camera. The principles of operation with a distorted array are discussed in Chapter 5, and the properties of the radiation pattern that result from retrodirective beamforming and the losses that can develop in array gain due to phase errors are described in Chapter 2. This chapter deals with the sources of these losses. Bounds are calculated on the necessary physical properties of retrodirective synchronizing sources. The sources include the passive reflector like the corner reflector, the active beacon, and radar ground clutter. Experimental one-dimensional radio camera images show how array gain can be degraded by an imperfect retrodirective source.

A retrodirective array samples the radiation field from a point source at a distance and adjusts the phase of the radiated wave at each element to be the complement of the measured signal phase [47, 39, 18]. This process, called phase conjugation, is discussed in Section 4.3. Adaptive beamforming is a retrodirective process. The source may be in the far or near field. It may be an active beacon or the echo from a passive reflector illuminated by a transmitter. The transmitter can be located in the neighborhood of the array or it may be one or more elements in the array or at a distance. All that is required for efficient operation is that radiation from the active source or passive target appear at the array as though from a point source in free space.

Retrodirective beamforming was introduced in the early days of satellite communication for the purpose of pointing a beam in the direction of a distant source without accurate knowledge of its direction [76], [38]. Modern communication satellite system techniques obviated the need for retrodirectivity and the technology did not fully mature.

Two Conditions

The point source requirement actually requires that two conditions be fulfilled. The first is on physical size, whereas the second is on echo or source strength in relation to the return signal from the remaining scatterers in the illuminated ground patch. Specifically, a source must appear to be confined in size so that only a single lobe of its radiation pattern illuminates the receiving array. Its signal strength must be larger than all other signals arriving simultaneously so that the phasefront of the received radiation is dominated by the beamforming source. An active beacon will always satisfy small size because its radiation will appear to emanate from a point. Similarly, the apparent size of a corner reflector will be less than its physical size because its radiation will appear to emanate from its center. A more complicated synchronizing source is a large scatterer, called a target of opportunity (TOO), immersed in a background of ground clutter. This source was used to form the Phoenixville map in Figure 5.17*a*.

Beamforming Versus Imaging

Figure 6.1 is useful in a discussion of both constraints for the target-in-clutter case. The range bin is modeled as a one-dimensional collection of scatterers of random back-scattering cross sections (illustrated by the sizes of the scatterers), located at random locations in the illuminated ground patch. At the right a large highlight like that due to a specular reflector is added. The received signal at each point in the array is the linear vector sum of the echoes from each of the scatterers. If no single reflector dominates (*left*), the return signal, as the vector sum of a large number of randomly phased echoes, represents a two-dimensional random walk process with the phase of the resultant uniformly distributed in the interval 0 to 2π. The correlation distance of the phase in the array is the order of the size d of the transmitting antenna, as shown in Figure 6.2, because the radiation pattern from the reflectors in the range bin consists of many lobes. The nominal lobe size is $\lambda/R\,\Delta\theta$ rad which, at the

FIGURE 6.1. The range bin is modeled as a one-dimensional collection of scatterers. (*a*) Small, randomly positioned scatterers, none dominant. (*b*) A single dominant reflector added.

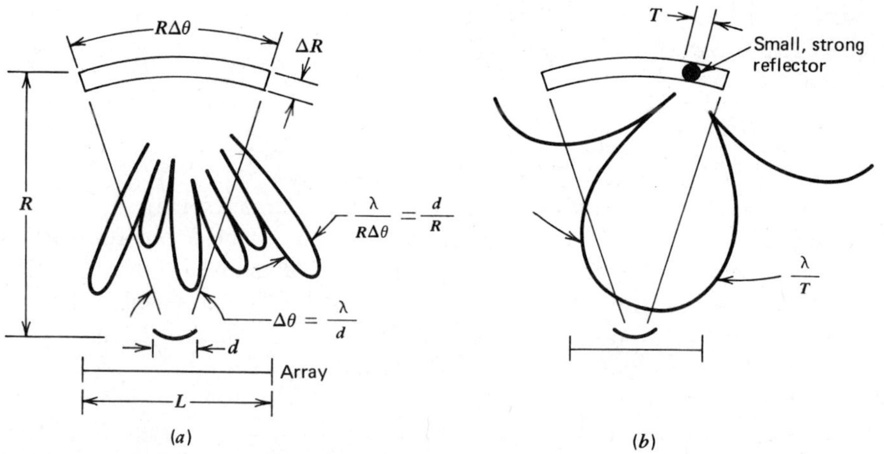

FIGURE 6.2. With no dominant reflector (a) the reradiation pattern consists of narrow lobes at the array. Strong, dominant reflector (b) produces a wide lobe.

array (a distance R away), has cross section $\lambda/\Delta\theta$. Because the transmitting beamwidth $\Delta\theta \simeq \lambda/d$, the correlation distance is about d. The phase pattern is illustrated in Figure 6.3 (*left*). It is clear that $\phi(x)$ is a random variable with correlation distance d. Therefore it is useless for adaptive beamforming. On the other hand, the fact that no single large scatterer dominates the phasefront means that this range bin is an interesting one to image.

Adaptive beamforming requires the condition shown in Figure 6.1 (*right*) in which a large specular reflector is present. The received signal is still the vector sum of all the echoes, but now the phasefront is dominated by the large scattering strength of the specular reflector. An adaptive beamforming system can self-cohere to this phasefront. Moreover, an excessively large echo is not required. Because the *phase* of the linear vector sum is a highly nonlinear combination of the amplitudes and phases of all the echoes, the large reflector need only exceed the collection of random scatterers by about 4 dB, which is shown in Section 6.2. This is an important distinction. When the phasefront is smooth, it is due primarily to a prominent reflector in the ground patch and

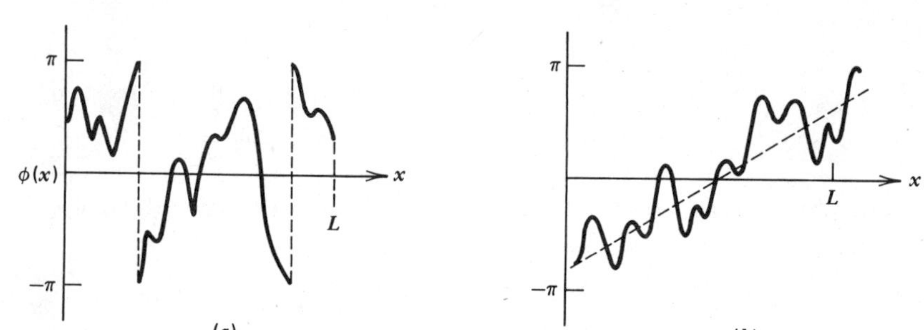

FIGURE 6.3. The phase of the reradiation pattern across the array: (a) good for imaging; (b) good for phase synchronizing.

self-cohering can take place. Following the retrodirective process, the array is prepared to image more complicated scenes.

6.2. LOSS IN ARRAY GAIN

Figure 6.4 shows two 1-dimensional radio camera images that illustrate the importance of a high-quality beamforming source. These one-dimensional angle scans were obtained with the time-shared cable receiver and the modified AN/APQ-102 radar transmitter used for the Phoenixville experiments (Section 5.7). The array was approximately linear, its length, 39m, and it consisted of 200 sample points located at random. The radar was an *X*-band ($\lambda = 3$ cm) set with 50-kW peak power and a range cell of 9 m. The adaptive beamforming source in the experiment in Figure 6.4*a* was a 1.2-m corner reflector 5.6 km from the radar; 31 m more distant was a 0.6-m corner reflector. The radio camera first self-focused on the 1.2-m reflector, then scanned in angle and range to the location of the smaller reflector, following which it scanned in angle across it to produce the image shown in Figure 6.4*a*. The procedure used is described in Sections 5.4 and 5.5. The ordinate is in amplitude units and the abscissa, in milliradians from the direction to the 1.2-m reflector. The cross

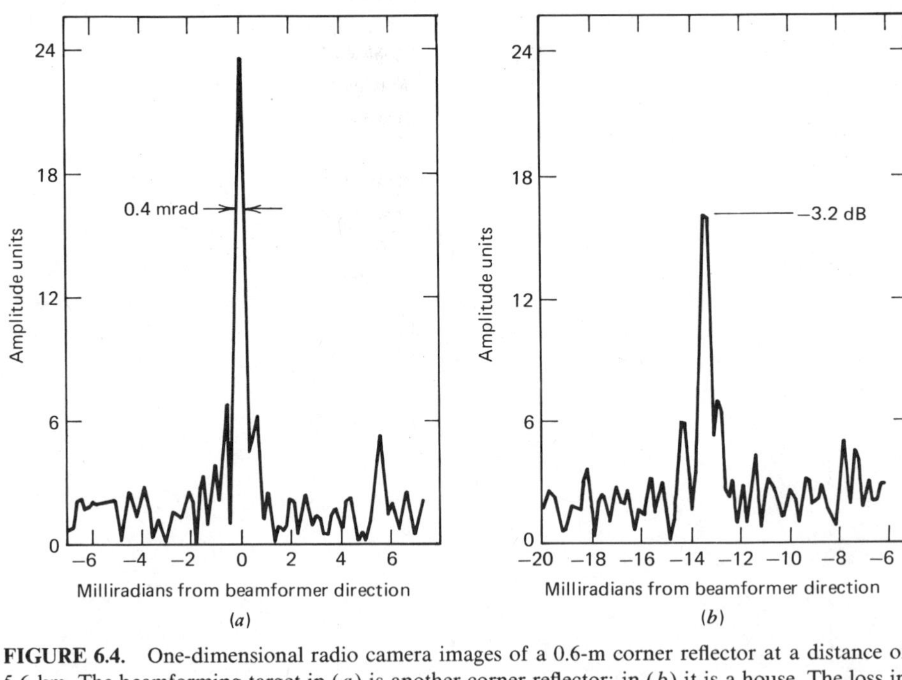

FIGURE 6.4. One-dimensional radio camera images of a 0.6-m corner reflector at a distance of 5.6 km. The beamforming target in (*a*) is another corner reflector; in (*b*) it is a house. The loss in array gain is 3.2 dB. (From B. D. Steinberg, *IEEE Trans. Antennas and Propagation*, AP-29, No. 5, September 1981.)

section of the imaged 0.6-m reflector is 0.4 mrad which is 2.2 m at the target range. The expected cross-range dimension based on the theory given in Chapter 2 is the same.

Figure 6.4*b* shows the same target imaged by the same equipment with the same array and imaging algorithm. The only difference is the beamforming source. The source in Figure 6.4*b* is a house located 13.5 mrad to the left of the target and 45 m from it in range. The resolving power of the instrument is not significantly altered but the array gain is reduced by 3.2 dB.

Degradation in Gain but Not in Resolution

It is predictable from random array theory that the gain will be seriously affected by the properties of the synchronizing source, whereas the beamwidth will not. As calculated in Chapter 2, array gain is sensitive to random phase errors across the array, according to

$$E\left[\frac{G}{G_0}\right] = e^{-\sigma_\phi^2} \tag{2.15}$$

where $E[\cdot]$ denotes expectation, G is array power gain, G_0 is the gain in the absence of errors, and σ_ϕ^2 is the phase-error variance across the array. Expressed in decibels the loss in gain is

$$\Delta G(\text{dB}) = 4.3\sigma_\phi^2 \tag{2.16}$$

On the basis of these equations, it may be deduced that the phase variance across the array after phase synchronization on the echoes from the house was 0.74 rad^2 or 0.86 rad rms. The nominal widths of all lobes (mainlobe and sidelobes) of a random array remain unchanged, irrespective of the phase errors. Thus, phase errors during the adaptive beamforming process reduce array gain but have no first-order effect on array beamwidth.

The Beam Is a Matched Spatial Filter

From a comparison of Figures 6.4*a* and 6.4*b*, it is evident that the house was not a fully satisfactory target of opportunity for adaptive phase synchronization of the distorted array. The reason is that the house is not a point reflector. The radiation power pattern formed by the adaptive retrodirective process is a replica of the source function or scene that produces the incident radiation field. Let the source or scene be at a distance R from the distorted array (Figure 6.5). Let y be an axis through the scene, perpendicular to the direction of phase synchronization of the array, called the z-axis. As before, define the reduced angular variable $u = \sin\theta = y/R$, where the scan angle θ is measured from the z-axis. The source or scene $s(u)$ produces a radiation field along the

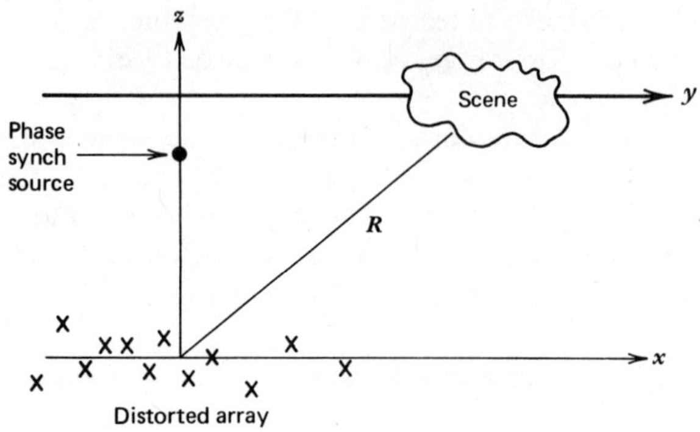

FIGURE 6.5. The scene to be imaged is at distance R from the array. The z-axis is in the direction of the synchronizing source. The x- and y-axes are in the same plane and are perpendicular to the z-axis.

x-axis in the array

$$S(x) = \int s(u) e^{jkxu} \, du \tag{6.1}$$

where the x-axis also is perpendicular to the z-axis. Now assume a phase conjugation operation such that the current excitation in the array along the x-axis is the complex conjugate $S^*(x)$ of (6.1). The radiation pattern becomes

$$f(u) = \int S^*(x) e^{jkxu} \, dx \tag{6.2}$$

which implies

$$S^*(x) = \int f(u) e^{-jkxu} \, du \tag{6.3}$$

by the properties of the Fourier transform. Equation 6.3 may be rewritten

$$S(x) = \int f^*(u) e^{jkxu} \, du \tag{6.4}$$

from which it is evident that the radiation pattern $f(u) = s^*(u)$ or $|f| = |s|$, thus validating the foregoing statement. In other words, the process of phase conjugation produces a radiation pattern that is a spatially matched filter to the source function of the radiation field. Now introduce a discrete sampling of the radiation field in the x-axis at locations x_i and let the adaptive circuits weight the N elements by $w_i = S^*(x_i)$. The radiation pattern

$$f_a(u) = \int_L \sum_{i=1}^{N} S^*(x_i) \delta(x - x_i) e^{jkxu} \, dx \tag{6.5}$$

is an approximation of (6.2) due to the discrete sampling in the aperture and to its finite extent L. In (6.5) $\delta(\cdot)$ is the Dirac delta function. Because the array is not solely in the x-axis, (6.5) is not an exact expression of the radiation pattern but an approximation in the angular neighborhood of the adaptive beamforming source.

Phase Weighting Usually Suffices

A further approximation may be made in the radio camera; that is, because the amplitude of the radiation field must be nearly constant for the source to approximate a point source, it is sufficient merely to phase-weight the elements in the array by the conjugate of the incident field and to ignore the amplitude variation across the array.

Required Target-to-Clutter Ratio

On the basis of the fact that the beam shape is an approximate spatial match of the reflectivity profile of the synchronizing target, we see that the axial gain of the adaptively formed beam will be reduced from its maximum possible value by the gain in the radiation pattern in directions toward the various scattering centers in the source region. The other scattering centers may be part of the synchronizing target, as in Figure 6.4*b*, or they may be clutter scatterers in the illuminated patch. To calculate the loss in gain it is necessary only to relate the phasefront distribution to the scatterer distribution and then deduce the gain loss from the phasefront distortion.

First, the scatterer distribution, whether on the beamforming target or in the illuminated clutter patch, may be presumed to be random, which is a sufficient condition to ensure that the phase perturbations in the phasefront of the reradiation are also random. Given this condition, the expected array gain in relation to the maximum possible gain is (2.15) and the loss in gain in decibels is (2.16). Next, assume that a tolerable loss in gain is 1 dB. From (2.16) the phase variance allowed in the phasefront is $\sigma_\phi^2 = 1/4.3 \text{ rad}^2$. Now assume that the target with radar cross section σ_T radiates as a point source and that the clutter cross section $\sigma_C < \sigma_T$. Clutter echoes received at some arbitrary antenna element in the array arrive with arbitrary phase α and amplitude ratio $a = \sigma_C^{1/2}/\sigma_T^{1/2} < 1$. The phase error

$$\delta\phi = \tan^{-1}\frac{a \sin \alpha}{1 + a \cos \alpha} \tag{6.6}$$

is a zero mean random variable. Equating its variance to $1/4.3$ leads to the condition

$$\sigma_T \geq 2.5\sigma_C \tag{6.7}$$

which guarantees that the loss in gain will not exceed tolerance. Thus the phase synchronizing source for the adaptive beamforming process need exceed the clutter by only 4 dB.

6.3. TYPES OF SYNCHRONIZING SOURCES

Figure 6.6 illustrates several types of synchronizing sources. Shown on the ground are a corner reflector (CR), which is a near-ideal beamforming source, and a large prominent TOO. Both are passive reflectors. Also shown is an active beacon (B) which can be airborne as well, as illustrated by the one carried in the remotely piloted vehicle (RPV). A mobile radio camera system must depend on TOOs, whereas active beacons or large passive targets like corner reflectors may be practical to implant for fixed installations. Implanted sources can be made to satisfy the required conditions of size and strength described later in this chapter. A condition more difficult to satisfy is the limited allowed phase distortion due to multipath and scattering. With care in siting of the adaptive beamformer, however, the scattered field can be held below the -6-dB tolerance derived in Chapter 7. The airborne beacon is problematical because fixed siting is not possible. Forward scatter from sea or ground might distort the beacon signal unless the vehicle altitude is chosen wisely. Even more uncertain is the TOO over which the system designer has no control. Nevertheless, as shown in Chapter 5, when the echo from the TOO reaches the array it is easily tested in the signal processor to determine whether it conforms to the required conditions. Only when the conditions are satisfied does the TOO echo have nearly constant amplitude across the array. The

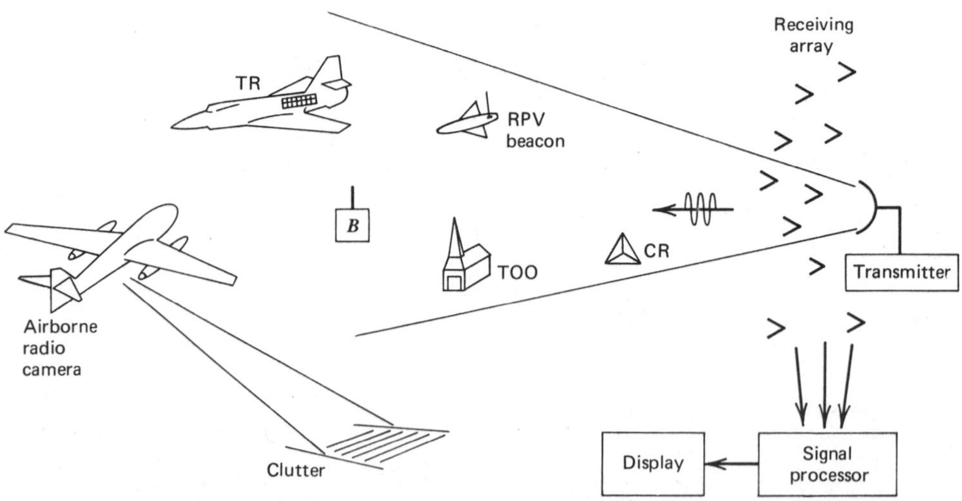

FIGURE 6.6. Types of synchronizing sources: passive reflectors, active beacons, and distributed clutter.

simple algorithm described earlier facilitates the search for a proper TOO. The same algorithm serves when an implanted corner reflector is the source.

The beamforming target also may be a reflecting surface on the target to be imaged (called a target reference), as illustrated by the large, specular reflecting surface of the airplane target. Denoted TR, this source introduces a basic problem. Physically, it is a confined area on the target (such as a portion of the fuselage momentarily oriented normal to the line of sight to the radar) that returns a specular reflection, or it may be a two-corner reflector formed by wing and fuselage oriented for a few tenths of a second to return an unusually strong echo. This "highlight" or strong return can serve as an efficient phase-synchronizing source for the array. The experiment based on radar echoes from aircraft, described in Section 5.8, showed that it is highly probable that the array can be made to self-cohere on a moving target. To do so the target echo must be dominated by the single, strong reflection. Under this condition, however, the target is hardly worth imaging. When this happens, the radio camera does not "see" the target; it sees only the point or small area from which the specular reflection derives. Hence there is an inherent contradiction in the use of a TR as a synchronizing source. On the one hand, it permits the self-cohering process to take place; on the other, it obscures the detail in the target image by dominating the reradiation from the target. This observation is reminiscent of the comment made near the end of Section 6.1.

As discussed in Section 5.8, the problem disappears when the target is in motion. Then the TR can be used successfully. Because the aspect angle of the target changes with time, a strong, specular return will sometimes reach the array, whereas at others no highlight will dominate. An object in motion will alternately act as a synchronizing source and a target worthy of being imaged. If the target's aspect angle does not change with time, however, the use of a TR is limited.

Last, the beamforming source can be distributed clutter echoes as illustrated for the airborne radio camera in Figure 6.6. These three sources (passive reflector, active beacon, and distributed clutter) are discussed separately in the sections that follow.

6.4. PASSIVE REFLECTOR

In Section 6.3 it was shown that the passive reflector must have a large enough radar cross section to allow its echo to dominate the phasefront of the radiation field illuminating the array. In this section it is shown that a second condition must be satisfied: its physical size must not be so large that its reradiation in the neighborhood of the array will differ significantly from a planar or spherical phasefront. These two conditions place bounds on the acceptable size of a passive target.

Figure 6.7 illustrates the second condition. A target is shown reradiating a field toward the array. Radiation from a target is always lobular. Unless the

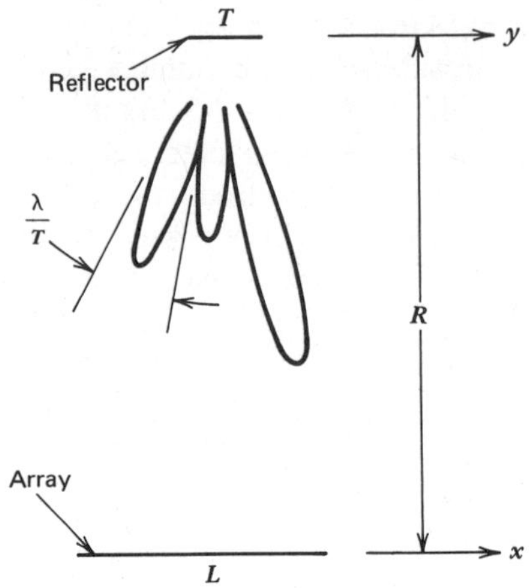

FIGURE 6.7. The reradiation pattern of the target is lobular.

lobes are large compared with the size of the array, the radiation will not be planar or spherical. Instead, the phasefront will vary in a random fashion similar to that shown in Figure 6.3 (*left*). Calculation of the condition for array size is based on the fact that a target of size T radiates lobes with a nominal angular spacing λ/T. This may be seen by considering structures of simple or known characteristics; for example, a flat-plate reflector of length T radiates a pattern with the angular characteristic $\sin(\pi Tu/\lambda)/(\pi Tu/\lambda)$, where $u = \sin(\theta - \theta_0)$, θ is the angle measured from the normal to the surface, and θ_0 is the direction of maximum reradiation. The width of the mainlobe is approximately λ/T, as are spacings between zero crossings in the remainder of the reradiation pattern. Similarly, if the target contains two prominent scatterers of equal strength spaced by T, the reradiation pattern will have the form $\cos(\pi Tu/\lambda)$. The zero-crossing interval is λ/T. If the scatterers are of unequal strength the radiation pattern will develop an additive constant but the angular modulation period will remain the same. The results will be similar if a target consists of many scatterers of random amplitudes and locations within the interval T. Let the scatterer distribution be a sample function of a random process characterized by $\sum_i a_i\delta(y - y_i)$, where a_i are the scattering amplitudes and y_i are their locations on the reflector. The radiation pattern $\mathcal{T}(u)$ is

$$\mathcal{T}(u) = \int_{-T/2}^{T/2} \sum a_i\delta(y - y_i)e^{jkyu}\,dy = \int_{-\infty}^{\infty} \text{rect}\left(\frac{y}{T}\right)\cdot\sum a_i\delta(y - y_i)e^{jkyu}\,dy$$

$$= \mathcal{F}\left\{\text{rect}\left(\frac{y}{T}\right)\cdot\sum a_i\delta(y - y_i)\right\} = \mathcal{F}\left\{\text{rect}\left(\frac{y}{T}\right)\right\} * \mathcal{F}\left\{\sum a_i\delta(y - y_i)\right\}$$

$$= \frac{\sin(\pi TL/\lambda)}{\pi TL/\lambda} * \sum a_i e^{jky_iu} \tag{6.8}$$

where $\text{rect}(y/T) = 1$ in the interval $[-T/2, T/2]$ and zero elsewhere, $\mathcal{F}\{\cdot\}$ is the Fourier transform, and $*$ denotes convolution. The second term is the underlying radiation pattern of the complex of scatterers; the first is due to the truncation of the random process and is the Fourier transform of the window function $\text{rect}(y/T)$ that represents the target. Because convolution is a spreading process, the lobe spacing of $\mathcal{T}(u)$ can be no smaller than those of $\sin(\pi Tu/\lambda)/(\pi Tu/\lambda)$, which is λ/T. All three examples indicate that λ/T is a typical value of lobe spacing.

A random radiation pattern with lobes nominally spaced by this value is sketched in Figure 6.7. Unless the prominent, central portion of a single lobe encompasses the entire array the phasefront condition will not be satisfied. The width of the lobe is about $\lambda/2T$ and its cross section at the array a distance R from the target is $\lambda R/2T$. Hence a minimum condition that must be satisfied for satisfactory beamforming is

$$\frac{\lambda R}{2T} > L \quad \text{or} \quad T < \frac{\lambda R}{2L} \tag{6.9}$$

To satisfy the first condition the radar cross section of the adaptive beamformer must exceed the combined cross sections of all the scatterers in the illuminated patch in order to dominate the phasefront. Figure 6.8 illustrates the geometry. The clutter cross section $\sigma_C = A_C\sigma_0$, where $A_C = R\,\Delta R\,\Delta\theta$ is the area of the patch illuminated by the transmitter, $\Delta\theta$ is the beamwidth, ΔR is the pulse length, and R is the distance from the transmitter; σ_0 is the normalized backscatter coefficient of the terrain. The radar cross section of the target $\sigma_T = A_T G$, where A_T is the projected target area illuminated by the transmitter and G is the gain or directivity of the target reradiation in the direction of the receiver. The peak gain of a flat-plate reflector is $4\pi A_T/\lambda^2$ and its maximum radar cross section is $\sigma_T = 4\pi A_T^2/\lambda^2$, which equals $4\pi T^4/\lambda^2$ for a square reflector of side T. The effective area of a corner reflector with a side of length T is that of an equilateral hexagon inscribed within an equilateral triangle of side T. Its area is $T^2/2\sqrt{3}$; its radar cross section is therefore approximately T^4/λ^2. Radar cross sections of other standard shapes are well documented [72].

By using the condition $\sigma_T \geq 2.5\sigma_C$ derived earlier (6.7) and expressing σ_T and σ_C in terms of radar and target parameters, the lower bound on target size is easily calculated for any shape; for example, the inequality

$$\frac{T^4}{\lambda^2} > 2.5\sigma_0 R\,\Delta R\,\Delta\theta \tag{6.10}$$

expresses a lower bound on corner reflector target size. Combining it with (6.9) yields

$$\left(2.5\lambda^2\sigma_0 R\,\Delta R\,\Delta\theta\right)^{1/4} < T < \frac{\lambda R}{2L} \tag{6.11}$$

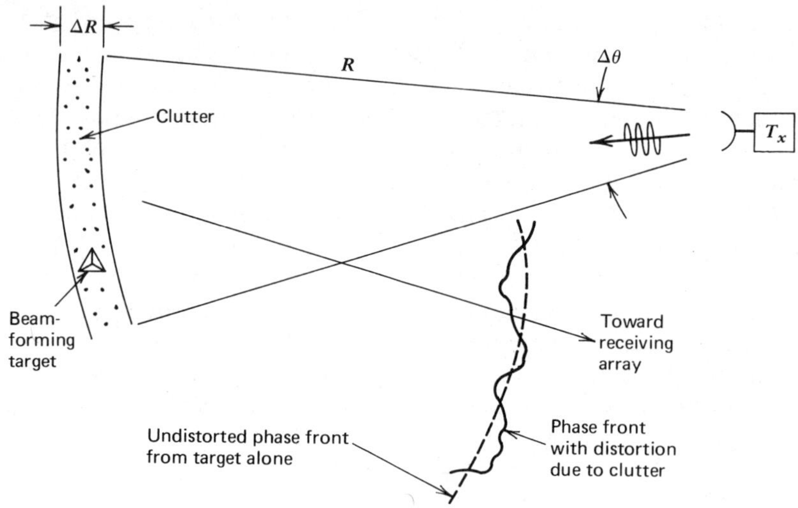

FIGURE 6.8. Clutter echoes distort the phasefront of the reradiation from the beamforming target.

Figure 6.9 expresses these bounds as a function of range R. It is evident that at short ranges the minimum allowed size exceeds the maximum, which means that adaptive beamforming cannot be accomplished with a passive reflector at distances of less than some minimum range. The minimum range is found by equating the bounds:

$$R_{\min} = \left(40L^4\lambda^{-2}\sigma_0\,\Delta R\,\Delta\theta\right)^{1/3} \tag{6.12}$$

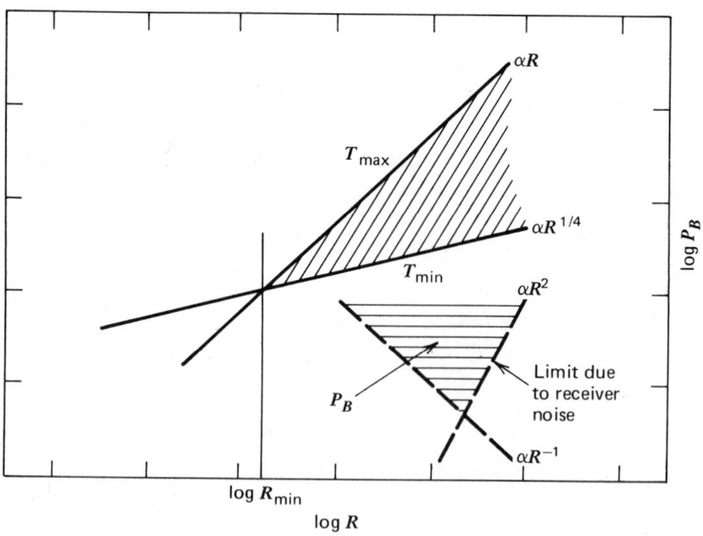

FIGURE 6.9. Range dependencies on size of passive reflector T and beacon power P_B. (From B. D. Steinberg, *IEEE Trans. Antennas and Propagation*, AP-30, No. 6, November 1982.)

TABLE 6.1. Maximum and Minimum Sizes of Corner Reflectors for Use as Sources for *X*-Band and *L*-Band Illustrations

Band	λ (m)	σ_0	ΔR (m)	$\Delta\theta$ (rad)	λ/L	$T_{\min}$ (m)	$T_{\max}$ (m)	$R_{\min}$ (km)
X	0.03	10^{-2}	5	1/20	1/2000	0.25	5	5.2
L	0.3	10^{-3}	20	1/20	1/2000	0.61	5	3.9

Table 6.1 illustrates two radars with beamwidths of 1/20 rad or approximately 3°. One has a 3-cm wavelength and a 5-m range resolution; the other has a 30-cm wavelength and a 20-m range resolution. In each case the design problem is to make the azimuthal resolution 100 times finer by the use of a large, distributed receiving phased-array. The adaptive beamformer is at a distance of 10 km in both cases. Values of $\sigma_0 = 10^{-2}$ and 10^{-3} are assumed for the two wavelengths. The table lists the maximum and minimum sizes of corner reflectors that satisfy the requirements described and the minimum range at which a passive reflector could be used.

6.5. ACTIVE BEACON

The second phase-synchronizing source discussed in this chapter is an active beacon that radiates a pulse sequence of power P_B at a distance R from the array. Isotropic radiation is assumed. The beacon may be triggered by the radar transmitter or by its own timing circuit. The beacon power density at the array is $P_B/4\pi R^2$. The clutter power density at the array is $P_T G_T \sigma_C/(4\pi R^2)^2$, where P_T is the radar transmitter power and G_T is the antenna gain. The clutter cross section σ_C is given in Section 6.4. A combination of these terms and the requirement based on (6.7) that the beacon signal exceed the clutter echo by a factor of 2.5 or more result in the following condition for beacon power:

$$P_B \geq \frac{P_T G_T \sigma_0 \,\Delta R\, \Delta\theta}{1.6\pi R} \tag{6.13}$$

Whereas the required passive target size grows with range (6.11), the minimum required beacon power is inversely proportional to distance. This surprising condition results because the beacon power density at the receiver suffers an inverse square propagation loss and the clutter power decreases with the cube of range. P_B cannot decrease indefinitely with range because the beacon signal must always exceed receiver noise. On the basis of the assumption that receiver noise is independent from antenna-element channel to channel, the received signal must exceed the noise by the same factor of 2.5 as it must exceed the clutter. The received beacon signal power is $P_B A_R/4\pi R^2$, where A_R is the effective area of the receiving array. Receiver noise power is $kTBF$, where k is Boltzmann's constant, T is receiver temperature, B is receiver bandwidth, and F is the system noise figure. A combination of these expressions leads to the second requirement for beacon power:

$$P_B \geq \frac{10\pi R^2 kTBF}{A_R} \tag{6.14}$$

These equations are plotted in Figure 6.9.

6.6. CLUTTER AS A SYNCHRONIZING SOURCE

It is evident from earlier discussion that without a prominent reflector in a clutter patch the beam pattern, following self-cohering, will approximate the angular clutter distribution weighted by the pattern of the illuminating beam. Its width will, therefore, be the same as that of the transmitter and no resolution improvement will result. When the radar system is airborne, however, a new factor that makes possible the development of a proper reference signal for the system from the clutter echo [85], [89] is introduced. Figure 6.10 describes an airborne radar moving with speed V and illuminating a clutter patch with beamwidth $\Delta\theta$ at distance R and at angle θ from the ground track. The receiving array of length L is assumed to be distributed on the airframe. Because the aircraft is nonrigid at microwave wavelengths, an external phase synchronizing source is necessary. Then $a \ll L$ is the aperture of the radar transmitter and the beamwidth $\Delta\theta = a/\lambda$. The clutter patch width $R\,\Delta\theta$ is too wide: it is $R\lambda/a$, whereas the proper width cannot exceed $\lambda R/2L$, as derived

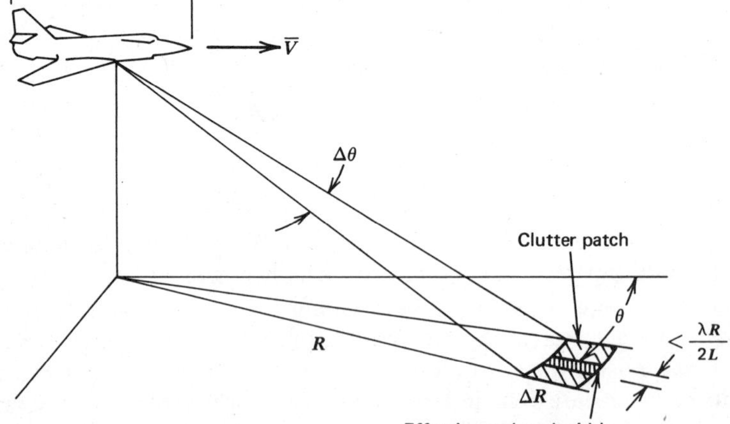

FIGURE 6.10. The clutter patch is too wide to be a synchronizing source. Narrowband filtering in the receiver reduces effective patch width to less than $\lambda R/2L$.

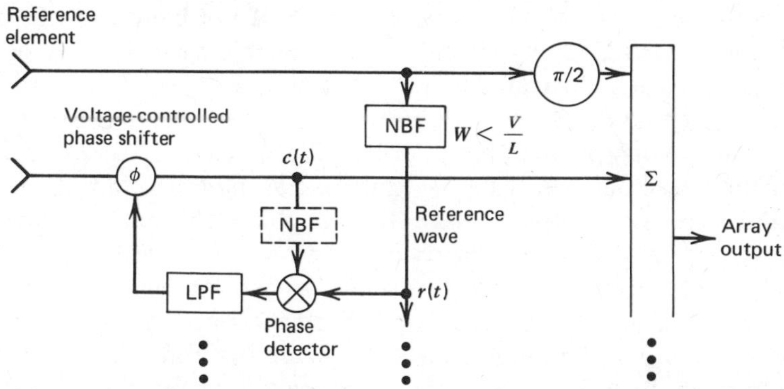

FIGURE 6.11. Reference element and one other element in airborne radio camera. All signals are phaselocked by a clutter-derived reference wave obtained from the narrowband filter.

earlier and as shown in Figure 6.10. By virtue of the aircraft motion echoes from scatterers within the ground patch may be distinguished from one another by their Doppler shifts; hence narrowband filtering of the received clutter echoes can extract the reflections from scatterers within a subpatch of the desired width. The output of this filter can be used as a phase synchronizing reference.

The Doppler shift of an echo from a scatter at angle θ from the ground track is

$$f_d \simeq \frac{2V\cos\theta}{\lambda} \tag{6.15}$$

A narrowband filter tuned to the mean Doppler shift (6.15) will pass echoes from the central region of the illuminated ground patch and reject the remainder. The edges of the desired subpatch are $\pm\lambda/4L'$ rad from the center ($L' = L\sin\theta$). Their Doppler shifts are $(2V/\lambda)\cos(\theta \pm \lambda/4L')$. Therefore to confine the response of the filter to echoes from the desired subpatch its bandwidth

$$\begin{aligned} W &\leq \frac{2V}{\lambda}\left[\cos\left(\theta - \frac{\lambda}{4L'}\right) - \cos\left(\theta + \frac{\lambda}{4L'}\right)\right] \\ &= \frac{4V}{\lambda}\sin\left(\frac{\lambda}{4L'}\right)\sin\theta \simeq \frac{V}{L'}\sin\theta = \frac{V}{L} \end{aligned} \tag{6.16}$$

Figure 6.11 shows how the reference signal would be used.[†] One element is

[†]A receiver chain that contains the usual circuits (amplifiers, mixer, and local oscillator) is implicit in each channel. Coherent detection is also implicit. The NBF is assumed to be preceded by a range gate so that the NBF will respond only to scatterers in the range interval $[R, R + \Delta R]$. Similarly, a range gate is assumed at the signal input port of the phase detector. The analysis that follows also pertains to wideband delay-line filters with narrow passbands at intervals of the pulse repetition frequency and to optical correlators.

chosen as the reference. A narrowband filter (NBF) of bandwidth W centered at the Doppler shift given by (6.15) delivers the reference signal to a bank of phase detectors, each associated with one antenna element. The clutter echoes received at each antenna element pass through a voltage-controlled phase shifter to the summer of the phased array. The signal also passes to the other input of the phase detector. The beat product is smoothed in a lowpass filter and applied as the control voltage to the phase shifter. The circuit is a phaselock loop which drives the two inputs to the phase detector into a quadrature relationship. The loop responds to those components of the element signal that correlate with the narrowband reference signal.

Figure 6.12 shows the clutter spectrum at an arbitrary element and after passage through the NBF. The clutter signal may be represented by the sum of M sinusoidal echoes of amplitudes a_i, Doppler shifts $f_d + f_i$, where $f_d = 2V\cos\theta/\lambda$ is the mean shift, and phases ϕ_i. The a_i, f_i, and ϕ_i are independent random variables. In addition, a phase shift Φ is due to the position of the element in the distorted array; Φ is the quantity to be corrected by the adaptive process. The equation of the clutter signal is

$$c(t) = \sum^{M} a_i \cos\left[(\omega_0 + \omega_d + \omega_i)t + \phi_i + \Phi\right] \tag{6.17}$$

where $\omega = 2\pi f$. The reference wave is the sum of $K < M$ echoes from the central portion of the ground patch. Its waveform is

$$r(t) = \sum^{K} a_j \cos\left[(\omega_0 + \omega_d + \omega_j)t + \phi_j\right] \tag{6.18}$$

The mixer output is the product $e(t) = c(t)r(t)$, only the low frequency terms

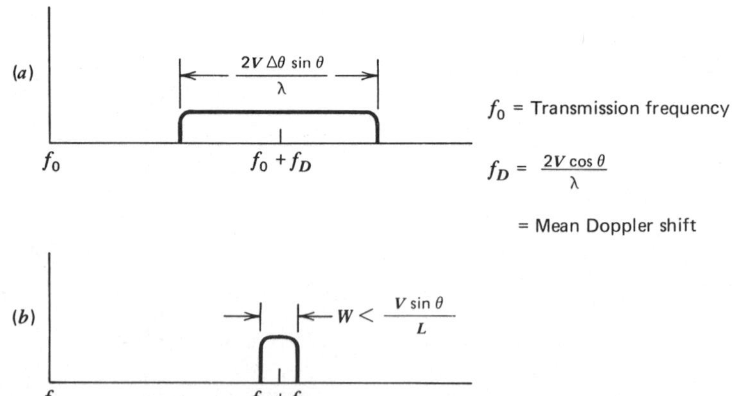

FIGURE 6.12. (a) Input clutter spectrum. (b) After narrowband filter in reference channel.

of which pass through the lowpass filter. Its output voltage is

$$v(t) = \frac{1}{2}\sum_{i}^{M>K}\sum_{j} a_i a_j \cos\left[(\omega_i - \omega_j)t + \phi_i - \phi_j + \Phi\right]$$

$$= \frac{1}{2}\cos\Phi \sum^{K} a_j^2 + \frac{1}{2}\sum_{i\neq j}^{M>K}\sum a_i a_j \cos\left[(\omega_i - \omega_j)t + \phi_i - \phi_j + \Phi\right] \tag{6.19}$$

The *DC* output is

$$\overline{v(t)} = \frac{K\overline{a^2}\cos\Phi}{2} \tag{6.20}$$

where $\overline{a^2}$ is the mean-square echo-strength of the scatterers. Thus the DC control voltage for the phase shifter is proportional to cos Φ. Because the loop drives this voltage to zero, the portion of the phase-shifted element signal within the passband of the NBF is brought into phase quadrature with the reference signal independent of the array distortion-induced phase error Φ. This procedure is performed in all array–element channels, with the result that their signals are cophased. The reference signal channel may be added, provided that it is shifted in phase by $\pi/2$. The sum is the array output. Adding the NBF (shown dashed in Figure 6.11) in the PLL improves its acquisition and tracking characteristics by reducing the phase noise in the loop. The steady-state, phase-error variance in the loop is proportional to the variance in (6.19), which is the power in the second term. That term consists of $K(M-1)$ sinusoids of average amplitude $\frac{1}{2}\overline{a_i a_j} = \frac{1}{2}\bar{a}_i\bar{a}_j = \frac{1}{2}\bar{a}^2$, assuming that the scatterers are statistically independent and have common statistics. Because the Doppler shifts and scatterer phases are independent from scatterer to scatterer, the average power in the second term is the sum of the component powers, each of which contributes $\frac{1}{2}\left(\frac{1}{2}\bar{a}^2\right)^2 = \bar{a}^4/8$. Hence the variance of (6.19) is

$$\sigma_v^2 = \frac{K(M-1)\bar{a}^4}{8} \tag{6.21}$$

By adding the dashed NBFs in the element channels the number of scatterers in the clutter signal (6.17) will drop from M to K, which reduces (6.21) to

$$\sigma_{v,NB}^2 = \frac{K(K-1)}{8}\bar{a}^4 \tag{6.22}$$

The reduction in phase-noise power is the factor $(M-1)/(K-1) \simeq M/K$

because the number of scatterers is large. This ratio is $2L\,\Delta\theta/\lambda$, which approximates the ratio of the clutter bandwidth to the filter bandwidth because random ground and sea clutter typically exhibit uniform angular distributions and the transformation (6.15) from angle to frequency shift is, for small angles, nearly constant. This phase-noise power reduction factor is the exact angular resolution improvement ratio, which can be very large; for example, consider an S-band radar ($\lambda = 10$ cm) with a 3° beamwidth ($\Delta\theta \simeq 1/20$ rad) aboard a 30-m aircraft. The resolution improvement factor is 30, which means that the reduction in the phase-noise power in the PLL can be as large as 15 dB when the NBF is added in the loop.

In addition to the proper filter bandwidth (6.16), the frequency selectivity of the filter must ensure that the clutter power of the echoes passing through the central region of the filter will exceed the remainder by at least the factor of 2.5, or 4 dB, calculated in (6.7). Let the clutter power-density spectrum be represented by $C(f)$ and the filter transfer function by $H(f)$, where the origin of the frequency variable f is taken as the mean Doppler shift f_d. Based on (6.15) and the assumption of a uniform angular distribution of scatterers, we find that $C(f)$ is proportional to the probability density function of the frequency variable in (6.15) and the weighting due to the antenna pattern. The pdf can be shown to be proportional to $[(V/\lambda)^2 - f^2]^{-1/2}$; the weighting function is the two-way antenna pattern transformed from the angle variable θ to the frequency variable f by (6.15). The required condition for $H(f)$ is given by

$$\frac{\int_0^{W/2} C(f)|H(f)|^2\,df}{\int_{W/2}^{\infty} C(f)|H(f)|^2\,df} \geq 2.5 \tag{6.23}$$

CHAPTER SEVEN

Tolerance Theory

7.1. INTRODUCTION

It was shown in Section 4.4 that when adaptive beamforming is used there is no loss of average array gain at the beamforming angle, even in the presence of phase errors in the beamforming signals. The loss is developed when scanning the radio camera away from θ_0, the direction of beamforming. Although different methods of adaptive beamforming exhibit detailed differences, the following discussion illustrates the general distinction between the closed-loop (adaptive) beamforming and the open-loop (nonadaptive) scanning functions.

Figure 7.1 is helpful in illustrating how phase errors degrade array gain during scanning. It is structured similarly to Figure 4.6 but emphasizes the primacy of phase by replacing the complex samples of the radiation field with their phases. The symbol $\hat{\phi}$ is used to distinguish between the measured phase, with error, and the proper or ideal phase ϕ. Let x in Figure 7.1 represent element number, which is monotonic with position in the array, and let the x-axis be approximately perpendicular to the target direction. Let θ represent target angular position and let the θ-axis be an arc of constant distance from the array. Next let the measured phase at point x, due to a target or source at angle θ in the field, be $\hat{\phi}(x, \theta) = \phi(x, \theta) + \delta\phi(x, \theta)$. The first term is the "geometric" phase (mod 2π) equal to $2\pi/\lambda$ times the distance between point x and the target. The second term is an error due to noise, multipath, element position error, and other sources of phase error. The amplitude of the received signal is $a(x, \theta)$.

Now assume a strong synchronizing source at $\theta = \theta_0$ radiating energy to the array. The retrodirective process focuses the array on the source. This procedure may be visualized (Figure 7.1a) by viewing $\hat{\phi}(x, \theta_0)$ as a column vector in a matrix $\hat{\phi}(x, \theta)$ in which x is the vertical index and θ is the horizontal index. To form a beam at $\theta = \theta_0$ it is necessary to rotate the phase of the received signal $a(x, \theta_0)\exp[j\hat{\phi}(x, \theta_0)]$ by the negative of $\hat{\phi}(x, \theta_0)$ (Figure 7.1*b*), and sum the phase-rotated vectors. In doing so all signals from the direction $\theta = \theta_0$ become cophased and may be added. This is the exact function performed by the phase shifters in a phased array.

In adaptive beamforming the phase rotation is accomplished by closed-loop self-phasing procedures. Once the phases are set adaptively signals arriving from any other direction θ are similarly phase-shifted by $-\hat{\phi}(x, \theta_0)$. The general entry in the table in Figure 7.1*b* becomes $\hat{\phi}(x, \theta) - \hat{\phi}(x, \theta_0)$. By assigning the subscript i to designate discrete element locations in x the vector sum

$$f(\theta) = \sum_{i=1}^{N} a(x_i, \theta)\exp\{j[\hat{\phi}(x_i, \theta) - \hat{\phi}(x_i, \theta_0)]\} \tag{7.1}$$

is the complex gain of the array in the direction θ. Of course, at $\theta = \theta_0$ the phase-rotated signals are cophased and (7.1) becomes

$$f(\theta_0) = \sum_{i=1}^{N} a(x_i, \theta_0)\exp\left\{ j[\hat{\phi}(x_i, \theta_0) - \hat{\phi}(x_i, \theta_0)] \right\} = \sum^{N} a(x_i, \theta_0) \tag{7.2}$$

In the absence of phase errors the radiation pattern (7.1) would have been

$$f_0(\theta) = \sum^{N} a(x_i, \theta)\exp\{j[\phi(x_i, \theta) - \phi(x_i, \theta_0)]\} \tag{7.3}$$

and its peak value

$$f_0(\theta_0) = \sum^{N} a(x_i, \theta_0) = f(\theta_0) \tag{7.4}$$

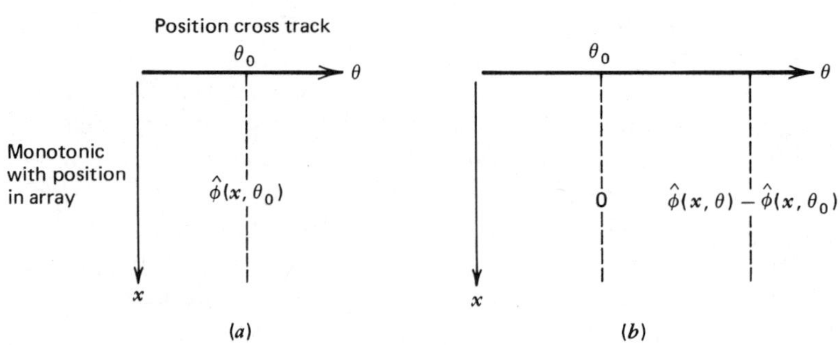

FIGURE 7.1. Beamforming procedure: $(a)\hat{\phi}(x, \theta)$ matrix; $\hat{\phi}(x, \theta_0)$ is the phase of the synchronizing signal received at point x in the array from a source at θ_0; (b) phase matrix after beamformation at $\theta = \theta_0$.

is the same as in (7.2), again showing that loss does not develop at the beamforming angle, even in the presence of phase errors in the beamforming signals.

The effects on scanning are different, however. The objective of the scanning operation is to make some other column vector in the matrix, say $\hat{\phi}(x, \theta_1)$, equal to zero. In doing so signals arriving from $\theta = \theta_1$ should, after processing, become cophased, and their vector sum should add to the sum of the amplitudes, as in (7.2). Errors in phase measurement, however, reduce the vector sum, which implies a loss in mainlobe gain.

This loss in gain is the dominant effect of phase-measurement error, provided that (a) the errors in phase measurement are random and independent and (b) the element placements are random. Beamwidth, beam shape, and pointing error effects then become nil. The sidelobe properties are already those of a random process; further random phase perturbations change the details of the side radiation pattern but not its statistics. Hence the effect of phase-measurement errors is to reduce the contrast in the scanned image (mainlobe to average sidelobe ratio) but not the resolution (proportional to beamwidth).

Resolution is influenced by a different factor. The cross section of the retrodirectively formed beam cannot be smaller than the radio size of the source of energy that phase-synchronizes the array. Hence the minimum cross-range dimension of the resolution cell is limited to the size of the beamforming source. This phenomenon is described in Chapter 6.

Equation (7.1) illustrates the adaptive beamforming function: the second term in the argument of the exponential zeros the phases at $\theta = \theta_0$. Scanning the beam after it has been adaptively focused at θ_0 requires a further phase correction. To scan the beam to an arbitrary position θ_1 requires that $\hat{\phi}(x, \theta_1) - \hat{\phi}(x, \theta_0)$ be subtracted from the phase of the signal in the matrix of Figure 7.1*b*. This is an open-loop process. The signal processor can only estimate the required phase correction from the geometry; it has no information regarding $\delta\phi$. In short, it applies as its estimate the geometric correction $\phi(x, \theta_1) - \phi(x, \theta_0)$. The difference between the desired phase shift and the actual open-loop phase shift is the residual phase error introduced by the scanning process. Its value may be written (dropping the subscript on beam direction θ)

$$\hat{\phi}(x, \theta) - \hat{\phi}(x, \theta_0) - [\phi(x, \theta) - \phi(x, \theta_0)] = \delta\phi(x, \theta) - \delta\phi(x, \theta_0) \neq 0 \tag{7.5}$$

The mean residual phase error across the array is assumed to be zero. The residual phase-error variance is the mean square error.

$$\begin{aligned} \sigma_\phi^2 &= E[\delta\phi(x, \theta) - \delta\phi(x, \theta_0)]^2 \\ &= E[\delta\phi(\theta)]^2 - 2E[\delta\phi(\theta)\,\delta\phi(\theta_0)] + E[\delta\phi(\theta_0)]^2 \end{aligned} \tag{7.6}$$

which reduces to

$$\sigma_\phi^2 = 2E\left[(\delta\phi)^2\right] \tag{7.7}$$

when the phase-error statistics are independent of source location and the correlation between errors from different directions is zero. In these expressions $E[\cdot]$ indicates the expected value or ensemble average.

It is shown in Chapter 2 that phase errors in the signals received across a receiving array decrease the main-beam power gain G under fairly general conditions according to

$$E[G] = G_0 \exp\left(-\sigma_\phi^2\right) \tag{2.15}$$

where G_0 is the gain in the absence of errors. Given K independent sources of phase error their variances add to form the total phase-error variance. Expressed in decibels, the expected change in gain is

$$10 \log E\left[\frac{G}{G_0}\right] = -4.3\sum_i^K \sigma_{\phi_i}^2 \text{ dB} \tag{7.8}$$

The *loss* in gain in decibels, called ΔG, is the negative of this quantity; that is,

$$\Delta G = 4.3\sum \sigma_{\phi_i}^2 \text{ dB} \tag{7.9}$$

From (7.9) it is seen that if the random phase error across an array is limited to $\frac{1}{2}$ rad rms the loss of gain will not exceed approximately 1 dB. As in the other chapters, this value is accepted as a nominal gain–loss tolerance.

It should be noted that the total phase variance in (7.8) and (7.9) is *not* the initial variance across the array. It is the residual variance following adaptive beamforming. The self-cohering process removes the variation due to the primary causes of phase error of the signal received across the array; these are random errors in element locations, random variations in the propagation paths to the various parts of the array, random perturbations in the phasefront due to multipath, receiver noise and, possibly, random phase shifts through the receivers in the array modules.

In the following sections four physical sources of phase distortion and their implications are discussed. They are measurement error, element position error, phase distortion due to multipath and scattering, and distortion in the propagation medium. All four cause losses in main-beam gain (therefore image contrast) when the beam is scanned away from the direction of phase synchronization. Among the four sources the first is not angle-dependent, although the three others are. Experimental test data accompany two of the theories.

7.2. PHASE-MEASUREMENT ERROR DUE TO RECEIVER NOISE

The simplest source of phase-measurement error is receiver noise. A signal of unit amplitude, shown in Figure 7.2, is contaminated by receiver noise of random amplitude n and random phase ψ, which is assumed to be uniformly distributed in the interval zero to 2π. The phase error

$$\delta\phi = \tan^{-1}\frac{n\sin\psi}{1 + n\cos\psi} \approx n\sin\psi \tag{7.10}$$

when $n \ll 1$. This weak-noise condition always will be satisfied in radar because detection of the beamforming signal must precede array synchronization; detection normally requires a minimum signal-to-noise power ratio of 20, or 13 dB. Hence the variance of $\delta\phi$ may be taken as $E[n^2]/2$. The phase error due to random receiver noise is unbiased and uncorrelated from array element to array element. From (7.7) therefore the receiver-noise contribution to σ_ϕ^2 is

$$\sigma_{\phi_N}^2 = 2E\left[(\delta\phi)^2\right] \approx E[n^2] = (\mathrm{SNR})^{-1} \tag{7.11}$$

and the dB loss in gain due to noise is

$$\Delta G_N = 4.3E[n^2] = \frac{4.3}{(\mathrm{SNR})}\ \mathrm{dB} \tag{7.12}$$

In these equations SNR is the power ratio of signal to noise. Because the minimum useful SNR in radar is approximately 20, the maximum expected loss is about 0.2 dB, which is very small. In general, receiver noise can be ignored as a source of loss in array gain when scanning the beam of the radio camera.

The same formula (7.12) pertains to all other uncorrelated sources of phase-measurement error, including noise in phaselock loops, calibration noise in phase detectors, and quantization noise in digital phase shifters and detectors.

7.3. ELEMENT POSITION ERRORS

Another cause of phase errors is due to position errors in the array elements. Although these errors have no effect on beamforming, they determine the

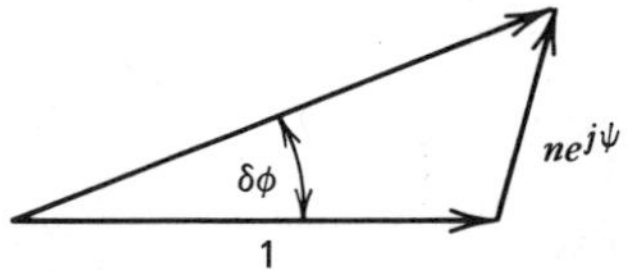

FIGURE 7.2. Phase measurement error due to noise.

extent in angle and range to which the beam can be scanned after self-cohering. Hence the field of view is limited by these errors. The geometric scan correction $\phi(x, \theta) - \phi(x, \theta_0)$ is in error if the target coordinates or element coordinates are not known accurately. The full theory of the limitation to scanning in angle is published in [101] and [85], and a simplified, one-dimensional theory is published in [83]. These sources indicate that the error that dominates the scanning loss is the error in estimation of element position transverse to the beamforming direction, which is nominally along the array. A review of the theory follows.

Two array elements are shown in Figure 7.3. One is a reference element against which phase measurements are made. The origin of an x, y-coordinate system is placed at this element. The second element is incorrectly assumed by the signal processor to be displaced by $(\delta x, \delta y)$ from the actual position (x, y).

During phase synchronization in the direction normal to the x-axis the adaptive beamformer correctly adds

$$\phi_0 = k(y - n\lambda) \tag{7.13}$$

to the phase of the signal received at (x, y), where n is the integer that reduces ϕ_0 to the interval $[0, 2\pi]$. The signal processor incorrectly assumes that the phase shift is

$$\phi_1 = k(y + \delta y - n\lambda) \tag{7.14}$$

To scan the adaptively formed beam through angle θ the phase shift should be

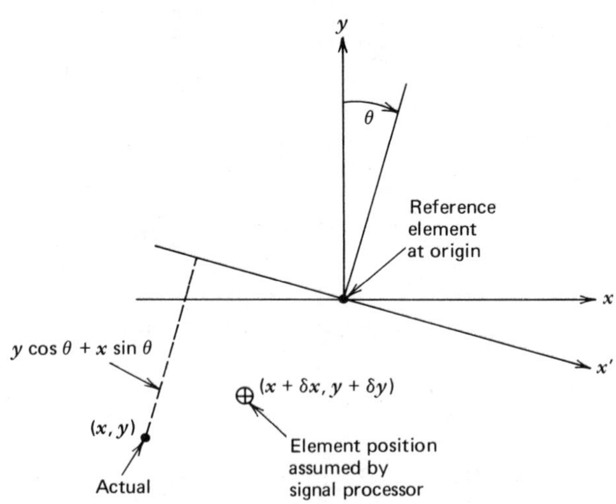

FIGURE 7.3. Placement error geometry.

changed from ϕ_0 to

$$\phi_2 = k(y\cos\theta + x\sin\theta - n\lambda) \tag{7.15}$$

The signal processor calculates that the new phase shift should be

$$\phi_3 = k[(y + \delta y)\cos\theta + (x + \delta x)\sin\theta - n\lambda] \tag{7.16}$$

It then adds $\phi_3 - \phi_1$ to the phase shift obtained during the adaptive phase-synchronization process. In doing so an open-loop, phase-shift error results:

$$\begin{aligned} \delta\phi &= (\phi_3 - \phi_1) + (\phi_2 - \phi_0) \\ &= k[\delta y(\cos\theta - 1) + \delta x\sin\theta] \\ &\simeq k\theta\,\delta x \qquad \theta \ll 1 \end{aligned} \tag{7.17}$$

Thus the errors along the direction to the target are secondary. It is shown in [101] and [85] that errors in estimation of target location are also secondary.

In (7.17) δx is the deviation between the x-coordinate of an element position and the value stored in the signal processor. Assuming that it is a zero-mean random variable with variance σ_x^2 the small-angle approximation to the phase-error variance, due to element position error, is given by

$$\sigma_{\phi_P}^2 \simeq k^2\sigma_x^2\theta^2 \tag{7.18}$$

According to (7.9) the loss in gain in decibels should be 4.3 times $\sigma_{\phi_P}^2$, namely

$$\Delta G_P = 4.3k^2\sigma_x^2\theta^2 \tag{7.19}$$

where the scan angle θ is measured in radians.

It should come as no surprise that the x-error dominates the residual phase error. Because the single beacon is in the y-direction, the self-synchronizing process compensates completely for position errors in that direction. Because no beacons occur in the orthogonal directions, such errors are uncorrected and remain part of the residual phase error.

This theory has been tested at the Valley Forge field site. A phase matrix (Figure 4.6*a*) was measured from the 27-m rooftop array described in Section 5.7. The experiment was conducted at L-band ($\lambda = 30$ cm). To make the measurements a transmitter was moved along the test track shown in Figure 5.7. The track is also visible in Figure 5.8. Phase was measured at each array element after every 2 ft of motion of the transmitter. The elements were randomly located, as shown (by the solid circles) in Figure 7.4. Fifteen elements were used, although the ninth element failed.

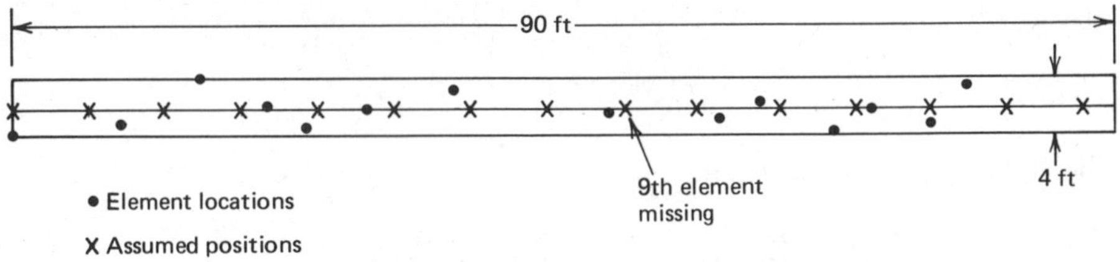

FIGURE 7.4. Simulated air-dropped array.

The element coordinates were known; this information, however, was not given to the signal processor. Instead, the processor was given the overall length of the array, the number of elements, and the fact that the ninth element failed. Having no further information about the positions of the elements, it assigned equally spaced locations (crosses) to them. The experiment thus simulated an air-dropped array in which a series of antenna elements was assumed to be dropped at approximately equal intervals from a moving aircraft.

Four experiments were conducted by selecting four of the locations of the transmitter on the test track as beamforming locations. The scanning algorithm was then applied and the loss in mainlobe gain was measured as a function of scan angle from the four beamforming directions. The resulting loss in gain with scan angle was not due to position errors alone, however; it was also due in part to multipath and scattering at the field site. According to (7.6), the measured loss is the sum of those losses due to the element position errors and to multipath and scattering. It was possible to measure the latter directly by the procedure described in Section 7.6. When this was done these losses could be subtracted from the total losses. The remainder are the experimental data in Figure 7.5.

Although the errors were in two dimensions, the dominant effect (according to the theory) was due to the position errors along the array. Therefore to check the theory it was necessary only to calculate or estimate σ_x^2 and apply it to (7.19). Figure 7.6 shows the position-error distribution in the direction of the array (the set of δ-functions on the abscissa) and also the staircase cumulative distribution function of the error, which is well approximated (points marked by $+$) by the cumulative distribution function of a Gaussian distribution with a variance of 11.1 ft^2. The Gaussian distribution may be thought of as an estimate of the population distribution of the random process for which the actual set of position errors is a sample function. By applying this value of the variance to (7.18) we obtain the solid curve of Figure 7.5. It is clear that the agreement is exceedingly good.

Two important interpretations follow from (7.19). This equation gives the loss in gain (and image contrast) as a function of scan angle. Given a system specification or a tolerance on the maximum allowed loss in image contrast $\Delta G_{\max}$, it is possible to calculate the permissible rms element position uncertainty in the array as a function of the maximum desired scan angle $\theta_{\max}$. The

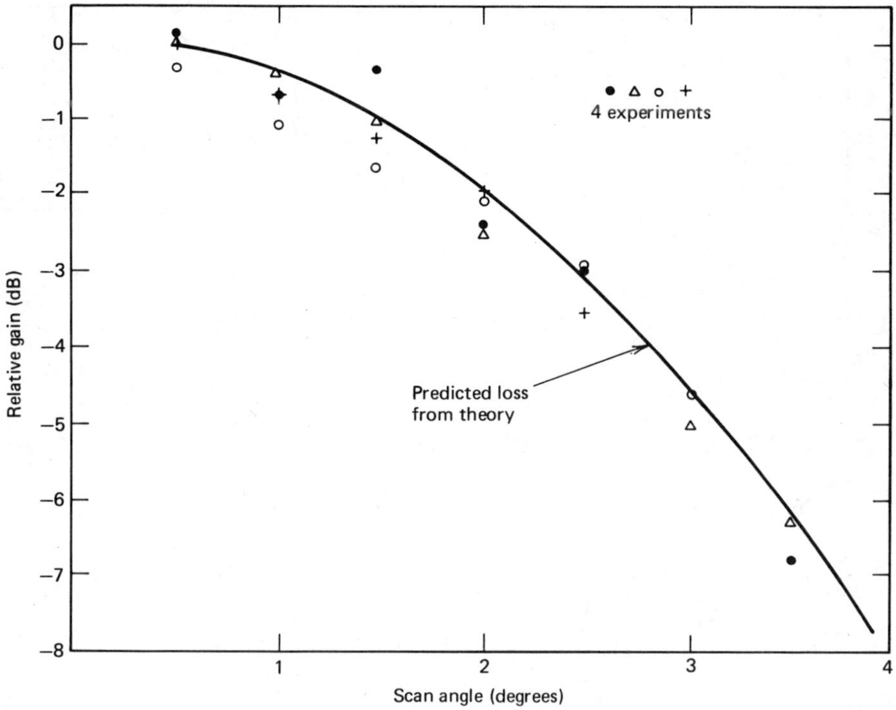

FIGURE 7.5. Effect of element position errors.

relation is

$$\Delta G_{\max} = 4.3k^2\sigma_x^2\theta_{\max}^2 \text{ dB} \tag{7.20}$$

Taking 1 dB as a nominal tolerance on loss of gain, the rms tolerance in element position becomes

$$\sigma_x = \left(\frac{\lambda}{4\pi}\right)\theta_{\max}^{-1} \tag{7.21}$$

The first factor on the right is the conventional tolerance of a diffraction limited aperture found in Section 2.4 [following (2.18)]. The second factor is the reciprocal of the maximum scan angle which, for target imaging, is always a small number. Hence the second factor is a large multiplier; for example, a huge target like the moon subtends only 10^{-2} rad at the earth. Most man-made targets are smaller still. Thus the maximum required scan angle for target imaging is 10^{-2} rad or smaller. Consequently the tolerance on position errors in a radio camera may be two or more orders of magnitude larger than in a conventionally designed phased array. It is this remarkable result that makes the radio camera a potentially attractive imaging instrument.

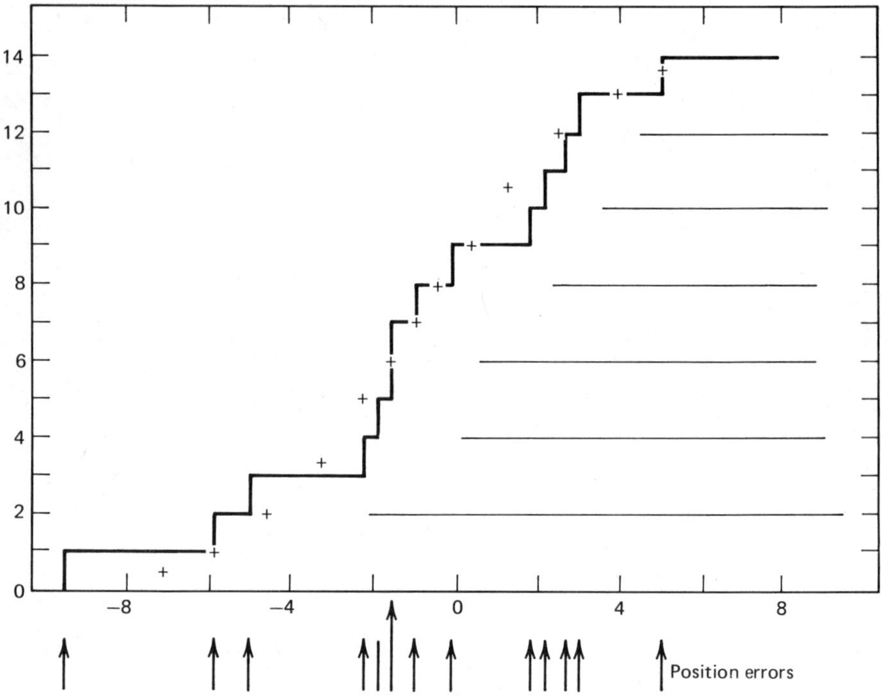

FIGURE 7.6. Cumulative distribution of element position errors and Gaussian fit ($\sigma^2 = 11.1$ ft^2).

7.4. FIELD OF VIEW

A second important conclusion drawn from (7.19) is that the useful field of view

$$\text{FOV} \triangleq 2\theta_{\max} = \frac{\lambda}{2\pi\sigma_x} \tag{7.22}$$

Within an angular sector of this size, centered on the direction of phase

TABLE 7.1. Field of View and Tolerance Liberalization Ratio Versus Rms Element Position Error

σ_x/λ	Tolerance Ratio	FOV (deg)
3	40	3.0
10	120	1.0
30	400	0.3

synchronization of the array, the loss in mainlobe gain or image contrast does not exceed 1 dB. This expression is believed to be a conservative estimate of FOV because the 1-dB degradation is experienced only at the edges; at half the distance to the edges the degradation is 1/4 dB. Table 7.1 lists the relation between rms position error in the array and FOV. It also shows the factor by which the position error tolerance is eased in relation to a nonadaptively formed array.

7.5. SCANNING IN THE NEAR FIELD OF THE ARRAY

The far-field analysis in Section 7.3 is easily extended to near-field scanning, where it is shown that the tolerance in scanning in the range direction is even larger than it is in the angular or cross-range direction. Figure 7.7 shows the reference element again at the origin of a rectangular coordinate system. The beamformer is at distance R_0 on the y-axis. The beam is to be scanned to x_T, y_T. The differential distance between an element at x, y and the reference element to the beamforming point is

$$R_0 - \sqrt{x^2 + (R_0 - y)^2} \simeq y - \frac{r^2}{2R_0} \tag{7.23}$$

The phase correction made by the adaptive signal processor is k times (7.23) modulo 2π. The differential distance to the scan point is

$$R_T - \sqrt{(x_T - x)^2 + (y_T - y)^2} \simeq \frac{x_T x + y_T y}{R_T} - \frac{r^2}{2R_T} \tag{7.24}$$

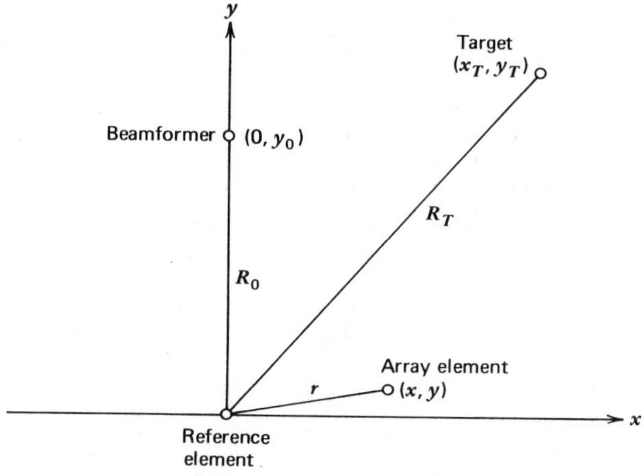

FIGURE 7.7. Near-field geometry.

The signal processor calculates the difference between (7.24) and (7.23)

$$y(\cos\theta - 1) + x\sin\theta + \frac{r^2}{2}\left(\frac{1}{R_0} - \frac{1}{R_T}\right) \tag{7.25}$$

and adds a phase shift to the element at x, y equal to k times this difference.

In the absence of errors, (7.25) times k is the correct phase shift for scanning. Let the coordinates be in error, however, by $\delta x, \delta y$. Then the calculated phase shifts are k times

$$(y + \delta y)(\cos\theta - 1) + (x + \delta x)\sin\theta + \frac{(x + \delta x)^2 + (y + \delta y)^2}{2}\left(\frac{1}{R_0} - \frac{1}{R_T}\right) \tag{7.26}$$

and the phase error is approximately k times

$$\delta y\,(\cos\theta - 1) + \delta x\sin\theta + (x\,\delta x + y\,\delta y)\frac{\Delta R}{R^2} \tag{7.27}$$

where quadratic position-error terms have been dropped and the differential distance $\Delta R = R_T - R_0$ is assumed to be small compared with either distance.

Comparing this equation to (7.17), we see that the last term introduces the near-field behavior of the system. The first term can be dropped as before. The mean square error, assuming independent, zero-mean coordinate errors $\delta x, \delta y$, is

$$u^2\sigma_x^2 + \frac{(\Delta R)^2}{R^4}\left(\overline{x^2}\sigma_x^2 + \overline{y^2}\sigma_y^2\right) \tag{7.28}$$

which, when the statistics of these errors are the same, simplifies to

$$\sigma_x^2\left[u^2 + \frac{(\Delta R)^2}{R^4}\left(\overline{x^2} + \overline{y^2}\right)\right] \tag{7.29}$$

$\overline{x^2}$ and $\overline{y^2}$ are the second central moments of the element distribution in the two coordinate directions. For uniform distributions over lengths L_x and L_y in the x- and y-directions, respectively, (7.29) becomes

$$\sigma_x^2\left[u^2 + \frac{(\Delta R)^2}{12R^4}\left(L_x^2 + L_y^2\right)\right] \tag{7.30}$$

which simplifies to

$$\sigma_x^2\left(u^2 + \frac{(\Delta R)^2 L_x^2}{12R^4}\right) \tag{7.31}$$

for a one-dimensional array and

$$\sigma_x^2\left(u^2 + \frac{(\Delta R)^2 L_x^2}{6R^4}\right) \tag{7.32}$$

for a square array.

Loss in array gain, in dB, is proportional to (7.30) from (2.16). The loss contributions are equal when the two terms in (7.30) are equal. Let the scan distance ΔS in the cross-range direction be defined by

$$u \triangleq \frac{\Delta S}{R}$$

and let the scan distance in range be ΔR as before. Thus equal losses in array gain due to element position errors develop when

$$u = \frac{\Delta R}{R^2}\sqrt{\frac{L_x^2 + L_y^2}{12}} \tag{7.33}$$

or

$$\frac{\Delta R}{\Delta S} = \frac{\sqrt{12}\,R}{L} \tag{7.34}$$

where $L = (L_x^2 + L_y^2)^{1/2}$ is identified with the size of the array. It is evident from (7.34) that, in general, $\Delta R \gg \Delta S$ because normally $R \gg L$. Thus the position-error limitation to scanning is most heavily felt in the cross-range or angle direction.

The Scanning Sector

A scanning sector may be defined in an analogous manner as the resolution cell or focal zone described in Chapter 2 (Fig. 2.8). The dimensions of the scanning sector are the transverse and longitudinal excursions from the focal point at which the mainlobe gain drops 3 dB due to the phase errors induced by the element position errors. The axial extremities of the scanning sector are found by equating each part of (7.30) to $3/4.3k^2$ [from (2.16)]. These values are

$$\sqrt{\frac{3}{4.3}}\,\frac{\lambda R}{\pi\sigma_x} \quad \text{and} \quad \frac{6\lambda R^2}{\sqrt{4.3}\,\pi\sigma_x L}$$

The corresponding values for the resolution cell in the near field are given by (2.41) and (2.42). The number of resolution cells in the scanning sector is the ratio of the areas of the scanning sector to the resolution cell, which is approximately the ratio of the products of their axial dimensions:

$$\text{number of resolution cells} \approx \frac{L^2}{30\sigma_x^2} \tag{7.35}$$

7.6. PHASEFRONT DISTORTION DUE TO MULTIPATH AND SCATTERING

Another source of phase error is multipath and scattering. Energy from the target may be scattered by reflectors located outside the direct path to the array and some of the scattered energy may arrive at the array. This situation is described in Figure 7.8. Energy also may be refracted to the array by refractive index variations in the propagation medium. HF propagation in the ionosphere and acoustic propagation in the ocean provide examples of multipath propagation. The direct effect of the scattered energy is distortion of the phasefront of the radiation across the array. In turn, this distortion leads to phase errors in adaptive beamforming.

It is desired to develop a theory that enables the designer to predict the effects of multipath and scattering on the imaging process. Because design is the objective, a theory to be useful may involve only a small number of general descriptors of the process. Examples are the scale of the scattering body or bodies and scatter strength statistics. On the other hand, a theory that requires a detailed description of the scattering process (number of scatterers, scatterer coordinates, and scattering coefficients) is useless for design purposes. A theory that involves only two properties of the scattering process and appears to check the experimental data is described [99].

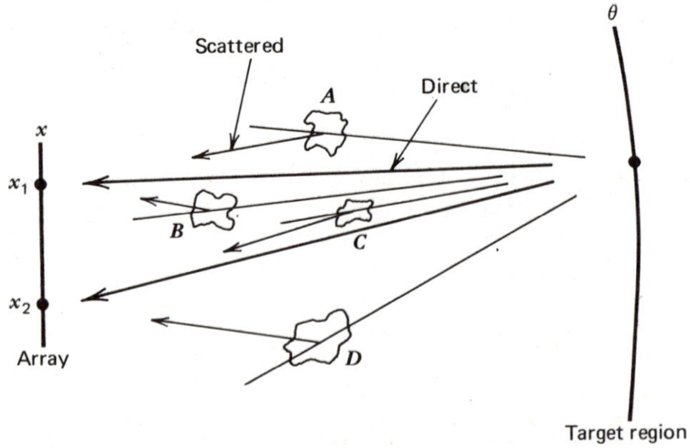

FIGURE 7.8. Scattering model.

A source in the target region, shown in Figure 7.8, radiates energy to positions x_1 and x_2 in the array along great circle paths; these energies constitute direct rays. In addition, scatterers A and B divert energy from other paths to point x_1, whereas scatterers C and D scatter energy to x_2. As a consequence, the signal phases at x_1 and x_2 differ from the signal phases that would exist in the absence of scattered energy. Because the signal processor of the adaptive array has no *a priori* means for calculating the phase changes due to the scattering, these phase alterations constitute phase errors that necessarily increase the phase-error variance σ_ϕ^2. From (7.6) the contribution to σ_ϕ^2 due to multipath and scattering is

$$\begin{aligned}\sigma_\phi^2 &= E[\delta\phi(x,\theta) - \delta\phi(x,\theta_0)]^2 \\ &= E[\delta\phi(x,\theta)]^2 - 2E[\delta\phi(x,\theta)\,\delta\phi(x,\theta_0)] + E[\delta\phi(x,\theta_0)]^2 \\ &= 2E[\delta\phi(x)]^2[1 - \rho(x,\Delta\theta)] \\ &= 2\sigma_S^2[1 - \rho(\Delta\theta)] \end{aligned} \tag{7.36}$$

where σ_S^2 is the initial phase-error variance across the array due to the scattered energy and $\rho(\Delta\theta)$ is the autocorrelation function of the phase error as a function of the beamsteering displacement $\Delta\theta = \theta - \theta_0$. Thus the residual phase-error variance is doubled because the residual phase error is the difference between two phase errors and is reduced according to the correlation between them.

Four assumptions are made in (7.36). The first is that the statistics of the phase at any position x in the array is independent of θ: thus

$$E[\delta\phi(x,\theta_i)]^2 = E[\delta\phi(x,\theta_j)]^2 \triangleq E[\delta\phi(x)]^2 \qquad \text{all } i, j \tag{7.37}$$

The second is that the covariance of the phase error is independent of θ and is dependent only on the spacing $\Delta\theta$. This condition is the equivalent of stationarity in time-series analysis. The third assumption is that the phase-error statistics are independent of x; therefore

$$E[\delta\phi(x_1)]^2 = E[\delta\phi(x_j)]^2 \triangleq E[(\delta\phi)^2] \qquad \text{all } i, j \tag{7.38}$$

Last, it is assumed that the phase error is unbiased (zero mean) such that $E[(\delta\phi)^2]$ is also the variance of σ_S^2. In addition, ergodicity is implicitly invoked in the assumption that measured spatial averages can replace ensemble averages.

σ_S^2 is related to the power ratio S of the scattered signal received at the array to the direct signal. By referring to Figure 7.2, we consider the "noise" vector in that figure to be the scattered "signal" with amplitude $S^{1/2}$, and ψ is a random phase variable uniformly distributed in $[0, 2\pi]$. The variance of $\delta\phi$, $E[\delta\phi^2]$, is

$$\sigma_S^2 = E\left[\tan^{-1}\frac{S^{1/2}\sin\psi}{1 + S^{1/2}\cos\psi}\right]^2 \tag{7.39}$$

which is approximately equal to $S/2$ when S is small. Making this substitution in (7.36) and redefining the angle θ to represent the scan angle from the direction of phase synchronization, we find that the loss in gain due to scattering becomes

$$\begin{aligned}\Delta G_S(\text{dB}) &= 4.3\sigma_\phi^2 \\ &= 4.3S[1 - \rho(\theta)]\end{aligned} \tag{7.40}$$

Equation (7.40), the desired result, gives the loss as a function of only two properties of the scattering process, the scatter strength and the correlation between phase errors induced by the scattering. Because $\rho(\theta)$ is an autocorrelation function, its value is unity when $\theta = 0$ and it should drop asymptotically to zero as θ gets large. The dB loss in gain should be zero at the origin and should grow asymptotically toward 4.3S dB. A distributed cloud of scatterers will yield a smoothly decaying correlation function, whereas $\rho(\theta)$ for one or a few discrete scatterers will exhibit undulations as it decays. The specific angular dependence is not particularly important, for in most cases the loss in correlation will occur within one or a few beamwidths from the direction of phase synchronization [48]. Hence a simple measure of the effect of scattering is obtained.

Equation 7.40 may be interpreted as defining a threshold of scattering strength $S_{\max}$, determined entirely by the tolerance or specification on allowed loss in gain $\Delta G_{\max}$(dB), below which scattering may be ignored. The relationship is $\Delta G_{\max} = 4.3S_{\max}$. For an allowed loss of 1 dB $S_{\max} \simeq \frac{1}{4}$ or -6 dB. Thus the intensity of the scattered field at the array can be as large as $\frac{1}{4}$ the direct field intensity without causing more than a 1-dB loss in image contrast.

The theory, appealing in its simplicity, requires a knowledge only of a single parameter, S, and a single function, $\rho(\theta)$; S can often be estimated at a field site from visual observation of the terrain and surrounding structures. Sometimes it is possible to estimate the most significant characteristic of $\rho(\theta)$, its correlation distance, which may be designated θ_c; θ_c is the argument of the correlation function by which $\rho(\theta)$ has dropped from unity (near the origin) to about one-half; θ_c is usually the nominal scattering angle of the scattering process. For natural terrain it is at most a few degrees.

Although the theory is attractive because it asks for only two gross characteristics of the scattering process rather than for the details, its validity is in question because of the spatial stationarity and ergodicity assumptions made in the derivation. Both assumptions are highly questionable. Only a test of the theory will disclose the extent of its applicability.

Data taken at the Valley Forge field site tend to validate the theory. In one experiment that involved strong forward scattering from the ground midway between the array and the beamforming source it was possible to estimate the value of S [68]. In that experiment the amplitude of the wave received from a source in the field was plotted as a function of the element number (monotonic with distance) in the array. This was done with the 27-m rooftop array at a wavelength of 30 cm; 20 elements were used. The source-to-array distance was 240 m. Figure 7.9 shows the measurements. The simple, undulatory variation of amplitude with position is suggestive of a direct ray, plus a single scattered or multipath ray. Figure 7.2 is a useful model for this situation. The main signal is the direct ray, whereas the "noise" is the scattered ray of amplitude $S^{1/2}$. The differential distance traveled by the two rays to an antenna element varies with the position of the element. Thus the phase that represents the scattered ray would be expected to rotate with position in the array. The magnitude of the sum of the direct and scattered ray would undulate, as in the experiment reported in Figure 7.9. The average peak-to-valley ratio, which is observed in the figure to be $2\frac{1}{2} : 1$, equals $(1 + S^{1/2})/(1 - S^{1/2})$. From these data the forward scatter coefficient $S^{1/2} \approx 0.43$. Hence the asymptotic loss should be less than 1 dB. In addition to the measurements shown in Figure 7.9, another set was made from another source in the field at a different location and the results were almost identical. On the basis of these observations we assume that $S \approx \frac{1}{5}$ typifies the scattering strength for this portion of the field site at the wavelength used.

A later experiment conducted at the same site completed the testing of the theory. Phase was measured at each element of a 16-element random array when a source was moved through an angle of approximately 6° [25]. The source was at a distance of 240 m and the 16 elements were distributed over the same 27-m section of the laboratory roof. The same area of the field site was used for which the estimated forward scattering coefficient was obtained.

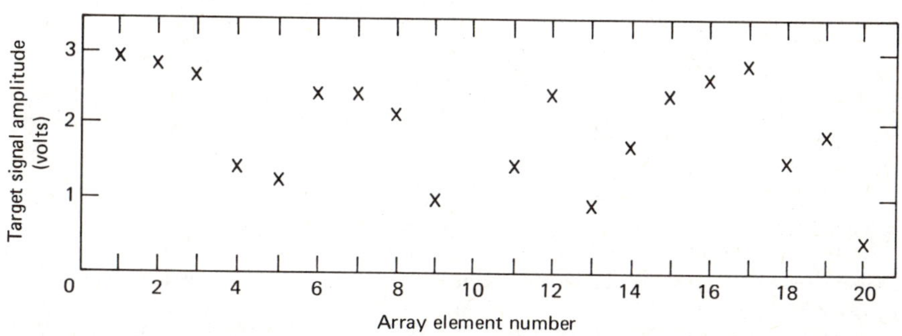

FIGURE 7.9. Signal amplitude versus element number. (Data from E. N. Powers, "Adaptive Arrays for Microwave Imaging," Ph.D. Dissertation, University of Pennsylvania, 1975.)

Four experiments were conducted. A typical result which showed the region in which the gain dropped with scan angle is represented by the curve labeled "experimental" in Figure 7.10. The direction of beamforming is at the left edge of the figure. The abscissa is the scan angle in degrees measured at the array. The general decrease in gain toward an asymptotic value of about 1 dB, as predicted by the theory, is observed. The undulations in the experimental results can be explained by discrete forward-scattering from the terrain [48]. The dashed curve is calculated from the theory and the agreement is excellent. Three additional experiments yielded similar results.

The dashed curve was obtained in the following manner. Perusal of the field site indicated that the dominant scattering was forward scatter from a region approximately midway between the radar and the target. The correlation function $\rho(\theta)$ was calculated for this case (see Section 7.7). Using the expression for the correlation function derived in that section, we determined the loss in array gain as a function of angle. Parameters of that expression are the reflection or the forward scattering coefficient of the terrain $S^{1/2}$ and the coordinates of the dominant reflecting portion of the terrain. A computer search was made for the set of three coordinates (scattering strength and location) for which the loss in gain with scan angle would most closely approximate the measured function of Figure 7.10. Constraints were imposed that both the scatter coefficient and scatterer location be realistic. The com-

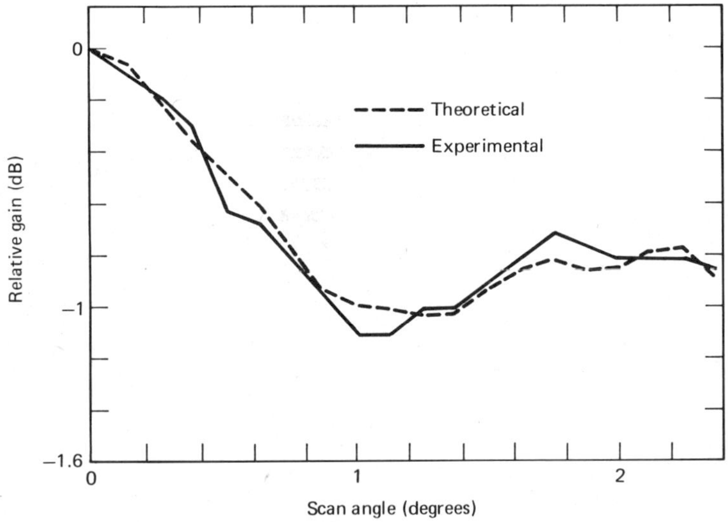

FIGURE 7.10. Main-beam gain versus scan angle following self-cohering beamforming at zero degrees. (Experimental data from Dorny and Meagher, Valley Forge Research Center Report No. 107, Moore School of Electrical Engineering, University of Pennsylvania, March 1977. Theoretical curve from E. Jadlovker (Yadin), "A Study of the Effect of Multipath and Scattering upon Scanning with a Radio Camera, " M. S. Thesis, Tel Aviv, 1978, and B. D. Steinberg and E. Yadin, "Effect of Multipath and Scattering on Array Gain of a Radio Camera, " submitted to IEEE Trans. AP.)

puter search that resulted in the excellent agreement shown in Figure 7.10 indicated a scatter strength of 0.42, which agreed with the calculation from the more direct measurement obtained from Figure 7.9, and a reflection location within a few meters of the midpath point. Thus the theory proved exceedingly successful at this field site, notwithstanding the theoretical inadequacies noted earlier.

7.7. MULTIPATH MODELS

Equation 7.40 gives the loss of gain as a function of the autocorrelation function of phase error in the array. In this section the autocorrelation function is calculated for two scattering models and for three scattering situations for each model.

Both are single-scatterer models. Realistically, many scatterers contribute to the reradiation. The scattered field, in general, is the linear vector superposition of the reflected energy from the several scatterers, plus the secondary reflections among them. A single-scatterer model, however, sometimes suffices if one of the scatterers is larger than the vector sum of the remainder. This is because the phase of the field at any point is a nonlinear function of the scattered components, even though the field itself is a linear sum:

$$\phi = \tan^{-1} \frac{\Sigma a_i \sin \psi_i}{\Sigma a_i \cos \psi_i} \tag{7.41}$$

In (7.41) a_i is the amplitude of the ith scattered component and ψ_i is $2\pi/\lambda$ times the distance to that scatterer. Because of the nonlinear operations in (7.41), ϕ can be dominated by the phase of a single large scatterer. This nonlinear phase-capture effect is assumed in the discussion that follows. The amplitude of the dominant scatterer is $S^{1/2}$.

The models are sketched in Figure 7.11. The first model is a single, fixed-point reflector. It could be a large rock or a prominent mound on the scattering surface. It may also be a shed or a small building. The second model is a specular surface in which the reflection point moves with target position and also with element position within the array. The three cases for each model are the following:

1. Scatterer close to or within the target area corresponds to the multiple scattering problem from a complex target in which energy reflected from one part of the target is reflected from another surface near the first surface before returning to the array.
2. Midpath scattering corresponds to forward-scatter surface reflections from horizontal or near-horizontal surfaces. It is typified by low-angle, forward-surface scattering in ground-to-ground radar. It is also common in forward scattering from the sea surface in air-to-air radar.

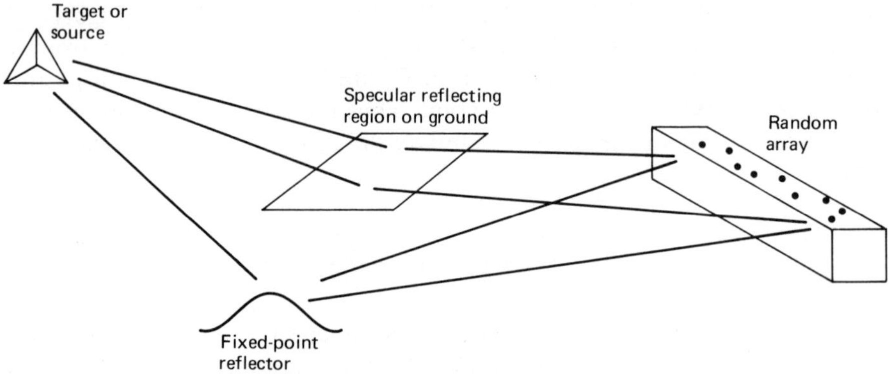

FIGURE 7.11. Two forward-scattering models.

3. Scatterers that are local to the array are typified by reflections from within the radome structure or by floors, walls and ceilings of structures on which antenna modules are located.

The Fixed-Point Reflector (FPR)

The geometry is sketched in Figure 7.12. Let the direct ray from target T to array point E have unity amplitude and let the amplitude of the reflected ray be $S^{1/2}$. The phase of the sum of the direct and the reflected energies differ from the phase of the direct ray by

$$\delta\phi = \tan^{-1}\left\{\frac{S^{1/2}\sin\left[(2\pi/\lambda)\left(\overline{TP}+\overline{EP}-\overline{TE}\right)\right]}{1+S^{1/2}\cos\left[(2\pi/\lambda)\left(\overline{TP}+\overline{EP}-\overline{TE}\right)\right]}\right\} \tag{7.42}$$

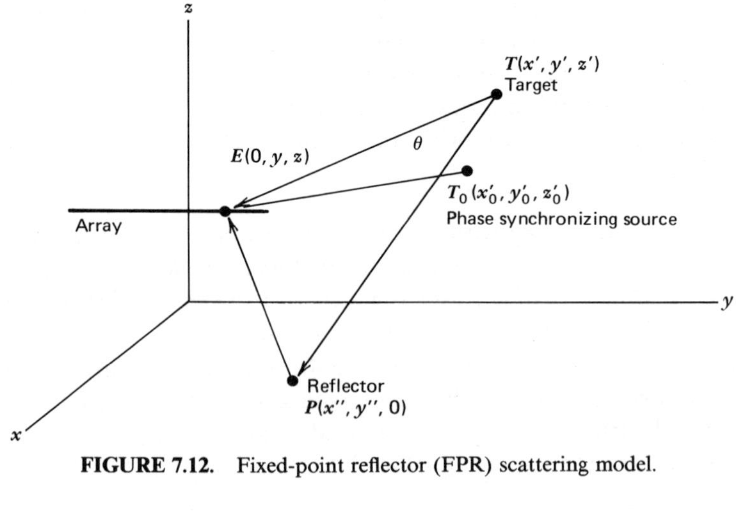

FIGURE 7.12. Fixed-point reflector (FPR) scattering model.

where $\overline{TP}$ is the length of the line segment that connects the target and the reflector. The phase error that affects the scanning performance is the difference between (7.42) and the phase error $\delta\phi_0$ in the direction of adaptive beamforming, which has the same form as (7.42); T_0 replaces T in that equation and T and T_0 are assumed to be at common altitudes. Their coordinates are related by $x' = x_0' - y_0'\theta$, $y' = y_0' + x_0'\theta$, $z' = z_0'$. The correlation function $\rho(\theta)$ in (7.40) is given by

$$\rho(\theta) = \frac{E[\delta\phi_1 \delta\phi_0] - E[\delta\phi_1]E[\delta\phi_0]}{\sqrt{E[\delta\phi_1^2] - (E[\delta\phi_1])^2}\sqrt{E[\delta\phi_0]^2 - (E[\delta\phi_0])^2}} \tag{7.43}$$

where the scan angle $\theta = \theta_1 - \theta_0$. $\rho(\theta)$ was computer-simulated by using (7.43). Expected values were estimated by averaging over the actual locations of the 16 elements in the array described in Section 7.6. The known locations of target, beamformer, and array elements were used to evaluate $\delta\phi_0$ and $\delta\phi_1$ with the formulation given by (7.42). The scatter coordinates (strength, position) used in (7.42) were found in a computer search for the best fit between the predicted and measured losses. One result is shown in Figure 7.10. The model accurately predicted the lengths of the correlation intervals and the general character of the loss versus scan angle in each of the four experiments.

Table 7.2 shows the best-fit scatter coordinates and the rms difference (σ_d) in dB between measured and predicted losses. The reflection point is located approximately midway between array and target area.

The reflector strength is reasonably consistent from experiment to experiment and is close to the approximate value (0.43) inferred by amplitude measurements across the array reported earlier. Thus it is evident that the scattering process at this field site can be modeled by a FPR for beam-scanning experiments following adaptive beamforming.

It is possible to develop a closed form approximation for $\rho(\theta)$ which closely matches the accurate expression given in (7.43). By accepting the assumptions inherent in the theory which lead to (7.40), that is, $E[\delta\phi_1] = E[\delta\phi_0] = 0$ and $E[\delta\phi_1]^2 = E[\delta\phi_0]^2 = \sigma_s^2 = S/2$, $\rho(\theta)$ reduces to

$$\rho(\theta) = \frac{E[\delta\phi_1 \delta\phi_0]}{S/2}$$

TABLE 7.2. Best-Fit Scatter Coordinates and Rms Loss Differences (dB)

Experiment	1	2	3	4
$S^{1/2}$	0.38	0.36	0.42	0.4
x''(ft)	328.5	337.5	301.5	381
y''(ft)	−47.5	−69.5	−45.5	−63
σ_d(dB)	0.130	0.065	0.060	0.075

Referring to (7.39) and for $S \le 0.5$, we can express $\delta\phi_1, \delta\phi_0$ as

$$\delta\phi_1 = S^{1/2}\sin\psi_1 \qquad \delta\phi_0 = S^{1/2}\sin\psi_0$$

where ψ_0, ψ_1 are the phases of the scattered signal associated with the beamforming direction θ_0 and the direction θ_1 respectively; $\delta\phi_1\delta\phi_0 = S\sin\psi_1\sin\psi_0 = (S/2)[\cos(\psi_1 - \psi_0) - \cos(\psi_1 + \psi_0)]$. Because ψ_1, ψ_0 are arbitrary values in the interval $[0, 2\pi]$ we can assume that $E[\cos(\psi_1 + \psi_0] = 0$. Therefore

$$\rho(\theta) = E\left[\cos(\psi_1 - \psi_0)\right] \tag{7.44}$$

From the geometry of Figure 7.12 ψ_1, ψ_0 and, therefore, $\psi_1 - \psi_0$ are given by

$$\psi_1 = \frac{2\pi}{\lambda}\left(\overline{TP} + \overline{EP} - \overline{TE}\right) \qquad \psi_0 = \frac{2\pi}{\lambda}\left(\overline{T_0P} + \overline{EP} - \overline{T_0E}\right)$$

$$\psi_1 - \psi_0 = \frac{2\pi}{\lambda}\left[\left(\overline{TP} - \overline{T_0P}\right) - \left(\overline{TE} - \overline{T_0E}\right)\right] \tag{7.45}$$

Let R_0 be the distance between the center point of the array and the phase synchronizing source T_0 and let R_S be the distance from T_0 to the reflector P; R_0, R_S can be written as

$$R_0 = \left[x_0'^2 + y_0'^2 + (z_0' - z)^2\right]^{1/2}$$

$$R_S = \left[(x_0' - x'')^2 + (y_0 - y'')^2 + z_0'^2\right]^{1/2}$$

$\overline{TE}, \overline{T_0E}$ are given by

$$\overline{T_0E} = \left[x_0'^2 + (y_0' - y)^2 + (z_0' - z)^2\right]^{1/2} = \left[R_0^2 - 2y_0'y + y^2\right]^{1/2}$$

$$\sim R_0\left[1 - \frac{2y_0'y + y^2}{2R_0^2}\right]$$

$$\overline{TE} = \left[x'^2 + (y' - y)^2 + (z' - z)^2\right]^{1/2}$$

Because $x' = x_0' - y_0'\theta$, $y' = y_0' + x_0'\theta$, $z' = z_0'$, $\overline{TE}$ can be written as

$$\overline{TE} = \overline{T_0E} + \frac{\left(x_0'^2 + y_0'^2\right)\theta^2}{2R_0} - \frac{yx_0'\theta}{R_0}$$

$\overline{TP}, \overline{T_0P}$ are given by

$$\overline{T_0P} = R_S$$

$$\overline{TP} = \left[(x_0' - x'' - y_0'\theta)^2 + (y_0' - y'' + x_0'\theta)^2 + z_0'^2 \right]^{1/2}$$

$$\sim \overline{T_0P} + \frac{2(y_0'x'' - x_0'y'')\theta + (x_0'^2 + y_0'^2)\theta^2}{2R_S}$$

Substitution of these approximations for $\overline{TP} - \overline{T_0P}, \overline{TE} - \overline{T_0E}$ into (7.45) and (7.44) yields

$$\rho(\theta) = E[\cos(k_1 + k_2 y)] \tag{7.46}$$

where

$$k_1 = \frac{2\pi}{\lambda R_s}(y_0'x'' - x_0'y'')\theta + \frac{\pi(x_0'^2 + y_0'^2)}{2\lambda}\left(\frac{1}{R_s} - \frac{1}{R_0}\right)\theta^2$$

$$k_2 = \frac{2\pi x_0'}{\lambda R_0}\theta$$

The expectation in (7.46) is with respect to element location y. Assuming that y is uniformly distributed between $-L/2$ and $L/2$, $\rho(\theta)$ is given by

$$\rho(\theta) = \frac{1}{L}\int_{-L/2}^{L/2} \cos(k_1 + k_2 y)\,\mathrm{d}y = \frac{\sin k_2 L/2}{k_2 L/2}\cos(k_1)$$

$$= \left[\frac{\sin(\pi x_0' L\theta/\lambda R_0)}{\pi x_0' L\theta/\lambda R_0}\right]$$

$$\times \cos\left\{\frac{\pi\theta}{\lambda R_S}\left[\left(1 - \frac{R_S}{R_0}\right)(x_0'^2 + y_0'^2)\theta + 2(x''y_0' - y''x_0')\right]\right\} \tag{7.47}$$

This closed form approximation was compared with the accurate expression for $\rho(\theta)$ given by (7.43) in several computer simulations. With various parameters the two curves obtained in each case were almost indistinguishable [48]. Figure 7.13 shows one pair of curves.

The Specular Reflector (SR)

The geometry for this case is shown in Figure 7.14. The phase error is given by

$$\delta\phi = \tan^{-1}\left\{\frac{S^{1/2}\sin[(4\pi/\lambda)zz'/\overline{TE}]}{1 + S^{1/2}\cos[(4\pi/\lambda)zz'/\overline{TE}]}\right\} \tag{7.48}$$

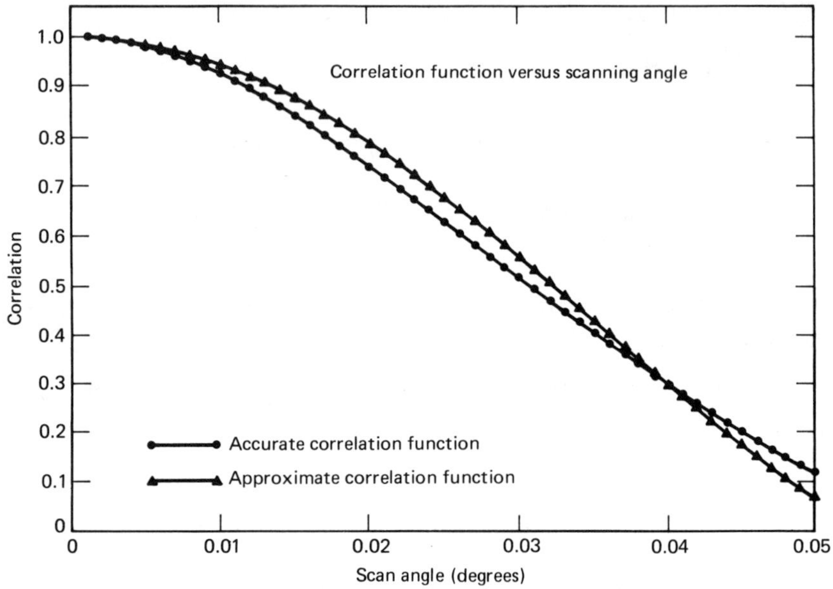

FIGURE 7.13. A computer simulation comparison between $\rho(\theta)$ calculated from (7.43) and the closed form approximation (7.47). A 100-element array, 1000 wavelengths long. The elements are randomly distributed along the array. The target is approximately 11,000 ft from the array at 26.5° from broadside. One fixed-point reflector is located at midpath. The scattered field is 8 dB below the direct field. (From E. Jadlovker (Yadin), "A Study of the Effect of Multipath and Scattering upon Scanning with a Radio Camera," M.S. Thesis, Tel Aviv, 1978.)

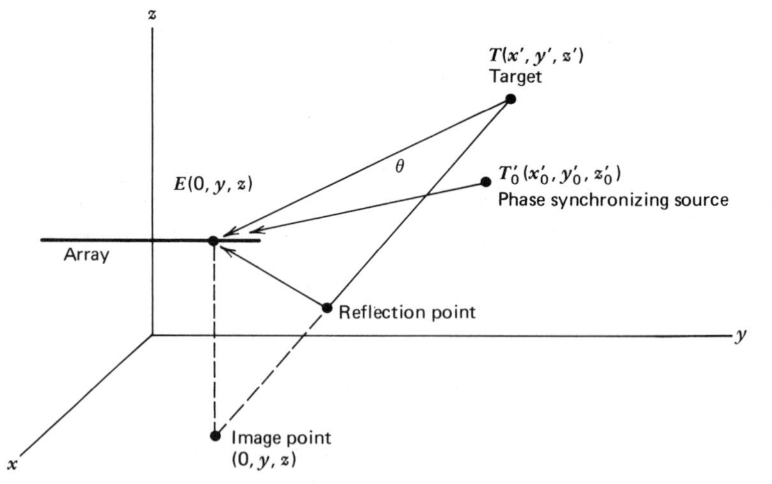

FIGURE 7.14. Specular reflector scattering model.

The closed form approximation for the correlation function $\rho(\theta)$ is derived similarly [48]:

$$\rho(\theta) = \frac{\sin\left(2\pi z z_0' x_0' L\theta/\lambda R_0^3\right)}{2\pi z z_0' x_0' L\theta/\lambda R_0^3} \cos\frac{2\pi z z_0' \theta^2\left(x_0'^2 + y_0'^2\right)}{\lambda R_0^3} \tag{7.49}$$

Computer simulation shows that (7.49) is in good agreement with $\rho(\theta)$ calculated from (7.43).

Extensions to Other Cases

Because a single point reflector model, when applied to the gain-loss theory, satisfactorily predicted the scanning loss or the field site, it is with some confidence that both models (FPR and SR) can be applied to three common and important cases. The first is typified by ground or sea surface reflections. The scattering region generally is somewhere between target and array, the exact location depending on their heights above the ground plane and, for the SR, the slope of the ground plane. The second case involves a reflecting object near the target or a large target structure with parts separated by more than a beamwidth. It can be above or below the target and can have arbitrary orientation and location. The third case involves reflectors in or near the array. A different reflector is assumed for *each antenna element*; that is, each element is influenced only by its immediate physical surroundings. This last case pertains to a large array with widely spaced elements in which the reflections may arise from the walls or floors or supports for the antenna modules. In each case the correlation function, which is the necessary ingredient for the calculation of the gain-loss, can be developed [48]. These functions are listed in Table 7.3.

It is evident from (7.40) that the loss in main-beam gain is small when the correlation function approaches unity (small scan angle) and is asymptotically large when the correlation drops toward zero. The scan angle interval in which the loss grows from zero almost to the maximum value (4.3S dB) may be called the correlation interval. This quantity, θ_c, and not the shape of $\rho(\theta)$, is the pertinent characteristic of the correlation function. It is given in Table 7.3 for the six cases studied.

The symbols used in Table 7.3 are the following: z is the elevation of the array; x', y', and z' are the target coordinates; R is the distance from array center to target; ΔR is the distance between the reference target and the reflection point; Δx, Δy, and Δz are the components of ΔR in the x, y, and z directions, respectively; $\Delta\theta = \lambda R/Lx'$ is the beamwidth of the array in the direction of the target; θ is the scan angle following adaptive beamforming; and A and B are statistical parameters that relate to the distribution of many reflectors near the array. It is assumed that each element receives scattered

TABLE 7.3. Correlation Function and Correlation Interval for Fixed-Point and Specular Reflectors

Scattering Model	Correlation Function $\rho(\theta)$	Correlation Interval (θ_c)
Many SRs near array	$\dfrac{\sin(\pi x'B\theta/\lambda R)}{\pi x'B\theta/\lambda}\,\dfrac{\sin(\pi y'A\theta/\lambda R)}{\pi y'A\theta/\lambda}$	$> N\Delta\theta$ for $x' > y'$ $> \dfrac{x'}{y'}N\Delta\theta$ for $x' < y'$
Many FPRs near array	Same as above	Same as above
Single specular ground plane approximately midpath	$\dfrac{\sin[(2\pi z'z/R^2)(\theta/\Delta\theta)]}{(2\pi z'z/R^2)(\theta/\Delta\theta)}\cos\left(\dfrac{2\pi z'z\theta^2}{\lambda R}\right)$	$\sqrt{\dfrac{\lambda R}{4z'z}}$
Single FPR approximately midpath	$\dfrac{\sin(\pi\theta/\Delta\theta)}{\pi\theta/\Delta\theta}\cos\left(\dfrac{\pi R}{\lambda}\theta^2\right)$	$\leq \Delta\theta$
Single vertical specular plane in target region	$\dfrac{\sin[\pi(1+\Delta x/\Delta R)(\theta/\Delta\theta)]}{\pi(1+\Delta x/\Delta R)(\theta/\Delta\theta)}\cos\left[\dfrac{2\pi}{\lambda}\dfrac{R}{\Delta R}\Delta y\theta\right]$	$\leq \Delta\theta$
Single FPR in target region	$\dfrac{\sin(\pi\theta/\Delta\theta)}{\pi\theta/\Delta\theta}\cos\dfrac{2\pi}{\lambda}\left[\dfrac{R^2}{2\Delta R}\theta^2+\left(\dfrac{x'\Delta y-y'\Delta x}{\Delta R}\right)\theta\right]$	$\leq \Delta\theta$

energy from a single reflector and that the reflector is different for each element. The distance between each element and its reflection point is a random variable uniformly distributed between $-A/2$ and $A/2$ in the x-direction and between $-B/2$ and $B/2$ in the y-direction.

Observations

The first line in the correlation interval column pertains to the forward scanning sector of the array, whereas the second line pertains to the end-fire region. The forward scanning sector is of dominant interest. In this region the correlation interval is more than N beamwidths, which means that scanning is essentially lossless in an angular sector $\pm N\Delta\theta$ when the multipath sources are located in or close to the array. This is true for both scattering models.

The midpath specular model is more complicated. The correlation interval is sensitive to array and target heights. It nearly vanishes when the array and target are airborne, and it can extend over the entire forward scanning sector for low-angle scattering.

In the last three cases decorrelation occurs within one beamwidth of the array; hence the asymptotic loss $4.3S$ dB is presumed to pertain everywhere for these cases.

7.8. MEDIUM DISTORTION

Although the electromagnetic medium is normally considered homogeneous at microwave frequencies, the same is not true at longer wavelengths (HF) or shorter wavelengths (optics); HF communications depend on the refractive properties of the ionosphere, which is a weak ionic plasma with spatial and temporal random properties. Similarly, spatial and temporal variations in the dielectric constant of the atmosphere at optical wavelengths limit the "seeing" conditions of optical telescopes. Long wavelength acoustic signals in the ocean experience refraction due to sound velocity variations with depth. Ultrasonic signals in biological tissue experience similar sound velocity variations, largely because of variations in density of the biological material. Because of the nonuniformities of the refractive index, no single medium-distortion model suffices. One problem has been addressed [93]: it is the case in which the propagation medium is statistically homogeneous and characterized by a spatially varying propagation speed $c = c_0 + \delta c$, where δc is a zero-mean random variable with correlation distance d. The spatial correlation coefficient of δc is assumed to be unity for distances less than d and zero beyond. Let a beamforming (or focal) point A and two array elements at a and b, separated a distance D, be in the medium (Figure 7.15). The distance from the array to A is $R = R_1 + R_2$. Divide the portion of the medium traversed by a ray into K cells of size d. The phase shift of a wave at frequency ω over a ray path from

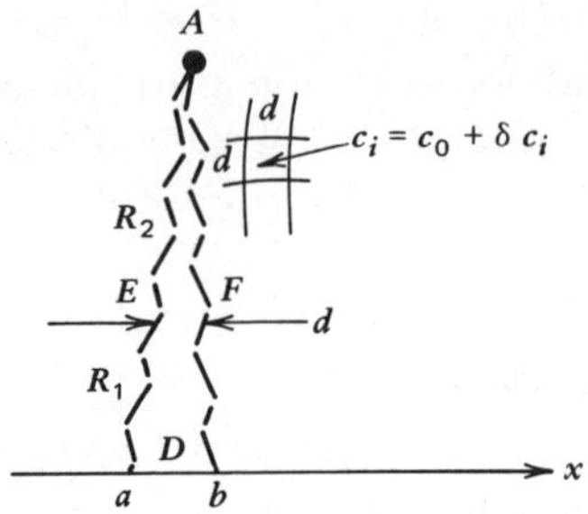

FIGURE 7.15. a and b are points in an array; A is the target or source.

the array to A is

$$\phi(R) = \sum_{i=1}^{K} \Delta\phi_i = \sum \frac{\omega d}{c_i} = \omega d \sum \frac{1}{c_0 + \delta c_i}$$

$$\simeq \frac{\omega d}{c_0} \sum \left(1 - \frac{\delta c_i}{c_0}\right) = \frac{\omega R}{c_0} - \frac{\omega d}{c_0^2} \sum \delta c_i \tag{7.50}$$

and

$$E\phi(R) = \frac{\omega R}{c_0} \triangleq \phi_0(R) \tag{7.51}$$

where K is the number of cells of correlation size d through which the ray passes and $R = Kd$. Also

$$\phi^2(R) = \phi_0^2(R) - \frac{2\phi_0(R)\omega d}{c_0^2} \sum \delta c_i + \left(\frac{\omega d}{c_0^2}\right)^2 \sum_i^K \sum_j^K \delta c_i \, \delta c_j \tag{7.52}$$

and

$$E\phi^2(R) = \phi_0^2(R) + \left(\frac{\omega d}{c_0^2}\right)^2 K \sigma_c^2 \tag{7.53}$$

where $\sigma_c^2 = E[\delta c]^2$. The variance of the phase shift

$$\sigma_\phi^2 = E\phi^2(R) - [E\phi(R)]^2 = \frac{\omega^2 R d \sigma_c^2}{c_0^4} \tag{7.54}$$

or, in terms of the wavenumber $k = \omega/c_0$,

$$\sigma_\phi^2 = k^2 R d \left(\frac{\sigma_c}{c_0}\right)^2 \tag{7.55}$$

Beamforming: Two-Element Array

In open-loop beamforming on point A the phase shifters (or time delays) at each element are set to accommodate the arrival time differences of a wave traveling from A to the array elements at constant speed c_0. Because of the variability of the medium, however, the phase shifts are in error.

Let ϕ_{Aa} be the phase shift of a ray traversing the path from A to a and let R_{Aa} be the best estimate of the length of that path. The phase error in the setting of the phase shifter at point a, when the array is to be focused at A, is $\delta\phi_{Aa} = \phi_{Aa} - kR_{Aa}$. Similarly, the phase error at point b is $\delta\phi_{Ab} = \phi_{Ab} - kR_{Ab}$. The net phase error is their difference and the variance of the net phase error may be calculated from (7.55).

Each propagation delay in the medium consists of two parts: one is the phase shift acquired over distance R_2 in which the ray paths from A to the elements at a and b are separated by less than d. In this region the phase difference is correctly calculated to a first order from the geometry; that is, $\phi_{AE} - kR_{AE} \simeq \phi_{AF} - kR_{AF}$. In the remaining portion the random components of the phase delays are uncorrelated. Thus the variance of the phase difference $\delta\phi_{Aa} - \delta\phi_{Ab}$ is twice (7.55) with $R = R_1$ in that equation. Substituting $R_1 = R(1 - d/D)$ gives

$$\sigma_\phi^2 = 2k^2R\left(1 - \frac{d}{D}\right)d\left(\frac{\sigma_c}{c_0}\right)^2 \tag{7.56}$$

The differential phase-error variance (7.56) determines the maximum allowed separation between the elements of the two-element array by its influence on the loss in array gain. Let the two phase errors at points a and b be $\delta\phi_a$ and $\delta\phi_b$, respectively. In following the beamforming phase shifting, the sum V of the element signals, each assumed to have unit magnitude, is $V = e^{j\delta\phi_a} + e^{j\delta\phi_b}$. The output power $VV^* = 2[1 + \cos(\delta\phi_a - \delta\phi_b]$. In the absence of phase errors the output power $V_0V_0^* = 4$. The expected value of the array gain G in relation to the design value G_0 is $EVV^*/V_0V_0^*$, or

$$E\left[\frac{G}{G_0}\right] = E\left[\frac{1 + \cos(\delta\phi_a - \delta\phi_b)}{2}\right]$$

which evaluates to

$$E\left[\frac{G}{G_0}\right] = \frac{1 + e^{-\sigma_\phi^2}}{2} \tag{7.57}$$

for independent Gaussian phase errors with common variance σ_ϕ^2. When the loss in array gain is small it may be approximated, in dB, by $\Delta G[\text{dB}] \simeq 2.17\sigma_\phi^2$. Allowing about 1 dB loss in array gain as a nominal tolerance, the phase variance may not exceed $\frac{1}{2}$ rad^2. Equating (7.56) to $\frac{1}{2}$ and solving for D yields the maximum allowed separation between the two elements:

$$D = \frac{d}{1 - \left[4k^2Rd(\sigma_c/c_0)^2\right]^{-1}}$$

$$= \frac{4k^2Rd^2(\sigma_c/c_0)^2}{4k^2Rd(\sigma_c/c_0)^2 - 1} \tag{7.58}$$

Note that $D \simeq d$ when $4k^2Rd(\sigma_c/c_0)^2 \gg 1$ and that D grows rapidly when that term approaches unity; D cannot be negative notwithstanding (7.58) and the fact that the denominator can be less than unity: that condition means that the phase variance is less than $\frac{1}{2}$ rad^2. For all practical problems D is the order of d, as shown by the following argument:

Consider the equality condition $4k^2Rd(\sigma_c/c_0)^2 = 1$, or (using $k = 2\pi/\lambda$)

$$\lambda = 4\pi\left(\frac{\sigma_c}{c_0}\right)(Rd)^{1/2} \tag{7.59}$$

Typically, σ_c/c_0 is a few percent in propagation media known to have perturbations in the refractive index; for example, HF propagation in the ionosphere, low-frequency acoustics in underwater sound propagation, and ultrasonics in biological tissue. By taking 5% as a nominal value (7.59) becomes

$$\lambda \approx 0.6\sqrt{Rd} \tag{7.60}$$

But d is never significantly less than λ on physical grounds and the simple ray theory used requires that $R \gg 1$. Hence the validity of (7.60) is denied and $D \approx d$. Thus the well known rule of thumb which states that the size of a two-element array should not exceed the correlation distance of the medium perturbation is obtained.

Adaptive Phasing

The situation is different when adaptive phasing or beamforming is used. In adaptive beamforming on a source or target at point A the phase shifter at an element (located, say, at b in Figure 7.16) is set by closed-loop circuits according to the measured phase difference between the wave received at that element and at a reference element (located, say, at a).

The phase shifter at element b is counterrotated by the measured phase difference $\Delta\phi_A = \phi_{Ab} - \phi_{Aa}$ and both waveforms are cophased. That phase-shift setting is the adaptive correction. Scanning the beam at a transverse distance S to B is an open-loop operation. Expected values of the phase delays are calculated from A and B to a and b and are added or subtracted to phases of

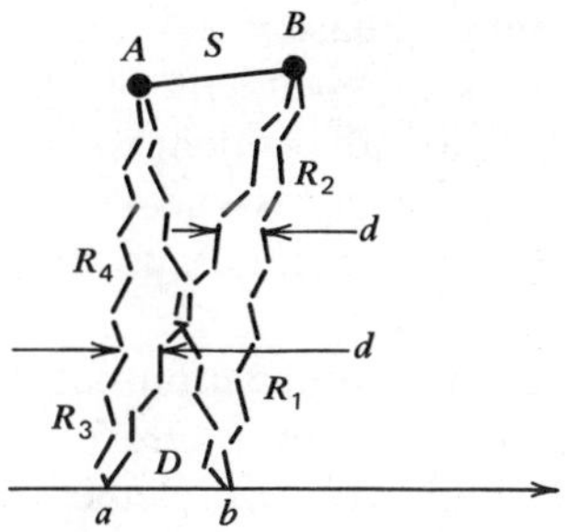

FIGURE 7.16. *a* and *b* are points in an adaptive beamforming array; *A* is the beamforming point; *S* is the scan distance.

the received signals to move the beam. Thus $k(R_{Ba} - R_{Aa})$ is added at element *a* and $k(R_{Bb} - R_{Ab})$ is added at *b*. The phase errors at the elements are $\delta\phi_{Bb} = (\phi_{Bb} - kR_{Bb}) - (\phi_{Ab} - kR_{Ab})$.

Again each component consists of two parts: one is the phase shift acquired over distance R_3 in which the ray paths from *A* and *B* to an array element are separated by less than *d*. In this region no errors that contribute to the phase-error difference $\delta\phi_{Bb} - \delta\phi_{Ba}$ occur. In the remaining part the phase delays from *A* and *B* are not equal and their random components are uncorrelated. The ray paths from *B*, however, experience correlated fluctuations over the distance R_2. Thus the variance of the phase error difference is four times (7.55) with $R = R_4 - R_2$ in that equation. Substituting as before, equating the variance to $\frac{1}{2}$ rad^2 and solving for *S* gives the limit to scanning:

$$S = \frac{d}{1 - d/D - \left(8k^2Rd(\sigma_c/c_0)^2\right)^{-1}} \tag{7.61}$$

The last denominator term, based on the discussion given earlier, can be ignored. Thus

$$S \simeq \frac{d}{1 - d/D} \tag{7.62}$$

However, $D \gg d$; this is because the purpose of adaptive beamforming is to overcome the limit to aperture size induced by the medium fluctuations and therefore would be used only when this condition is experienced. Hence $S \approx d$ and is not a function of the element separation. Thus adaptive beamforming removes the limit to array size normally imposed by a perturbed medium. Instead the correlation distance *d* imposes a limit to the maximum scan distance *S* or maximum scan angle S/R which, by following the preceding argument, is $S \approx d$. The total transverse scanning distance is twice this amount.

N-Element Array

Extension of the problem to the *N*-element array can be done according to the following logic. The output voltage from an *N*-element array is the linear phase-weighted sum of the individual transducer outputs. Phase weighting is applied to achieve beam scanning. The output sum is independent of the method of addition; that is, the output voltage is the same independent of the feed structure for the array, as, for example, whether the array is end-fed or driven by a corporate-type feed or whether all elements are fed simultaneously from a single power distribution point.

The analysis of the effects of phase errors due to varying propagation paths is most readily discussed in terms of the feed structure sketched in Figure 7.17. Let the number of tiers be n where $N = 2^n$. Pairs of phase-weighted elements are spaced and added. Their outputs are added further by pairs. Each layer can be viewed as the input to an array of one-half the number of elements of the preceding layer. It is evident that the expected output power at each node of a tier will be the same. Let the signal voltages approaching the output terminal be V_{B1} and V_{B2}. The output power is $|V_{B1} + V_{B2}|^2$ and its expected value is $2\overline{|V_B|^2}[1 + \exp(-\sigma_{\phi B}^2)]$, where $\overline{|V_B|^2}$ is the expected power at each B node and $\sigma_{\phi B}^2$ is the variance of the phase error at each B node. In terms of the A signal, the expected output power can be written as $4\overline{|V_A|^2}[1 + \exp(-\sigma_{\phi A}^2)][1 + \exp(-\sigma_{\phi B}^2)]$.

In general, the expected output power can be written as

$$E[G] = \overline{|V_1|^2} \prod_{i=1}^{n} 2(1 + e^{-\sigma_{\phi i}^2})$$

where $\overline{|V_1|^2}$ is the expected output power of a node at the first level and $\sigma_{\phi i}$ is the variance of the phase error at the ith level. Let G_0 be the gain of the array in the absence of noise; $G_0 = N^2\overline{|V_1|^2}$ and $N = 2^n$. Therefore

$$\frac{E[G]}{G_0} = \prod_{i=1}^{n} \frac{(1 + e^{-\sigma_{\phi i}^2})}{2} \tag{7.63}$$

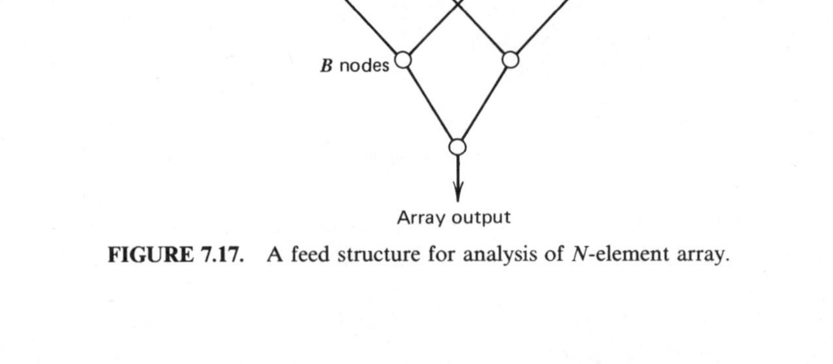

FIGURE 7.17. A feed structure for analysis of N-element array.

The expected decrease in gain, in dB, is

$$\Delta G\,(\text{dB}) = 10 \sum_{i=1}^{n} \log\left(\frac{1 + e^{-\sigma_{\phi i}^2}}{2}\right)$$

$$\approx 2.17 \Sigma \sigma_i^2 \qquad (7.64)$$

for small phase-error variances σ_i^2. Restriction of the decrease to no more than 1 dB requires that all contributing terms be less than 1 dB because each one is positive. Consider the first level alone. According to the earlier theory, equating it to the 1-dB loss limits the spacing between the elements that drive the same node at the second level to d. Therefore $(N/2) \times [L/(N-1)] = d$ or $L = 2d(1 - 1/N)$, which for large N is approximately $2d$.

Scanning in Range

The geometry of scanning in range in the near field is shown in Figure 7.18. Let the array be adaptively focused at point A at distance R. To scan the beam a distance ΔR to point B geometrically calculated phase shifts are added to the adaptively set values. The phase errors $\delta\phi_{Ba}$ and $\delta\phi_{Bb}$ were given earlier. As before, one portion of each term is zero, representing ray paths separated by no more than d. Consider the separation between ray paths from B to the arrows at distance R from the array. Assume that that spacing does not exceed $2d$. Then the paths from A to the array and the corresponding portions of the paths from B are spaced by no more than d, in which case no uncorrected phase errors due to the portions from the arrows to the array contribute to the phase-error variance. Further, the result of the earlier section showed that the phase-error variance due to the remainder of the paths, from B to the

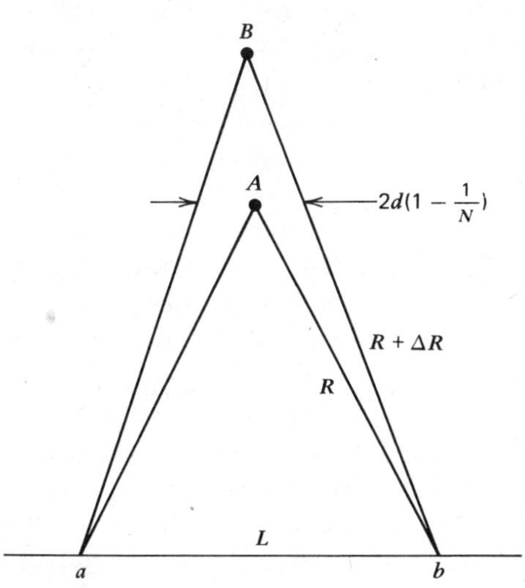

FIGURE 7.18. Range scanning to point B following adaptive beamforming in point A; a and b are points on the array.

arrows, will be not be excessive if the path separation at the arrows does not exceed $2d(1 - 1/N)$. Thus the maximum path separation at distance R is the smaller value or $2d(1 - 1/N)$. Given an array length L and the range scanning geometry of Figure 7.18,

$$\frac{\Delta R}{R + \Delta R} = \frac{2d(1 - 1/N)}{L}.$$

from which

$$\Delta R = \frac{2dR(1 - 1/N)}{L - 2d(1 - 1/N)} \simeq \frac{2dR}{L}(1 - 1/N) \qquad (7.65)$$

is the maximum distance, in range, that the array may be scanned from the adaptive beamforming point. The total range-scanning interval is twice this amount.

Observations

By using an open loop or nonadaptive beamforming array we find that the nominal useful array size is twice the correlation distance d and $\lambda/2d$ is approximately the minimum angular beamwidth. The cell size at range R is $\lambda R/2d$. Angular scanning is unlimited. The number of angular resolution elements per radian is $N_R = 2d/\lambda$. Open-loop beamforming is most effective for low-angular resolution and wide-angle scanning.

With a closed-loop adaptive beamforming array the array size L can be much larger than d. Array gain increases by the ratio $L/2d$, provided that the elements in the enlarged array have about the same spacing as before. The transverse cell size at range R is $\lambda R/L$. The width of the sector to be scanned, however, is limited to $S \approx 2d$. The number of transverse resolution cells is $N_S = 2dL/\lambda R$. Closed-loop beamforming is most effective for high angular resolution imaging within a confined sector.

When the adaptive beamforming source is in the near field scanning in the longitudinal or range direction is also possible. The range dimension of the scanning sector is $4dR(1 - 1/N)/L$.

Figure 7.19 sketches the range-angle scanning sector in which point A is the location of the source or target on which the array self-focuses. The beam may then be refocused by open-loop beam-steering procedures anywhere within the sector shown. It is a thin sliver of width $2d$, oriented in the range or longitudinal direction, and length $4dR(1 - 1/N)/L$.

The range resolution is determined by the waveform of the radiation. In active radar or sonar systems it normally is proportional to the reciprocal of the radiated bandwidth W expressed in units of distance; that is, $c_0/2W$. If, however, the bandwidth is small and the array is focused in the near field, the range resolution may be determined instead by the depth of field of the array;

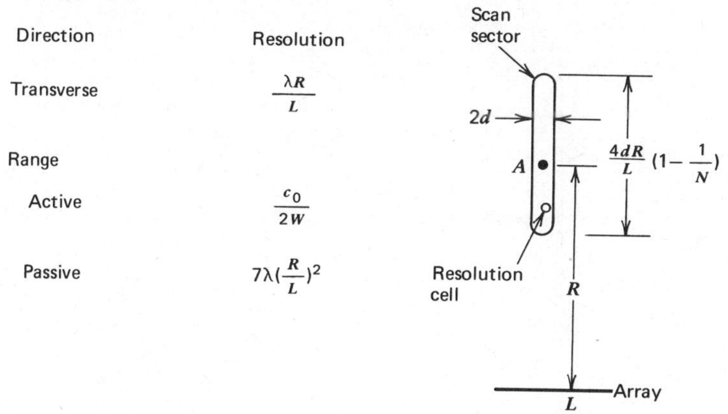

FIGURE 7.19. Dimensions of scanning sector following adaptive beamforming at point A.

given a uniform distribution of elements in the array, the 3-dB range extent of its focal spot is approximately $7\lambda(R/L)^2$. The range resolution with an active system is the smaller of these values. For narrowband radar or for passive sonar operation, on the other hand, the range resolution is determined by the depth of field. Thus a large array not only enhances the azimuthal resolution but can also provide useful resolving power in the range direction even for unmodulated CW signals.

The number of resolution cells in the scanning sector is approximately $16d^2W/\lambda c_0$ for a wideband active system and approximately $(dL/\lambda R)^2$ for a narrow-band active or passive system; λ in the first expression is the midband wavelength.

To scan a still larger sector more than one beamforming source or target is required; their spacings need be no more than $2d$ in azimuth or $4dR(1 - 1/N)/L$ in range. The sources can be active beacons or active sources of opportunity, or they may be passive reflectors. When a field of view is required such that more than four beacons would be needed the self-survey technique of Section 4.4 is preferable.

7.9. COMBINED EFFECTS OF THE PHASE ERRORS ON MAINLOBE GAIN

Because the gain-losses are additive in the logarithm, the combined effect of the first three sources of loss is given by (7.12) + (7.19) + (7.40):

$$\text{total loss in mainlobe gain (db)} = 4.3\left\{\frac{1}{\text{SNR}} + k^2\sigma_x^2\theta^2 + S[1 - \rho(\theta)]\right\} \tag{7.66}$$

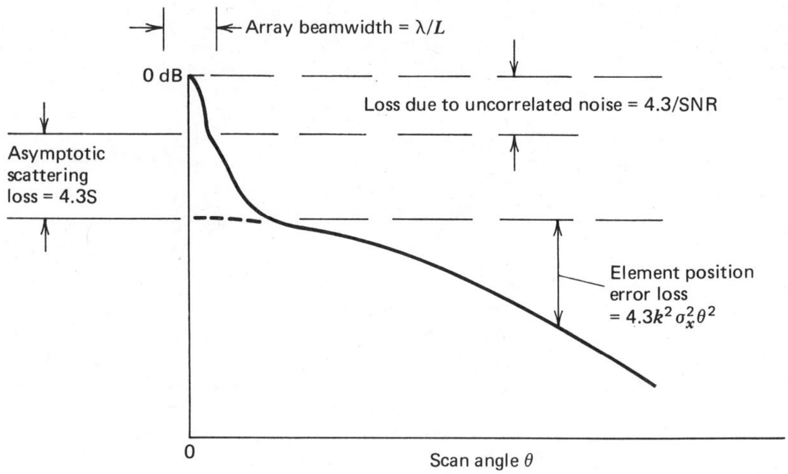

FIGURE 7.20. Relative array gain versus scan angle showing effects of various sources of phase error.

which is illustrated in Fig. 7.20. The uncorrelated errors, for example, from receiver noise, cause an immediate fixed loss as the scanning process begins. Scattering losses develop within a correlation interval of the scattering process. The dominant angular dependence is due to element position errors, which add a quadratically increasing loss.

Because the scattering loss asymptotes beyond some small angular scan from the phase synchronizing direction, (7.66) may be simplified to

$$\text{total loss (dB)} \cong 4.3\left(\frac{1}{\text{SNR}} + S + k^2\sigma_x^2\theta^2\right) \tag{7.67}$$

7.10. SUMMARY

Phase synchronization of the array (called adaptive beamforming, self-phasing, self-focusing, or self-cohering) is a closed-loop process that compensates for uncertainty in element locations and other sources of apparent and real phasefront distortion. No loss is suffered in mainlobe gain.

Scanning or imaging is an open-loop procedure. Hence phase distortion and phase measurement errors, although correctly compensated during phase synchronization of the array along the direction to the reference source, cause losses in array gain as the beam scans away from the direction of phase synchronization. Beamwidth, beam shape, and sidelobe statistics are relatively unaffected by the scanning process. During scanning phase errors decrease the contrast in the scanned image (ratio of mainlobe gain to average sidelobe level) but not the resolution (which is proportional to beamwidth).

Receiver noise and all other uncorrelated internal sources of phase-measurement error, such as noise in phaselock loops and quantization noise in phase shifters or detectors, reduce mainlobe gain (in dB) by 4.3/SNR, where SNR is the signal-to-noise power ratio that results from all the internal noise sources. Because SNR in radar generally exceeds 20 (13 dB) to satisfy detection requirements, noise is not a limiting factor.

Multipath and scattering also distort the phasefront of the synchronizing wave. The effect is a loss in mainlobe gain, which increases with scan angle and reaches a maximum value (in dB) of $4.3S$, where S is the power ratio of scattered-to-direct signal strength.

Position errors of the elements in the array also induce an angle-dependent loss of mainlobe gain. The loss (in dB) is quadratic and equal to $4.3k^2\sigma_x^2\theta^2$; $k = 2\pi/\lambda$ is the wavenumber, σ_x^2 is the position-error variance at right angles to the direction of phase synchronization, and θ is the scan angle (in radians) from the beamforming direction. From this formula it is deduced that the position-error tolerance may be two or more orders of magnitude larger for a radio camera than for a conventionally designed phased array.

The gain-losses (or decreases in image contrast) are additive in dB. The total loss is approximately

$$\text{total loss (dB)} = 4.3\left(\frac{1}{\text{SNR}} + S + k^2\sigma_x^2\theta^2\right) \tag{7.67}$$

The useful field of view of the radio camera is estimated to be greater than $\lambda/2\pi\sigma_x$ rad. At the edges of the FOV the loss in image contrast is 1 dB.

Losses due to medium distortion are not easily characterized in a single model. One analysis of a perturbed, infinite medium with correlation distance d shows that a nonadaptive array is limited in size to the order of the correlation distance. Adaptive beamforming eliminates this size constraint on the array. However, the scanning distance is limited to an amount proportional to the correlation distance and the proportionality constant is a function of the scanning direction.

CHAPTER EIGHT

Transmit-Receive Array

8.1. INTRODUCTION

The radio camera is described in Chapter 1 as a bistatic instrument. The transmitter shown there plays the function of the sun in optics by broadly illuminating the target region with microwave energy. The reradiated energy is operated on by the adaptively organized receiving array; however, the same array can be used for transmission, and there are two reasons for doing so. The first has to do with sidelobes. It is learned from Chapters 2 and 3 that a reduction in the sidelobe level of a random array is a matter of paramount importance in any radio camera design. Transmission through the same array is the most effective means of reducing sidelobes, for then the radiation pattern is the square of the one-way radiation pattern. The average sidelobe power level reduces to N^{-2} in relation to the mainlobe power, whereas the peak sidelobe will rise 20 to 25 dB above the average. This represents a significant reduction in sidelobe level when N is large.

A second reason for interest in the transmit–receive array for use in the radio camera is the growing availability of transmit–receive, solid-state modules, which include a transmitting source, antenna, circulator, local oscillator, mixer, IF amplifier, and second detector. With this front-end equipment becoming available, it is natural to inquire about the practicality of a radio camera with a transmit–receive capability.

Five distinct problem areas can be addressed to achieve a satisfactory design:

1. Initial illumination for phase-cohering.
2. The phase-synchronizing source.
3. Phase conjugation.
4. Generation and distribution of a phase reference.
5. The nature of the transmitter.

8.2. INITIALIZATION OF THE ARRAY

The large adaptive array must first self-cohere as a receiving system to the radiation from a synchronizing source. The synchronizing wave may be the radiation from an active beacon or the reradiation of a large reflector illuminated by a microwave transmitter. After phase synchronization on reception the system must organize the array as a transmitter and focus it retrodirectively on the source. Figure 8.1 shows the components of this system.

Achievement of this objective is not a simple task. There is a logical difficulty in its accomplishment plus a host of practical obstacles. The logical difficulty is the following: the system requires a radiator to illuminate the target and a large receiving aperture, the size of which implies *a priori* unknown geometric distortion in the array that the self-cohering process corrects. The correction is an electrical compensation in the receiving system and not a physical or geometric correction of the distorted array; therefore the array is not suitable for transmission prior to adaptive beamforming. Adaptive beamforming cannot be accomplished before illumination of the target by the

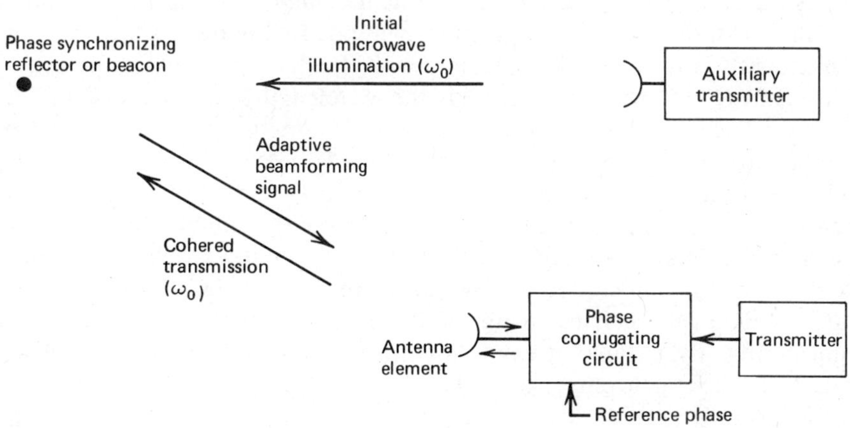

FIGURE 8.1. Components of adaptive beamforming transmit-receive array using an auxiliary transmitter for phase synchronization. The lower part of the figure shows one module of the large array.

transmitter, yet without transmission there is no target illumination and no adaptive beamforming and imaging.

To break this conundrum the adaptive beamforming system must be initialized before the array is used as a spatially coherent transmitter. Phase-synchronized, narrow-beam radiation may then take place. Thus two separate processes must be accomplished. The terms acquisition and tracking may be applied to them, reminiscent of the target acquisition and tracking functions performed by radar systems. Another term for acquisition is initialization. To accomplish acquisition the array must react to target reflections from a radiation field that may or may not have been produced by the same array. The self-cohering process operates on these target reflections to phase-synchronize the system. After initial synchronization the same array is used for coherent transmission. Subsequent reflections from the target keep the array synchronized.

The problem, then, falls into two parts: initialization and tracking. The solutions, or rather the approaches to the solutions, are different in kind. The tracking problem is primarily a technical one, whereas the solution to the acquisition problem is influenced by the nature of the system and the constraints placed on it. The tracking problem is relatively easy to solve once phase synchronization of the transmitter has been accomplished: when the transmitter beam is cohered the receiving array must follow the rules described in Chapter 5 and the transmitting array must be updated from time to time. This chapter deals with the remaining problem, that of acquisition or initial phase synchronization of the array.

8.3. SENSITIVITY TO SYSTEM CONFIGURATION

Figures 8.2 and 8.3 illustrate how the system configuration can impose limitations or constraints on the acquisition method. In Figure 8.2 an array of N independently positioned elements is shown. Each element has radiating arca A_e and gain $G_e = 4\pi A_e/\lambda^2$, where λ is the wavelength. The total aperture area $A = NA_e$. The available transmitting power P is divided equally among the N elements each of which is driven by P_e W; thus $P = NP_e$.

Now let it be desired to self-cohere on a target of radar cross section σ_T at a range R and let it be assumed that the total power P available for transmission is sufficient to do so, provided that the array was initially synchronized. In this case the power P transmitted through aperture A and reflected from the target returns a signal to each receiving module strong enough to permit adaptive beamforming. That signal strength is easily calculated. The effective radiated power from the synchronized array is

$$PG = \frac{4\pi PA}{\lambda^2} \tag{8.1}$$

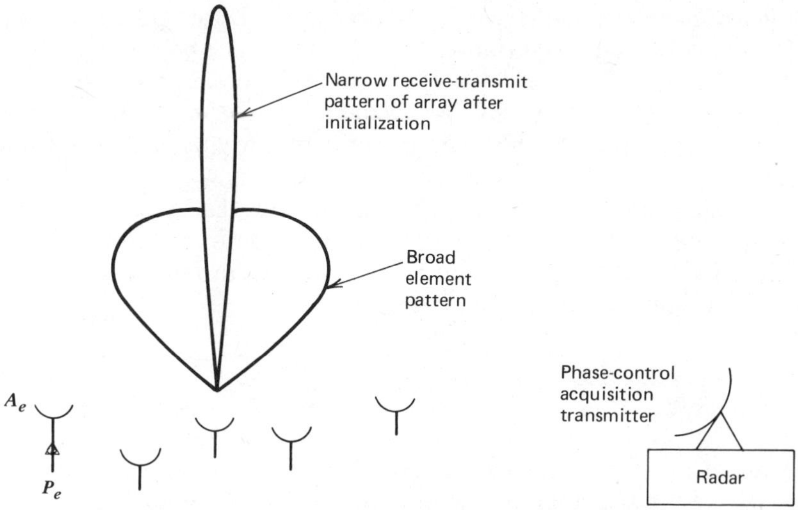

FIGURE 8.2. Array of N independently positioned elements.

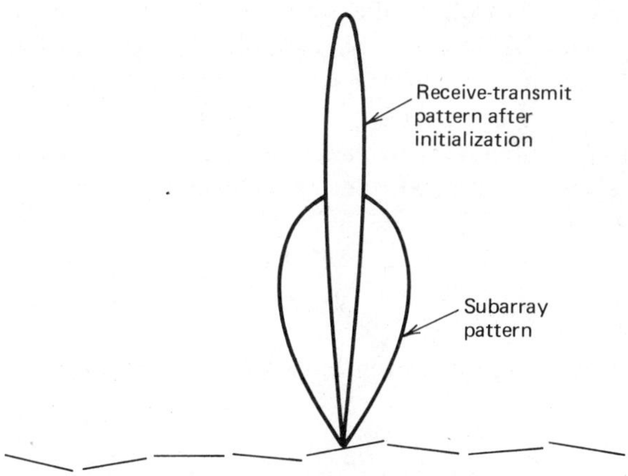

FIGURE 8.3. M edge-connected rigid subarrays each having N/M elements.

At the target distance the power density

$$W = \frac{PG}{4\pi R^2} = \frac{PA}{\lambda^2 R^2} = \frac{N^2 P_e A_e}{\lambda^2 R^2} \tag{8.2}$$

The reradiated power in the direction of the array is (8.2) times σ_T, which when divided by $4\pi R^2$ is the reflected power density at the array. The signal power S received by a single antenna module is the power density times the element area. Thus

$$S = \frac{N^2 P_e A_e^2 \sigma_T}{4\pi \lambda^2 R^4} \tag{8.3}$$

which is assumed to be sufficient for self-cohering. Before phase synchronization, however, the individual transmitters do not radiate coherently. Hence the effective radiated power is less than (8.1) and the received power per element is less than (8.3). The difference may be several tens of decibels. The system could be overdesigned to provide this margin, but then it would be overpriced for its function.

Therein lies the difficulty, for initialization of the system does require considerable power margin. One way to try to beamform the array in Figure 8.2 is to radiate P_e, the available power per element, through a single element. The received signal power at each element

$$S_1 = \frac{P_e A_e^2 \sigma_T}{4\pi\lambda^2 R^4} = \frac{S}{N^2} \tag{8.4}$$

is a factor of N^2 smaller than S. Because N is likely to be the order of 10^3 or larger, the received signal power can be 60 dB or more below the level necessary.

This troublesome matter can be alleviated by a factor of N by one of two methods. The first is to radiate the total power NP_e from a single element, which requires that the element be capable of handling the total power. Because N will be the order of 10^3, a separate high-power transmitting element and high-power source are needed. A much more desirable procedure is to radiate P_e from each of the N elements, even though they are not cophased. The average received signal power per element becomes

$$S_{\text{NON}} = \frac{NP_e A_e^2 \sigma_T}{4\pi\lambda^2 R^4} = \frac{S}{N} \tag{8.5}$$

This is seen by examining the power density at the target due to radiation from a single element:

$$W_1 = \frac{P_e G_e}{4\pi R^2} = \frac{P_e A_e}{\lambda^2 R^2} \tag{8.6}$$

If N elements radiate simultaneously with random phases, the expected power density at the target is NW_1, which leads to (8.5). Thus the synchronized transmitting array has a power gain N times larger than the same array before phase synchronization; the latter, in turn, provides a phase-synchronizing signal at the array N times larger than when a single element is used for transmission. This reduces the required margin from the order of 60 dB to 30.

Now let the array consist of M rigid subarrays, each having N/M elements. Figure 8.3 shows an edge-connected example of this array. Assume that the subarray is small enough physically that its radiation pattern encompasses the entire sector to be scanned by the radio camera. The N/M elements in each subarray can be combined into a single coherent aperture of area A_{sub} =

NA_e/M through which power $P_{sub} = NP_e/M$ is radiated. This radiation from a single subarray would result in an element signal

$$S_{N/M} = \left(\frac{N}{M}\right)^2 \frac{P_e A_e^2 \sigma_T}{4\pi\lambda^2 R^4} = \frac{S}{M^2} \tag{8.7}$$

which is larger than (8.4) by the factor $(N/M)^2$. Noncoherent radiation from all M subarrays raises this level by a factor of M to

$$S_{\text{NON }N/M} = \frac{S}{M} \tag{8.8}$$

These equations are summarized in Table 8.1, which demonstrates the strong dependence of the available signal power for the adaptive-beamforming acquisition function on the configuration of the system.

8.4. PRIMARY OR INITIAL ILLUMINATOR

The upper two lines in Table 8.1 pertain to Figure 8.2. At best, the acquisition beamforming signal is 30 dB below the level of the phase control signal during tracking. The system designer has several alternatives for overcoming this deficiency, five of which are described:

The first alternative is to overdesign the transmitter by the necessary margin of N, which is the relative gain of the synchronized array to the nonsynchronized. This is a costly choice.

Another possibility is the use of a separate high-gain illuminator for initial synchronization, after which power is transferred to the array. Figures 8.1 and

TABLE 8.1. Signal Power Available for Adaptive Beamforming Acquisition Function Depends Upon System Configuration

Transmission	Relative Received Signal Power for Initial Adaptive Beamforming[a]		
Single element	1	−60 dB	
N noncoherent	N	−30	
N coherent	N^2	0	← Adaptive beamforming tracking mode
$\frac{N}{M}$ coherent	$\left(\frac{N}{M}\right)^2$	−40	
$\frac{N}{M}$ coherent, M subarrays noncoherent	$\frac{N^2}{M}$	−20	

[a] In last column $N = 1000$ elements, $M = 100$ subarrays.

8.2 show a radar transmitter being used for this purpose. This is a sound approach when an auxiliary transmitter is available. The auxiliary radar transmitter might be a nearby radar operating in the same frequency band but not at the same frequency. The auxiliary transmitter in Figure 8.1 radiates at frequency ω_0' to a large TOO. Target reflections at that frequency phase synchronize the receiving system. As described in Chapter 5, phase conjugation of the received waveforms results in a receiving beam focused at the source. The transmission of the phase conjugated waves, however, is at $\omega_0 \neq \omega_0'$, which somewhat offsets the transmitting beam [18]. Instead of a passive source, the auxiliary radar could actuate an active beacon that would radiate to the array at the design frequency ω_0 to self-cohere the array and avoid the angular offset or "squint" (discussed in Section 8.7).

An auxiliary radar transmitter in a different frequency band could also be used to initiate a beacon signal or induce TOO reflections. No angular offset results if the beacon reradiates at ω_0. Otherwise the beam angle offset on transmission is too large and must be prevented by special phase-conjugation circuits like those described by Chernoff in [18]. An important advantage is obtained in the use of a lower frequency for initial synchronization with a TOO, for then the lobes of the reradiation pattern are widened by the frequency ratio, thereby easing the size-limitation tolerance on the synchronizing source found in Chapter 6.

A rigid subarray of the full system can be used for coherent transmission of a broad beam to deliver the initializing signal to the beacon or TOO.

The fifth alternative is to drive the entire array noncoherently before self-phasing, in which case the average power density is N^{-1} times the full power density of the system after the array is synchronized. Although this loss is large, it can be compensated by the use of a beacon or by initializing on the reflected signal from a nearby target. The squint problem is avoided because synchronization is at the system frequency ω_0. The method is effective with a TOO reflection because of the R^{-4} dependence of received signal power on target distance. To overcome a 30 dB initializing disadvantage ($N = 1000$) the reference reflection must be located no farther away than 18% of the maximum distance at which it could be placed if the transmitters were cohered. This distance reduces to 10% when $N = 10^4$. This technique, appropriate for an airborne system with ground or sea clutter for the reference target (see Section 6.6 and Chapter 9), is illustrated in Figure 8.4. Two modules of the array are shown, each of which consists of an antenna element, circulator (or any other diplexer), receiver, phase-stable reference oscillation common to the entire array, transmitter phase shifter, mixer, local oscillator cohered in frequency to the reference wave, and pulsed power amplifier. This technique, and those in the remainder of the chapter, follow [90].

To initialize the system each module radiates an RF pulse with a common system frequency ω_0 and random phase. The instantaneous transmission phase of the wave from the ith module may be designated $\omega_0 t + \phi_i$. This wave arrives at the synchronizing source with phase $\omega_0 t + \phi_i - \phi_{s_i}$, where ϕ_{s_i} is the phase

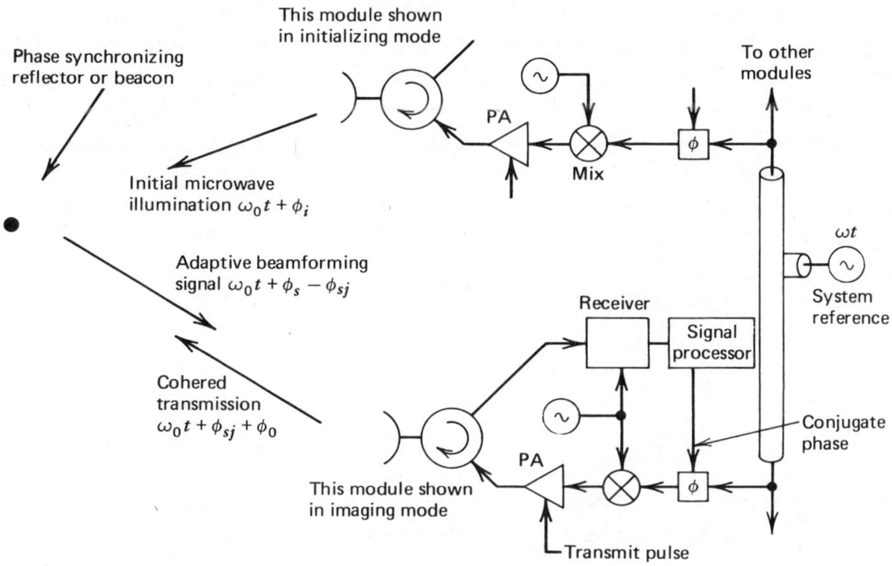

FIGURE 8.4. Adaptive beamforming transmit-receive array without auxiliary transmitter.

delay from the module. The combined illumination at the reflector or beacon is $\Sigma_{i=1}^{N} a_i \exp j(\omega_0 t + \phi_i - \phi_{s_i})$, the instantaneous phase of which is $\omega_0 t + \Phi_s$, the subscript s meaning source. The source wave is returned to the array with a different phase delay ϕ_{s_j} at each module. Thus the signal phase received by the jth module is $\omega_0 t + \Phi_s - \phi_{s_j}$.

To focus the transmitting beam on the source the radiation from the jth module must be changed from $\omega_0 t + \Phi_s - \phi_{s_j}$ to $\omega_0 t + \phi_{s_j} + \Phi_0$, where Φ_0 is an arbitrary phase constant across the array. This is the phase-conjugation process needed to achieve retrodirectivity. Methods for its accomplishment are discussed in Section 8.6.

8.5. PHASE-SYNCHRONIZING SIGNAL

The synchronizing source may be active or passive. A microwave transmitter is needed to initiate the synchronizing signal in both cases by actuating an active beacon or by causing a strong reflection from a passive source. In principle, a remote CW source could be used. However, being on all the time, it would make target detection more difficult.

Unless the total available transmitting power is large enough to overcome the initial transmitting inefficiency, the power density of the initial illumination will be considerably lower than that of the full transmitting array after it is self-cohered. To compensate for this loss in signal strength, a passive synchronizing source must be larger or closer to the array than is necessary for a receive-only system. In Section 8.3 it was shown that the magnitude of the expected power loss is between N^2 and M (N = total number of array elements; M = number of rigid subarrays). These factors can be related as follows: let P = total available transmitter power, G = effective transmitter antenna gain, σ_T = radar cross section of the reference target, R = range to the reference target, and S = received signal power. Then $S \propto PG\sigma_T/R^4$. Now let primed quantities pertain to the transmit–receive system before adaptive beamforming of the array and let unprimed quantities pertain to adaptive beamforming with the receive-only array. The received signal powers S and S' needed to synchronize the systems may be assumed equal. Equating them leads to

$$\frac{PG\sigma_T}{R^4} = \frac{P'G'\sigma_T'}{(R')^4} \tag{8.9}$$

The gain ratio G/G' is between N^2 and M. Hence the relation between the various quantities becomes

$$M \lesssim \frac{P'\sigma_T'}{P\sigma_T}\left(\frac{R}{R'}\right)^4 \lesssim N^2 \tag{8.10}$$

In most design problems an auxiliary transmitter will not be available for initial synchronization of the array, in which case $P' = P$. This leaves the cross section and the range of the reference source as the available parameters. Both may be needed to overcome the loss in initial transmitting gain.

In some situations range to the reference source is not an available parameter as, for example, when the reference source is in the same range bin as the target. In this case initial synchronization will take longer at best and may fail entirely. An alternate, initializing procedure, such as the use of a beacon, is needed.

An equation equivalent to (8.10) for the beacon is casily written: σ_T is replaced by beacon power P_B and the range dependence drops from quartic to quadratic:

$$M \lesssim \frac{P'P_B'}{PP_B}\left(\frac{R}{R'}\right)^2 \lesssim N^2 \tag{8.11}$$

8.6. PHASE CONJUGATION TECHNIQUES

Retrodirectivity requires phase conjugation at each element, which in turn, demands symmetry with respect to some reference. As discussed earlier in Section 4.3, if $\phi_R(x; \theta_0)$ is the phase of the received wave at position x in the array due to a synchronizing source at angle θ_0 and $\phi_T(x; \theta_0)$ is the transmitted phase variation needed to achieve retrodirectivity, the required relation is $\phi_T(x; \theta_0) = -\phi_R(x; \phi_0)$ plus an arbitrary constant. This equation implies the existence of some reference phase from which ϕ_T and ϕ_R may be measured. The symmetry can exist in more than one domain. Spatial symmetry is used in one of the oldest forms of retrodirective array, the Van Atta array [76], [38]. Symmetrical sidebands in the frequency domain is another. A third is paired symmetrical phase shifters.

Dependence on spatial symmetry is inappropriate for a large system such as an airport-sized or airborne array distributed about the airframe. In these circumstances the array may be presumed to be distorted, random, and highly thinned.

The other two techniques are practical. The use of symmetrical sidebands is illustrated in Figure 8.5. Let a signal characterized by the real (or imaginary) part of $\exp[j(\omega_0 t + \phi)]$ be received and heterodyned (mixed) with a local oscillator at frequency ω_{LO}. The mixer products are $\exp\{j[(\omega_{LO} + \omega_0)t + \phi]\}$ and $\exp\{j[(\omega_{LO} - \omega_0)t - \phi]\}$. The sidebands are symmetrically displaced about the local oscillator frequency and the lower sideband has the desired phase. Further, if the local oscillator frequency is made exactly twice the frequency of the received signal, the lower sideband is $\exp[j(\omega_0 t - \phi)]$.

In Figure 8.5 the input signal at frequency ω_0 and phase ϕ (indicated by the instantaneous phase $\omega_0 t + \phi$) is passed through a circulator in which it mixes with the second harmonic at an arbitrary phase ϕ_0. The difference frequency output of the mixer, at ω_0, is passed by the bandpass filter to the amplifier. The phase of this signal is $\omega_0 t + \phi_0 - \phi$. This signal is amplified and radiated. Provided that ϕ_0 is constant across the array the radiation is retrodirective.

This circuit has two limitations. First, the down-converting mixer is a source of trouble because of the harmonic relation between the signals at its inputs. Because the circuit is nonlinear, the second harmonic of the fundamental frequency at its input will be generated. A current due to the second harmonic will flow in the source impedance of the second-harmonic input circuit, thereby altering the phase of the reference signal at $2\omega_0$. Also direct feedthrough of the input signal to the output will alter the phase of the output signal. Either the

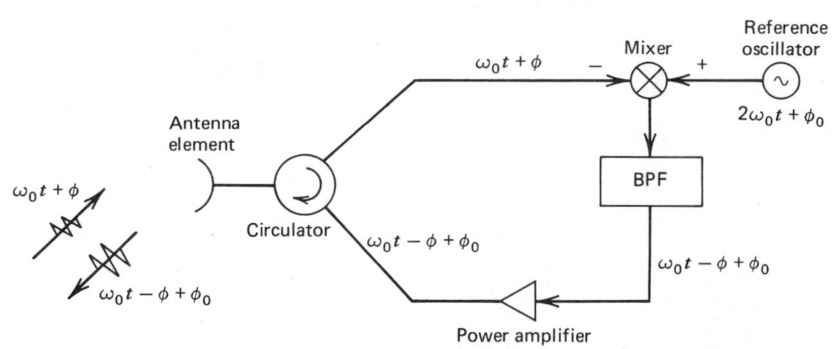

FIGURE 8.5. Phase conjugation schematic.

frequency of the reference signal must be different from $2\omega_0$ to avoid these troubles or the mixer must be carefully balanced; otherwise second harmonic generation or input-output leakage will affect the phase of the radiated wave.

The second problem is that this circuit transmits an amplified replica (with conjugated phase) of the received radar trace concurrent in time with the received signal. The desired transmission, however, is an RF pulse (with conjugated phase) that occurs at a later time. This means that the echo received from the synchronizing source must be sampled and its phase extracted, conjugated, and applied to an oscillation at the echo frequency, which is gated at the appropriate time, amplified, and radiated.

Figures 8.6 to 8.10 show several circuits for accomplishing this task. They differ in their means of storage of the phase information (as the phase of a coherent oscillator, e.g., as in a phaselock loop (PLL) [31], [102], [54], [52], [56], or as a digital number in the signal processor of the system) according to whether control of the phase shifter through which the signal to be transmitted is passed is open- or closed-loop and according to the choices of components; for example, analog versus digital phase shifters.

Figure 8.6 shows a means of phase-conjugating when a digital signal processor is used to combine the signals received from the distributed array elements. The received signal is shown as an echo pulse with instantaneous phase $\omega_0 t + \phi$. It is heterodyned to IF by a local oscillator at frequency ω_{LO} with some arbitrary phase α. The LO is assumed to be cohered to the reference wave at the intermediate frequency ω. Phase β of the reference is assumed to be constant across the array. The IF pulse, with instantaneous phase $\omega t + \phi - \alpha$, is delivered to the signal processor, which measures the phase in relation to the phase of some reference element in the array whose phase is arbitrarily identified as zero phase. Because all signals entering the signal processor experience the same phase offset α, the local oscillator phase cancels out. The

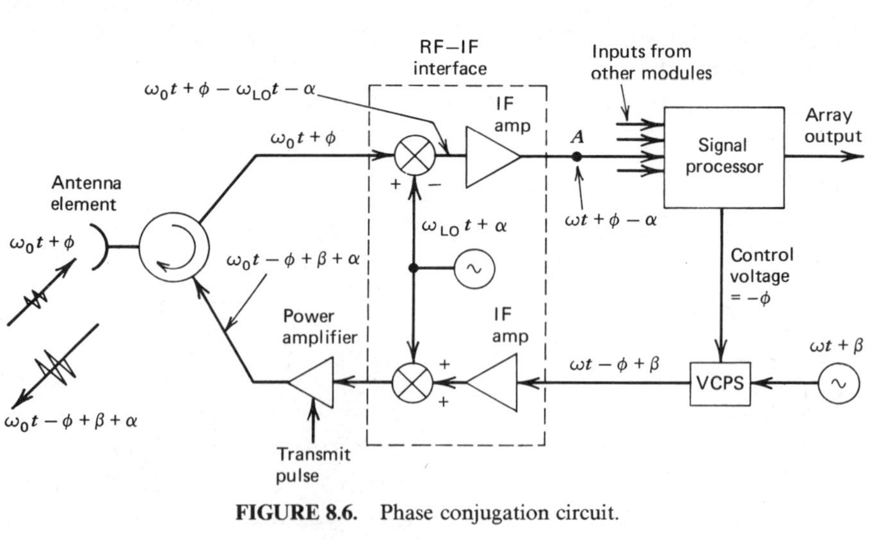

FIGURE 8.6. Phase conjugation circuit.

negative of the measured signal phase ϕ is delivered as a control voltage to a voltage-controlled phase shifter (VCPS) in the reference signal path. The output, with instantaneous phase $\omega t - \phi + \beta$, is up-converted to form the transmitted wave. The transmitted phase is $\omega_0 - \phi + \beta + \alpha$. Provided that β and α are constants across the array, the transmitted wave is the phase conjugate of the input signal.

The accuracy to which the signal processor measures ϕ (or $-\phi$) is limited by noise and multipath. The *precision* with which this measurement is made and held for delivery to the phase-conjugation circuit can be made arbitrarily fine; it is determined by how many significant figures or bits are used in the measurement. Thus the quality of the delivered value $-\phi$ need be no poorer than that of the measured $+\phi$.

Errors develop when the phase shifts in the system are not tuned out. Let the phase shift from the antenna through the circulator, receiver, and signal processor to the VCPS be δ. The output of the phase shifter then becomes $\omega t - \phi - \delta + \beta$. Let the phase shift in the transmitting chain from VCPS through the transmitter and circulator to the antenna be η. The phase of the radiated wave becomes $\omega_0 t - \phi - \delta + \beta + \alpha + \eta$. This phase is in error by $-\delta + \eta$. The variance in this error (in square radians) across the array, multipled by $10 \log e$, is the expected loss in array gain in decibels (2.16). As noted earlier, $\frac{1}{4}$ rad^2 phase-error variance leads to a loss in mainlobe gain of 1 dB. Thus each module must be carefully tuned to balance the phase shifts in the receiving and transmitting chains.

Another source of error is in the phase reference. Maintaining constant phase in the reference wave across the array is difficult. The variance in the phase of the reference wave also reduces transmitter gain according to (2.16). This problem is discussed in Section 8.8.

The circuit in Figure 8.6 stores the signal phase in the signal processor and uses open-loop phase control. The next circuit (Figure 8.7) retains open-loop phase control but remembers the signal phase in a PLL. This circuit also demonstrates the use of paired symmetrical phase shifters. The receiver chain is the same as in Figure 8.6 to point A, where the circuit branches. The received signal continues to the signal processor as before. It also is applied to the input port of the phase detector in a PLL in which the controlled element is again a VCPS rather than the more common voltage-controlled oscillator (VCO). The control voltage in the loop drives the phase of the signal at the VCPS to be in quadrature with the input IF signal. As in the earlier system, the signal through the VCPS is the reference oscillation at ω with arbitrary phase β. Hence the loop drives the VCPS to a phase shift $\phi - \alpha - \beta - \pi/2$. Ganged to the VCPS is a matched phase shifter with opposite phase. Its phase is $-\phi + \alpha + \beta + \pi/2$. Although the phase of the VCPS is set in a closed loop, the paired phase shifter is set open-loop. Carefully matched analog phase shifters are required; otherwise digital phase shifters must be used.

Digital phase shifters generally are preferable. The number of discrete phase-shift components required is easily calculated. If m is the number of

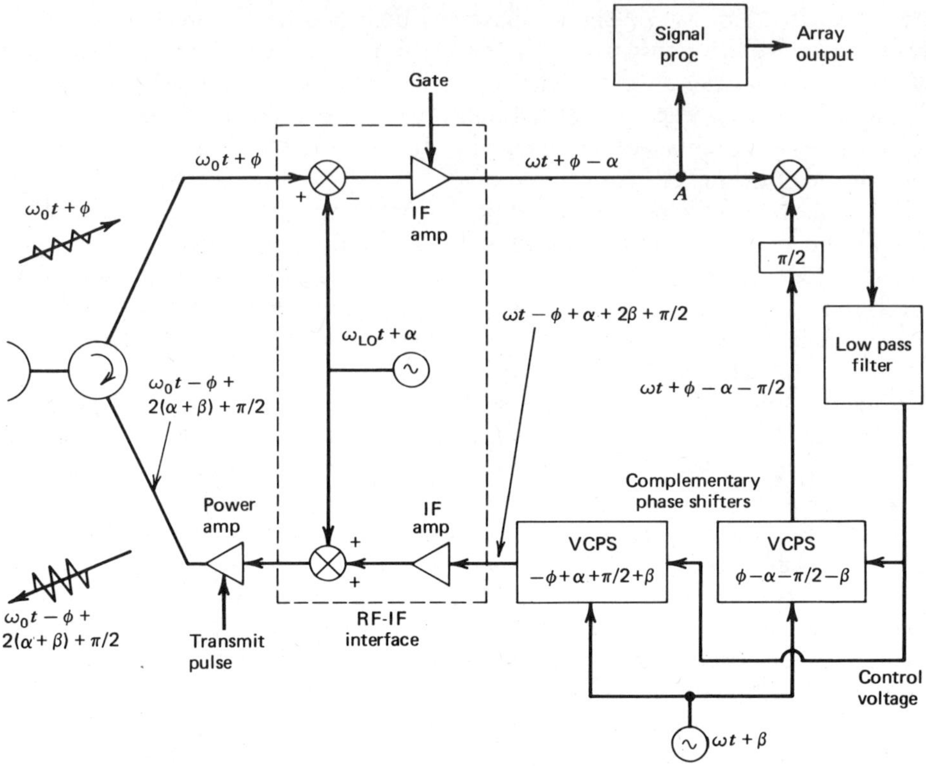

FIGURE 8.7. Phase conjugation circuit.

quantization bits and M is the number of levels of quantization, the relation between them is given by $M = 2^m$. The loss in array gain as a function of the number of quantization bits [85] is

$$\text{loss in gain (dB)} = 20 \log \frac{\sin(\pi/2^m)}{\pi/2^m} \tag{8.12}$$

Two bits limit the loss to less than 1 dB, three bits, to less than 1/4 dB, and four bits, to 0.06 dB. The side radiation pattern is also affected by phase-quantization errors, although the statistics in the sidelobe region are not. This is because the element positions are randomized and lead to Rayleigh sidelobe statistics. Further phase errors do not noticeably increase the sidelobe levels; they only reduce the mainlobe gain.

Returning to Figure 8.7, we see that the input to the paired phase shifters is $\omega t + \beta$; its output, therefore, is $\omega t - \phi + \alpha + 2\beta + \pi/2$. This IF signal is heterodyned to RF via the mixer and the local oscillator at $\omega_{LO} t + \alpha$. The upper sideband at ω_0 is selected by the power amplifier to provide a signal at the desired frequency with a phase $\omega_0 t - \phi + 2(\alpha + \beta) + \pi/2$. This signal is

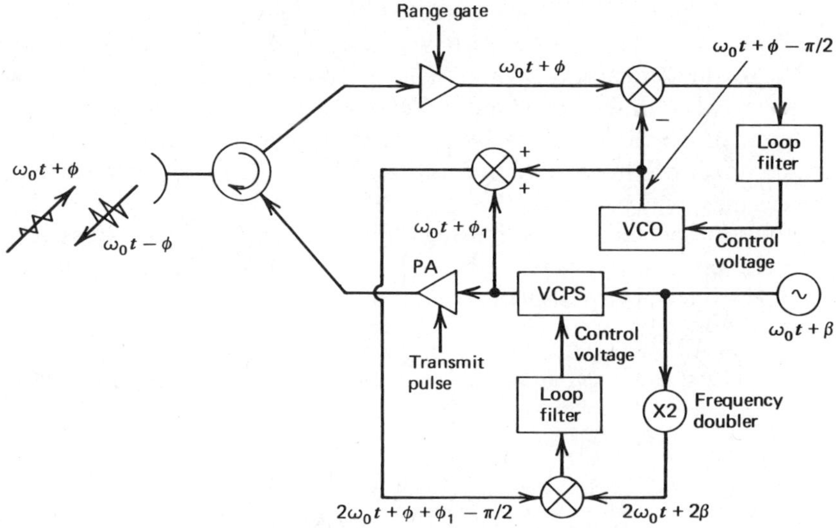

FIGURE 8.8. Phase conjugation circuit.

gated in the power amplifier by the transmit-pulse waveform from the radar synchronizer and radiated. Provided that $\alpha + \beta$ is a constant across the array, the high power transmitted pulse has the desired phase.

Figure 8.8 shows a phase-conjugating circuit (RF–IF heterodyning circuits not shown) which replaces the open-loop ganged phase shifter in Figure 8.7 with a phase-shift circuit under closed-loop control.† The circuit is a PLL. The received TOO echo at $\omega_0 t + \phi$ is applied to the PLL. The VCO phase, after the loop is locked, is $\omega_0 t + \phi - \pi/2$. It is applied as one input to an up-converter. The other input, also at ω_0, is delivered by a voltage-controlled phase shifter driven by the reference oscillation. Call its phase ϕ_1. The up-converter output is at the second harmonic frequency $2\omega_0$. Its instantaneous phase is $2\omega_0 t + \phi + \phi_1 - \pi/2$. Assume that adequate balance is achieved in the up-converter to avoid second harmonic feed through. This signal is compared in a phase detector with the second harmonic of the reference. The beat product is integrated in the loop filter and applied as the control voltage to the VCPS. Convergence requires that $\phi + \phi_1 = 2\beta$ or $\phi_1 = -\phi + 2\beta$. Thus the instantaneous phase of the CW output of the VCPS is $\omega_0 t - \phi + 2\beta$, which is the correct frequency and phase, provided that β is constant across the array. Hence the drive for the gated power amplifier is taken from this point.

The circuit in Figure 8.9 introduces another method of phase conjugation. The received echo pulse from the TOO is gated as before. The input to the phase-conjugating network is switch S_1, connected as shown. The echo at $\omega_0 t + \phi$ passes through a quadrature hybrid that delivers pulses at $\omega_0 t + \phi$ and

† This circuit and the next two were suggested by Professor Y. Bar-Ness of Drexel University while a Visiting Professor at the University of Pennsylvania from Tel Aviv University.

$\omega_0 t + \phi - \pi/2$. These signals are weighted by real-gain controls w_1 and w_2 and added. The sum, $w_1\cos(\omega_0 t + \phi) + w_2\cos(\omega_0 t + \phi - \pi/2)$, passes through switch S_2 to the comparator, where it is subtracted from the reference wave $\cos(\omega_0 t + \beta)$. The difference is fed back to the mixers of two correlators, the other inputs of which are driven by the quadrature outputs of the hybrid. The integrated mixer products drive the real weights w_1 and w_2 to those values that cause the sum waveform to equal $\cos(\omega_0 t + \beta)$.

The portion of the circuit between the switches is used extensively in adaptive nulling and interference cancellation problems. The closed loops set the weights w_1 and w_2 to solve the equation

$$w_1\cos(\omega_0 t + \phi) + w_2\cos\left(\omega_0 t + \phi - \frac{\pi}{2}\right) = \cos(\omega_0 t + \beta) \qquad (8.13)$$

The solution is

$$\tan(\phi - \beta) = \frac{w_2}{w_1} \qquad w_1^2 + w_2^2 = 1 \qquad (8.14)$$

When the loops converge, the circuit between the two switches transforms the input $\omega_0 t + \phi$ to the output $\omega_0 t + \beta$. In short, the transfer function of the circuit at ω_0 is $\exp[j(\beta - \phi)]$, which means that the circuit is a phase shifter with phase shift $\beta - \phi$. Following loop convergence, the weights are frozen and both switches are thrown to their lower positions. The reference wave with phase $\omega_0 t + \beta$ then passes through the circuit and emerges with phase $\omega_0 t - \phi + 2\beta$. It is amplified, gated, and radiated.

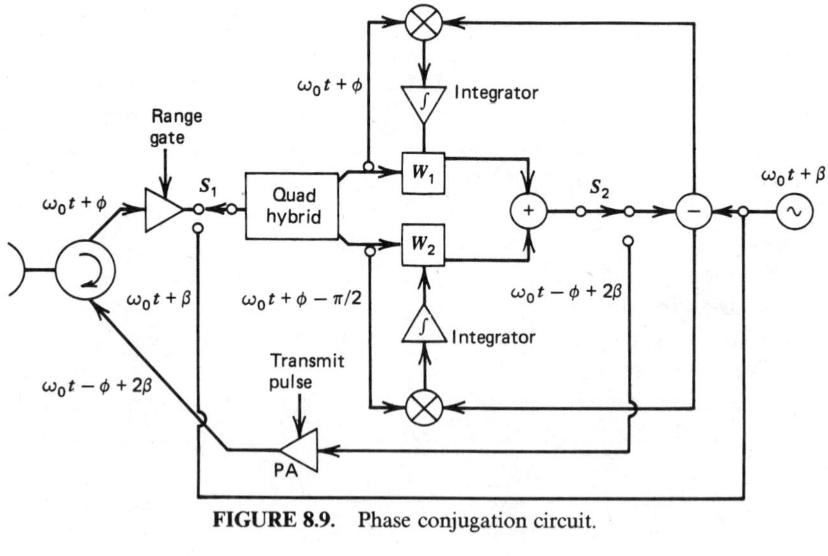

FIGURE 8.9. Phase conjugation circuit.

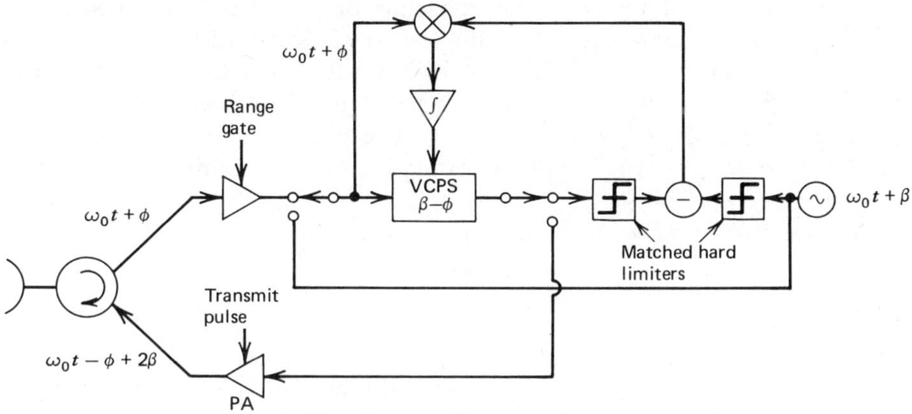

FIGURE 8.10. Phase conjugation circuit.

Figure 8.10 shows another version of the preceding circuit in which the signals entering the comparator are hard-limited in carefully matched limiters. Given that their amplitudes are matched, it is necessary only to shift the phase of the input echo by $\beta - \phi$ to zero the comparator output. Only a single cancellation loop is needed because there is only a single parameter to be varied. A latching phase shifter like the digital phase shifter discussed earlier is required. After loop convergence is completed the phase shift is frozen, the switches are thrown, and $\omega_0 t - \phi + 2\beta$ is radiated.

8.7. ERRORS IN PHASE CONJUGATION

Two types predominate. The first is a random phase shift due to mistunings in open-loop portions of phase-conjugating networks. The second is a linear phase shift due to a frequency offset. The first constitutes a random variation in phase across the array, the effect of which is loss in mainlobe gain. Assuming a 1-dB total loss budget for the entire system, the allowed phase error is about $\frac{1}{2}$ rad rms. It is evident that the allowed random error in phase conjugation is smaller still. Frequency offset, which has been ignored in the preceding section, may occur in two ways. First, the initializing microwave illuminator may have a somewhat different frequency from the transmitting array. Second, the need for isolation between low-level incoming signals and high-level outgoing signals may force a frequency offset. The effect is the same in both cases.†

†Only the first of these two reasons for an offset frequency is pertinent to the distributed radar array because after the array is phase synchronized it will radiate at the same frequency as the echoes (except for Doppler), as in ordinary radar. The second reason given above for a frequency offset pertains to a retrodirective array that operates under the CW or quasi-CW conditions common in point-to-point communication systems. It is possible for the system at each end of the propagation link to cohere its array to the signal radiated by the other. A frequency offset between

The magnitude of the frequency offset is determined by the manner in which the phase is conjugated. The simplest way is to adjust the phase shift at the initial frequency and accept the error that results. Let the reflecting source be at angle $\theta_0 = \sin^{-1} u_0$ from the array normal and the initializing (self-cohering on reception) frequency be $\omega' = k'c$. The phase of the wave across the array is $\phi_R = k'xu$. Let $i_0(x)$ be the excitation across the transmitting array. Let the conjugated phase be $-k'xu_0$ and let the wavenumber of the signal transmitted by the array be k. The radiation pattern becomes

$$f(u; u_0) = \int i_0(x) e^{-jk'xu_0} e^{jkxu}\, dx$$

$$= \int i_0(x) \exp\left[jkx\left(u - \frac{k'}{k} u_0 \right)\right] dx \tag{8.15}$$

Note that the beam no longer points to u_0 but to $k'u_0/k$. The error or displacement

$$\Delta u_0 = u_0\left(1 - \frac{k'}{k}\right) = u_0\left(1 - \frac{\omega'}{\omega}\right) \tag{8.16}$$

is called the squint angle. Equation (8.16) can be rewritten

$$\left|\frac{\Delta u_0}{u_0}\right| = \left|\frac{\Delta \omega}{\omega}\right| \tag{8.17}$$

to indicate that the magnitude of the fractional change in the beamsteering angle equals the magnitude of the fractional change in the frequency. The largest typical value of $\Delta\omega$ is the receiver bandwidth. Rarely will the angular displacement exceed one or a few percent of the scan angle. Such scale distortion will be unimportant unless the beam displacement exceeds the beamwidth of the large array and no synchronizing source (e.g., a large TOO) resides within the transmitting beam. Figure 8.11 illustrates this case. A phase conjugating circuit devoid of squint is required if this problem is anticipated [18].

The reason why angle distortion arises is that the phase is measured and conjugated at one frequency but radiation takes place at another. The error is eliminated if the phase shift resulting from the conjugation process is correct at the new frequency. Then the radiation pattern of the retrodirective array is

$$f(u; u_0) = \int i_0(x) e^{jk'x(u-u_0)}\, dx = f[k'(u - u_0)] \tag{8.18}$$

---

the signals passing in the two directions is good engineering design that minimizes the cross-talk and intermodulation between them. A frequency offset is useful in the phase-synchronizing subsystem of the Solar Power Satellite [33], [73], a remarkable plan for the delivery of solar-generated electrical power from geosynchronous orbit to earth. The technique for operating with a frequency offset is described in [18].

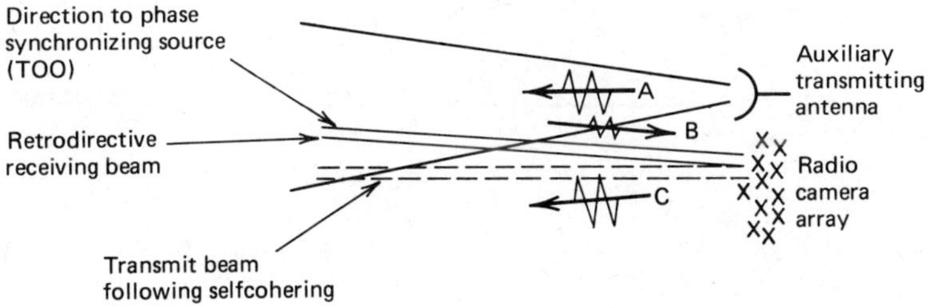

FIGURE 8.11. Beam squint due to frequency offset. (*a*) Radiation from an auxiliary transmitter at ω'. (*b*) A weak echo from the target of opportunity at ω'. (*c*) Coherent transmission from a distorted array at ω. Transmit beam displaced by frequency offset.

The argument of the function is $k'(u - u_0)$. The beamsteering angle, therefore, is u_0. Hence the frequency change is no longer reflected in an angular displacement. The sole effect is a change in the angular scale, measured from u_0, by a factor k'/k. This scale change is of no consequence in adaptive beamforming.

8.8. PHASE REFERENCE

A reference oscillation with constant phase $\omega t + \beta$ is required in every module in the array. This signal must be derived from an oscillator arbitrarily located in the array and delivered to each module by a circuit or subsystem. A frequency-stable and phase-stable oscillator as well as a frequency synthesizer capable of generating the local oscillator waveform are assumed.

Cables of equal and constant lengths can deliver the reference wave from the source to each module. This is a practical technique when the array is compact and the modules are contiguous. It becomes impractical when the array is large and distributed. Furthermore, in an open-loop system differential phase changes between cables due, for example, to temperature differences or mismatches at connections are passed directly as phase errors to the modules.

Circuits have been devised to deliver the phase reference from source to module or from module to module. The major impetus to date has been design work for the Solar Power Satellite. Figure 8.12 illustrates a method due to Lindsey [55] which consists of two distinct circuits separated by a cable with arbitrary phase delay Δ. A reference signal at frequency 2ω and phase β enters the left circuit where its frequency is halved, and passes to the right circuit via the cable.

The upper left circuit is a PLL. Its VCO phase is $\omega t + \phi_0$, where, for the moment, ϕ_0 is an arbitrary value. Oscillator output is taken from the loop and passed via the first diplexer (shown as a circulator) to the cable, which delivers $\omega t + \phi_0 - \Delta$ to the right-hand circuit. There the cable-delivered signal is doubled in frequency to provide the reference $2\omega + 2\phi_0 - 2\Delta$ for the second

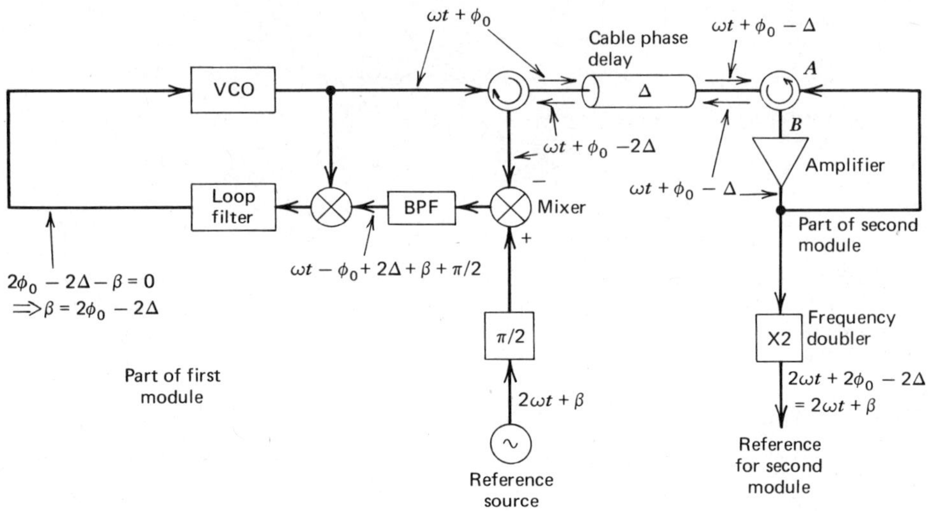

FIGURE 8.12. Frequency and phase-reference delivery circuit. (From W. C. Lindsey and A. V. Kantak, "Automatic Phase Control in Solar Power Satellite Systems," NASA/JSC, February 1978.)

module. In this way both modules are driven by the same reference frequency. It is shown below that they are also driven to the same phase.

Prior to frequency doubling in the second module, the amplified wave at $\omega t + \phi_0 - \Delta$ is fed back to the second diplexer and returned to the first module. Its phase is further retarded by the cable delay Δ. It is passed by the circulator to the mixer, where it is heterodyned by the 90° phase shifted reference wave. The lower sideband is selected by the bandpass filter (BPF). Its output is the input to the phase detector of the PLL. The instantaneous phase of this wave is $\omega t - \phi_0 + 2\Delta + \beta + \pi/2$. The other input is $\omega t + \phi_0$, delivered by the VCO. The low-pass filtered output of the phase detector is zero when the loop drives the phase difference between the inputs to 90°. Hence $2\phi_0 - 2\Delta - \beta = 0$ or $\beta = 2\phi_0 - 2\Delta$, which is the condition sought. Thus these modules have the same reference phase $2\omega t + \beta$ independent of the cable length between them. Other modules are fed in the same manner.

This extremely clever circuit is subject to several phase-error sources. First, the nonlinear mixer will generate harmonics of the input signal. The second harmonic will add to the reference source at 2ω to produce a net reference signal with altered phase. In addition, feed through the mixer at the fundamental frequency will alter the net phase of the signal delivered by the bandpass filter to the phase detector of the PLL.

A second source of phase errors is the phase shifts through all the non-closed-loop-controlled portions of the circuit. The circulators, the bandpass filter, and the signal-return loop are examples. This is a tuning-type problem common to all the preceding circuits.

Third, the signal-return loop has a special problem. Unless there is sufficient isolation in the circulator from ports A to B, the loop will oscillate. The amplifier is needed to overcome the signal losses in the cable in both directions. Hence the signal level delivered back to the circulator as port A is larger by the gain of the amplifier than the signal delivered by the circulator at port B. The ideal circulator (or other diplexer) provides zero coupling between ports A and B; the practical circulator has limited isolation. To avoid the danger of oscillation the isolation must exceed the amplifier gain, which, in turn, must at least equal the two-way cable loss. Hence the maximum allowed cable loss is limited by the isolation available in practical circulators.

A modification to the circuit avoids the more serious of these problems. Figure 8.13 shows that the reference source frequency is m times the VCO frequency and that the frequency of the return signal is $n\omega$. The VCO signal $\omega t + \phi_0$ is again delivered by cable to the next module, amplified, and returned. The return signal is frequency-multiplied by the factor n, delayed $n\Delta$ by the cable, and mixed with the reference in the down converter. The phase of the output of the BPF is $(m - n)\omega t + \beta - n\phi_0 + (2n)\Delta + \pi/2$.

The VCO output is multiplied in frequency by $(m - n)$ to equate the frequencies of the phase-detector inputs. The loop drives these signals into quadrature, resulting in the phase equation $m\phi_0 - \beta - (2n)\Delta = 0$, which implies that the instantaneous phase of the reference source $m\omega t + \beta = m\omega t + m\phi_0 - (2n)\Delta$.

The output to the next module is derived from a frequency multiplication by the factor m of the signal delivered by the cable; its phase is $m\omega t + m\phi_0 - m\Delta$. The only condition required to equate the last two expressions is $m = 2n$. When this condition is met the desired phase reference is transferred from the

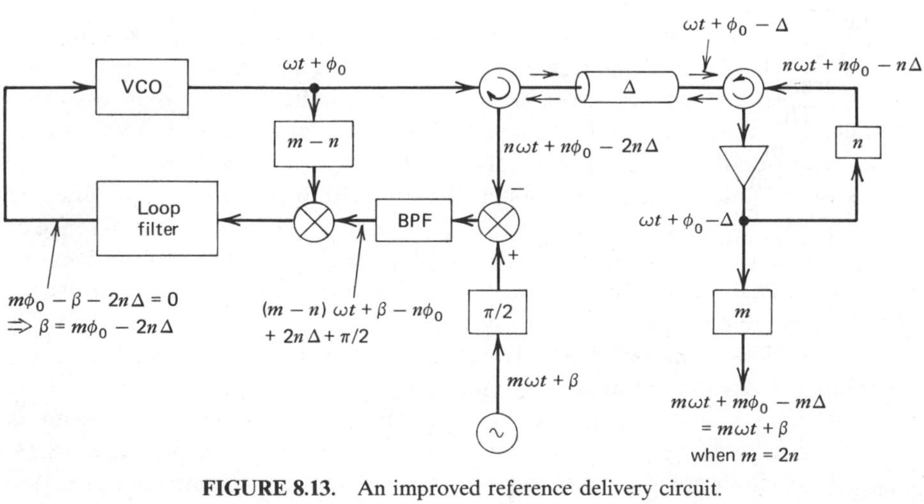

FIGURE 8.13. An improved reference delivery circuit.

first to the second module. The frequency multiplication which follows the VCO is the factor n.

System frequencies are determined by the choice of a convenient VCO frequency ω and the desired frequency offset $\omega(n-1)$; n must be small to avoid exceeding the bandwidths of the circulators and the delay line; n need not be integral. Frequency multiplication by ratios of integers is easy to obtain when modern frequency synthesizer techniques are used. Thus n can be made close to unity. A value of $4/3$ permits an adequate frequency separation between the input and output of the mixer, although not requiring excessive bandwidth of the components. The reference phase is delivered at $7\omega/3$ when n equals this value.

8.9. EXTERNAL CALIBRATION

Random, uncompensated phase shifts in components throughout the module and the reference-delivery circuit will always exist and will degrade mainlobe gain as indicated in Section 8.6. Periodic calibration or tuning is necessary to limit the loss in mainlobe gain to an acceptable level. Calibration can be manual or automatic in a small system but must be automatic in a system with a large array.

A calibration transceiver outside the array can be used for ground-based and shipboard installations. The design in Figure 8.14 includes a procedure for correcting the phase of the reference wave at each module and thereby removes the need for the circuit shown in Figure 8.13. One module of a distorted array is shown with the calibration transceiver. Each module has a phase-conjugating circuit (PCC) and a reference wave. Assume that the reference wave is stable in frequency and constant across the array but that the phase varies from module to module. Call the phase of the reference to the ith module β_i or its total instantaneous phase $\omega t + \beta_i$.

The calibration unit is external to the array and in the general direction toward which the antenna elements are pointed. Its front end consists of a pulsed transmitter and receiver and is similar to the front end of an array module. The reference oscillation in the array is delivered to it by cable or radio. The broadcast reference technique described in Section 5.9 is suitable and is the technique illustrated. When the calibration transmitter is connected in the SYNC mode (switches thrown to S) radiation of the system reference at some arbitrary phase from some arbitrary antenna within the array drives the PLL to a frequency ω'. The phase $\omega' t$ of the VCO in the calibration transceiver is the reference phase of the entire system.

After the loop is locked the VCO output is heterodyned to RF and radiated. Its frequency is ω_0, the radiation frequency of the entire system. Because there is some phase shift α from VCO to antenna, the radiated waveform is characterized by $\omega_0 t + \alpha$, as indicated next to the symbol **1** in the Figure 8.14. The system calibrates one module at a time. The system controller (i.e., the

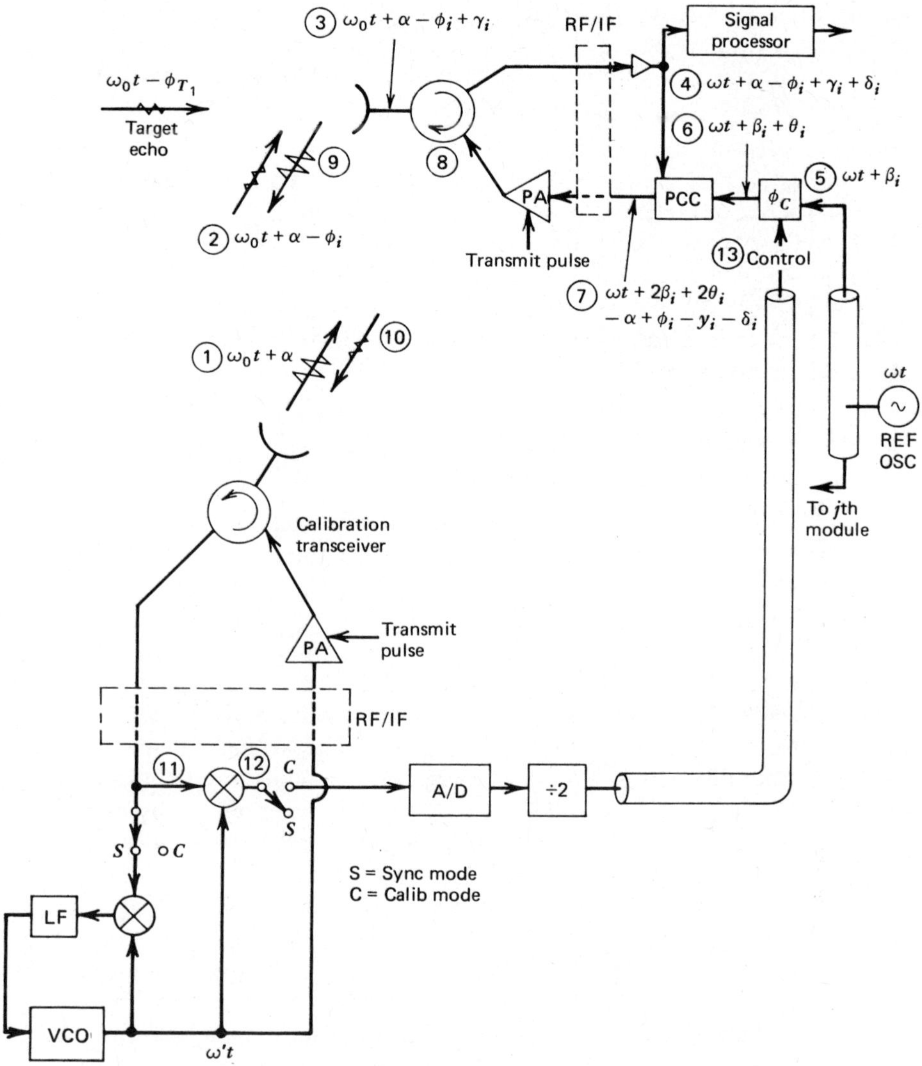

FIGURE 8.14. A transmit-receive array with external calibration subsystem.

central computer) turns on each module in sequence. On arrival of the radiated signal **1** at the antenna of the ith module **2** the phase is $\omega_0 t + \alpha - \phi_i$, where ϕ_i is the propagation phase delay. The phase shift through the antenna in the direction toward the calibration system is γ_i. Because the phase of the element pattern may be different in the direction to the target, γ_i is explicitly retained in this description or system operation. Thus the signal phase after the antenna is $\omega_0 t + \alpha - \phi_i + \gamma_i$. There is also a phase shift δ_i to the input **4** to the PCC, where the phase is $\omega t + \alpha - \phi_i + \gamma_i + \delta_i$.

The primary reference oscillator of the system runs freely at frequency ω delivering $\omega t + \beta_i$ to the input **5** to a digital phase shifter ϕ_c. The initial phase shift is some arbitrary value θ_i, making the input reference phase **6** to the PCC $\omega t + \beta_i + \theta_i$. The output **7** of the PCC is

$$\text{phase } \mathbf{7} = \omega t + \beta_i + \theta_i - [(\alpha - \phi_i + \gamma_i + \delta_i) - (\beta_i + \theta_i)]$$

$$= \omega t + 2\beta_i + 2\theta_i - \alpha + \phi_i - \gamma_i - \delta_i$$

After the wave is heterodyned to RF and passed through the transmitter its phase **8** is increased by η_i to

$$\text{phase } \mathbf{8} = \omega t + 2\beta_i + 2\theta_i - \alpha + \phi_i - \gamma_i - \delta_i + \eta_i$$

γ_i is added through the antenna **9** and the propagation delay ϕ_i is contributed on arrival at the calibration system **10**. There the instantaneous phase is $\omega t + 2\beta_i + 2\theta_i - \alpha - \delta_i + \eta_i$. It is evident that the propagation delay and antenna element phase shifts have dropped out.

This wave is received with the switches thrown to C, which places the calibration transceiver in the CALIB mode. The phase shift from antenna to phase detector **11** is μ. Hence the phase detector output **12** is a video voltage proportional to $-2\beta_i - 2\theta_i + \alpha + \delta_i - \eta_i - \mu$. This voltage is converted to a digital number, divided by two, and delivered, **13**, to the digital phase shifter in the module, where it changes the phase through that component by $\Delta\phi_c = -\beta_i - \theta_i + \frac{1}{2}(\alpha + \delta_i - \eta_i - \mu)$. The signal reentering the PCC as a phase reference therefore is $\omega t + \frac{1}{2}(\alpha + \delta_i - n_i - \mu)$.

This is the desired reference phase: the random initial phase θ_i of the phase shifter and the random reference oscillation phase β_i have been removed, the circuit phase shifters δ_i and η_i cancel out during system operation (see below), and the remainder is a constant across the array.

Now let a target echo $\omega_0 t - \phi_{T_i}$ arrive at the ith antenna element. The IF signal delivered to the PCC is $\omega t - \phi_{T_i} + \gamma_{T_i} + \delta_i$. The phase-conjugated output is $\omega t + \frac{1}{2}(\alpha + \delta_i - \eta_i - \mu) - [(-\phi_{T_i} + \gamma_{T_i} + \delta_i) - \frac{1}{2}(\alpha + \delta_i - \eta_i - \mu)] = \omega t + \alpha - \eta_i - \mu + \phi_{T_i} - \gamma_{T_i}$, and the signal radiated in the direction of the target is $\omega_0 t + \alpha - \mu + \phi_{T_i}$, which is exactly the conjugated phase plus an arbitrary constant.

The only circuit in the system not under closed-loop control is the phase-measuring and phase-control branch **12** to **13** from calibrator to module. Gain and bias errors can develop in this circuit. Let the phase detector gain be in error by the factor K and let a bias M develop in its output circuit. Then the phase shift in the digital phase shifter is changed by $\Delta\phi_c = K[-\beta_i - \theta_i + \frac{1}{2}(\alpha + \delta_i - \eta_i - \mu)] + M$. The reference for the PCC becomes $\omega t + \beta_i + \theta_i + \Delta\phi_c \triangleq \omega t + \psi$. The PCC output is $\omega t + 2\psi + \phi_{T_i} - \gamma_{T_i} - \delta_i$. After passage through the transmitter and antenna the radiated wave toward the target is

$$\omega_0 t + 2\psi + \phi_{T_i} - \delta_i + \eta_i$$

$$= \omega_0 t + \underbrace{2M + K(\alpha - \mu)}_{\text{constant}} + \underbrace{\phi_{T_i}}_{\substack{\text{conjugate}\\ \text{phase}}} + \underset{\substack{\uparrow\\ \text{fractional}\\ \text{gain error}\\ \text{in PD}}}{(1 - K)} \underbrace{(2\beta_i + 2\theta_i + \eta_i + \delta_i)}_{\text{random variables}} \tag{8.19}$$

The first four terms are the conjugated phase plus an arbitrary constant. The last term represents a phase error in transmission from the ith module, which goes to zero when the gain error goes to zero. The maximum effect is easily calculated. β, θ, η, and δ may be assumed to be random variables independent of the errors in the other modules. Their sum is a random phase error. The magnitude of the net phase error Φ_i is a function of the fractional gain error K of the phase detector. Note that the phase error is not a function of the bias error of the phase detector.

Using the theory of mainlobe gain-loss found in Chapter 2, a tolerance can be calculated for the phase-detector gain error. The random phase error is, at worst, uniformly distributed in the interval $[-\pi, \pi]$.† The variance of a uniform distribution is one-twelfth the square of the length of the interval.‡ Hence

$$\sigma_\Phi^2 \leq \frac{(1 - K)^2 (2\pi)^2}{12} \tag{8.20}$$

which must not exceed $\frac{1}{4}$ rad^2 if the loss in mainlobe gain is to be limited to 1 dB. Taking this value as the tolerance in gain loss, we find the allowed fractional gain error in the phase detector by equating (8.20) to 0.25, which yields $K = 0.724$. In other words, the gain can change by 27% without causing more than 1 dB loss in system performance. This is a relatively easy tolerance to maintain.

†For example, the probability density function of the modulo-2π sum of two random variables, each uniformly distributed in a 2π interval, is also uniform in the interval. This case corresponds to the equality condition in (8.20). If the pdfs are clustered near the center of the interval, the pdf of the sum also is clustered, leading to the strict inequality in (8.20). The random variables in (8.19) will generally correspond to this case. If the pdfs are lower in the central region than at the edges, the inequality could reverse. There is no physical basis for assuming that this situation will occur in this system.

‡Let $w(x) = 1/L$, $|x| \leq L/2 = 0$, elsewhere. Then $\sigma_x^2 = 1/L \int_{-L/2}^{L/2} x^2\,dx = L^2/12$.

8.10. TRANSMITTERS FOR ADAPTIVE TRANSMIT-RECEIVE ARRAYS

Although most radars operate with conventional microwave sources and amplifiers, such as magnetrons, klystrons and TWTs, the unusual nature of the radio camera transmitter warrants looking farther afield. The solid-state microwave transmitter is becoming practical and is the preferred device. Transmitters consist of low-power oscillators (which may be VCOs) that drive solid-state power amplifiers. Both silicon and gallium arsenide devices are available. The relativistic electron beam tube is more distant. These devices are in the gyrotron family [34].

The nominal peak power levels available from these microwave sources cover a remarkably large range, as listed in Table 8.2.

The choice of transmitter for the transmit-receive module depends largely on the power requirements. Consider (8.2), which shows that the required power per element is inversely proportional to the square of the number of elements. Because the power required in radar is compatible with the (conventional) devices designed for radar (10^5–10^6 W), the power per element needed for a large transmit-receive array is $(10^5–10^6)/N^2$, which is the order of watts. Hence the solid-state device is the natural component.

On the other hand, when the array is organized into rigid subarrays (as in Figure 8.3) it is practical to use a single transmitter per subarray. If the number of subarrays with which the system is composed is small (e.g., 10) and the required peak power is large ($> 10^6$ W), the conventional devices are more appropriate. Clearly the gyrotron, which is a remarkable invention, is inappropriate for a radar-range, large-array detection and/or imaging system.

8.11. SUMMARY

The logical requirements for a self-adaptive, nonrigid, distributed radar antenna array are discussed. A transmitter is required to illuminate a target, the reflections from which are received by the elements in the array. The target must reradiate a nearly spherical wavefront. The phases of the received echoes are used to set the phase shifts in the antenna elements to allow a receiving

TABLE 8.2. Range of Peak Power Available from Microwave Transmitting Devices

Microwave Device	Peak Power
Solid state	1–100 W
Conventional sources	10^5–10^6 W
Gyrotron	10^8–10^9 W

beam to be focused on the target. The same phase information permits setting the transmission phase shifts.

Once the transmitting beam is formed and focused on the target the initializing illumination is no longer required. The beam is scanned by modifying the phase shifts in the same manner used in a conventional phased array.

Phase conjugation of the received wave at every element is necessary to achieve focused transmission. Two primary circuit and system choices are made in the design of phase-conjugating networks. The first is between analog and digital circuits. The second is between open- and closed-loop control of the phase shift. The bases for these choices are discussed in this chapter and several circuits are given. The phase conjugating circuit at each antenna element requires a reference wave of constant frequency and fixed phase. Methods of distributing the phase reference across the array are described.

Coherent, focused transmission of power P_e from each of N antenna elements increases the power density on a target by the factor N^2 in relation to transmission of P_e from a single element. Because of the quadratic dependence on N, the required power per element is small; for example, the order of watts. Hence a separate solid-state transmitter per element is appropriate for most designs.

CHAPTER NINE

Airborne Radio Camera

9.1. GENERAL CONSIDERATIONS

An airborne radio camera is illustrated in Figure 1.3. The array is shown distributed throughout the surface of the aircraft. Because the airframe is flexible and the skin vibrates, it is necessary to phase-synchronize the array from an external source; hence imaging would follow some form of adaptive beamforming and all the discussions of the earlier chapters would be relevant.

As an example for the need for self-cohering, the rms uncertainty in the position of an arbitrary point on the fuselage of a Boeing 707 when in flight due to airframe twisting is 4 cm [49]. The loss in gain of the mainlobe due to this distortion is, in decibels, 4.3 σ_ϕ^2, where $\sigma_\phi = (2\pi/\lambda)(4\text{ cm})$. The loss at VHF is negligible. At L-band ($\lambda = 30$ cm) it is approximately 3 dB. The loss rises to 10 dB at S-band ($\lambda = 10$ cm). The mainlobe totally disappears at X-band ($\lambda = 3$ cm), where the loss, calculated from the expression above, is more than 300 dB.

The airborne radio camera differs fundamentally from the ground-based in two major ways. First, the external source used for adaptive beamforming cannot be a single, fixed source: it must move with the aircraft or the system must be capable of obtaining new and different sources of phase synchronization as often as it is required to do so. Second, the aperture size is limited by the size of the airframe and is *not* determined by the resolving power requirements. In other words, the available resolution is *given* by the airframe size and the smallest wavelength that the expected propagation conditions will permit and is not a design parameter of the system. Because the airframe is at most tens of meters and not the hundreds to ten thousands discussed in the ground-based examples in Chapter 1, the resolving power would be much more limited. It may still be useful, however; it is shown in Table 1.1 that the potential resolving power of a microwave system as large as a Boeing 707 airplane is comparable to human vision. Hence the reduced beamwidth offered by the aircraft size enhances the potential of the system for target counting and classification. Nevertheless, as a consequence of the limited size of the aperture, other attributes of the system become more relevant than in ground-based radio camera design. One performance parameter is signal-to-noise ratio (SNR) or its equivalent, radar detection range. Second, improved detection performance can be traded for a reduction in transmitter power for a given performance to permit the successful design of a low-probability-of-intercept radar. A third important advantage has to do with protection against jammers that are close to the axis of the beam. Rejection of jamming energy by low sidelobe design or adaptive techniques, such as coherent sidelobe cancellation and adaptive nulling, [43], [1], [103], [13], [29], [2] are useful techniques. Mainlobe jamming, however, is not suppressed without hazarding the suppression of target return. Increasing the size of the aperture reduces the width of the mainlobe and thereby reduces the minimum angular separation between target and jammer at which sidelobe suppression techniques can operate.

In Chapter 6 it is shown that ground clutter, the radar echoes from the earth's surface, can be used to phase-synchronize an airborne radio camera. A Doppler filtering technique is described, akin to that used in synthetic aperture radar, for deriving a reference wave from the ground clutter, notwithstanding the fact that the clutter echoes are returned to the aircraft from a wide angle. The derivation of the required filter characteristics is made in the context of Chapter 6, which concerned the development of ideal or nearly ideal phase-synchronizing sources. A proper question remains whether lesser requirements could be imposed on the system design and still permit adaptive beamforming of an airborne, nonrigid array on clutter echoes. A second question is whether phase synchronization on clutter can be accomplished when the Doppler echoes are not directly identifiable with bearing angle; examples are wind-driven ground and sea clutter. A third question concerns the potential usefulness of an airborne radio camera. Clearly, the size of conventional aircraft precludes creating an airborne microwave array of 10^4 or 10^5 wavelengths. Hence the utility of the techniques discussed in this chapter must relate to attributes other than high angular resolution. These are the questions addressed in the sections that follow.

9.2. MINIMUM AMPLITUDE VARIANCE TEST

The first question raised was whether a simple adaptive beamforming procedure might also work, although perhaps producing a less ideal reference wave than the Doppler technique described in Section 6.6. The basic procedure described in Chapter 5 is tempting to apply to the airborne problem on the assumption that at some unpredictable range in the echo trace there will be a

TOO suitably small in size but sufficiently large in radar cross section to act as a proper phase-synchronizing source. This hypothesis has been tested and the results are promising.

The algorithm consists of three parts. First, the variance of the amplitudes of the group of echoes from each range bin, measured at each element in the array, is calculated. The range bin with the lowest echo variance, when normalized to the average echo power in that range bin, is selected as the reference range for the system. Second, the processor either multiplies the complex sample at each array element from each range trace by the complex conjugate of the echo at the reference range or, more simply, merely phase rotates the received echoes by the phases of the complex conjugates of the signals at the reference range. This is the adaptive part of the process. Third, the processor applies linear phase weighting across the array to scan the adaptively formed beam electronically in angle.

The object of the first step is to find a target or a clutter patch whose reradiation most closely approximates that of the point source. The object of the second step is to self-cohere the array on that target. The object of the third step is to scan the beam in angle to the left and right of that target.

This algorithm is simple to implement. It requires neither special filtering of the radar echoes nor complicated signal processing. The algorithm was tested with airborne radar data [49], [96], [98]. The experiment used recorded data obtained from the Naval Research Laboratory. An eight-element UHF phased array was flown at 200 knots at an altitude of 15,000 to 20,000 ft over the southeastern portion of the United States. The transmitting pattern was Dolph-Chebyshev-weighted for −24-dB sidelobes. The received radar echoes were separately recorded on each of the 8 channels in a high-quality digital recording system. Demodulation was coherent and the inphase and quadrature video channels were recorded with 10-bit precision. Figure 9.1 is a functional sketch.

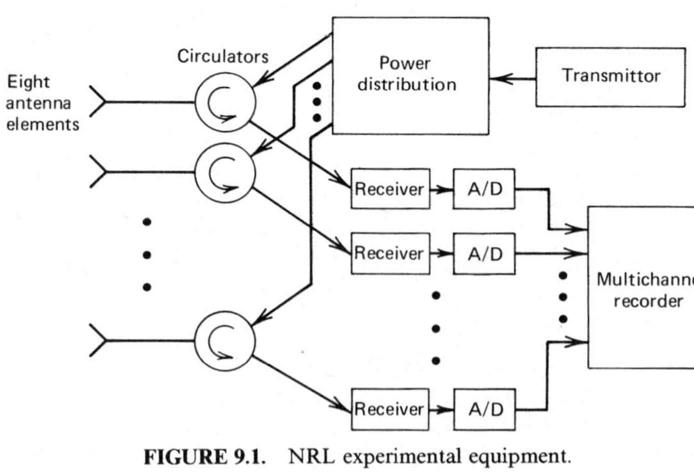

FIGURE 9.1. NRL experimental equipment.

Although a radio camera array would be expected to be flexible and distorted, the NRL antenna was rigid with one-half wavelength element spacing. Hence its proper performance was predictable and could be compared to the performance of the system when the adaptive algorithm was applied; that is, two beamforming algorithms were applied to the multichannel recorded data. The first was the conventional linearly phase-weighted, beamforming procedure. The second was the self-cohering algorithm described above. The beam patterns that resulted are shown in Figure 9.2.

The solid curve in Figure 9.2 shows the pattern that resulted from the adaptive process. It was calculated from the measured echoes at the reference range and the known geometry of the array. The beam formed by the adaptive procedure is directed toward the dominant scattering center in the illuminated ground patch in the range bin selected by the signal processor for adaptive beamforming. Hence the origin will usually differ from the bearing of the transmitter array. In this experiment the difference was 0.07 rad. Because $L = (8 - 1)\lambda/2$, $\Delta\theta \sim \lambda/L = 2/7$, and the beam displacement is $\sim 0.25\lambda/L$. The bearing of the transmitting array was fixed at 0.6° from the flight direction. The dashed curve shows the receiving pattern of a uniformly weighted eight-element array associated with conventional nonadaptive beamforming. The origin of that pattern which is the bearing of the transmitting array has been shifted by 0.07 rad to compare the adaptive and nonadaptive patterns. The results, which are similar, indicate that the adaptive process worked well. The smallest normalized echo variance observed in the range trace used for this experiment was 0.0026.

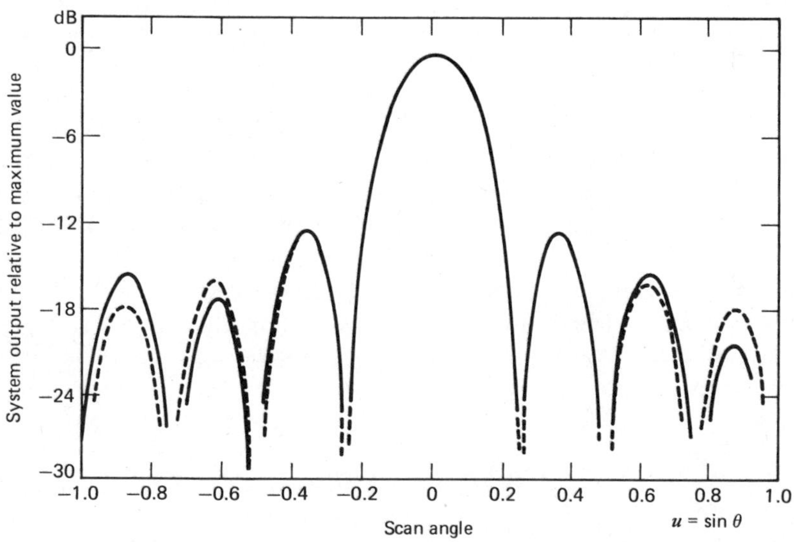

FIGURE 9.2. Comparison of patterns formed adaptively (solid) and nonadaptively. The beam direction is near the ground track of the aircraft. (From B. D. Steinberg and E. Yadin, *Proc. IEEE Letters*, Vol. 70, No. 1, January 1982.)

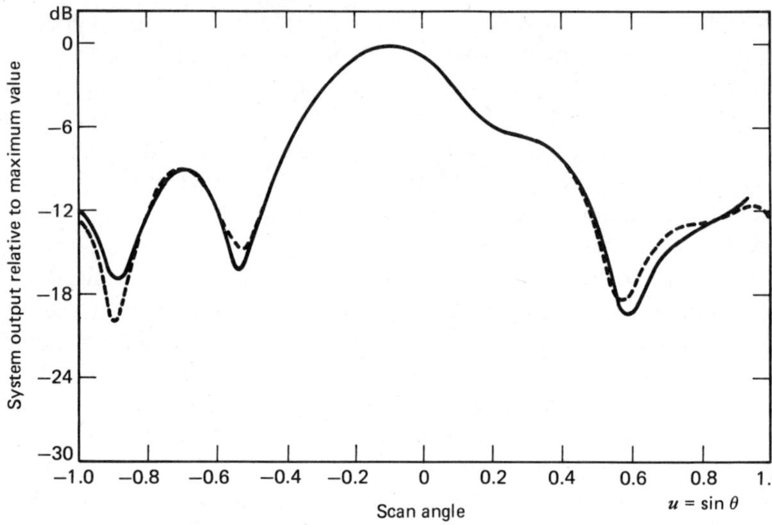

FIGURE 9.3. Comparison of scanned outputs of adaptively (solid) and nonadaptively formed beams. (From B. D. Steinberg and E. Yadin, *Proc. IEEE Letters*. Vol. 70, No. 1, January 1982.)

An empirical study was made of the data to determine the largest normalized echo amplitude variance that provided satisfactory adaptive beamforming results. Considering the radiation pattern acceptable when the first sidelobe had no more than doubled (risen to -10 dB), we found that the normalized echo amplitude variance can be as large as 0.02.

The largest echo variance observed in the range traces examined was 0.77. Adaptive beamforming on such a range bin is not fruitful. The high variance implies a complicated echo profile in the ground patch which therefore makes such a range bin interesting to image. It is possible to self-cohere the array on the echoes from a low-variance bin and scan the adaptively formed beam at the range of the high-variance bin as well as at all other ranges. Figure 9.3 is an example in which the beam is self-cohered at the range bin used in Figure 9.2 and then scanned at the range bin with the 0.77 variance (solid curve). The response of the rigid, electronically scanned phased array is shown as a dashed line. Again the excellent agreement indicates that this simple algorithm has promise of being suitable for self-cohering an airborne phased array on ground clutter.

9.3. DOPPLER-AIDED SYSTEM

Although the experiment just reported yielded highly satisfactory results, the technique contains an inherent difficulty. The problem is that an illuminated ground patch, because of its large size, is highly likely to contain more than one large radar target. When this happens the reradiation from the range bin is inappropriate for use as a phase-synchronizing field. That the cross-range dimension of the illuminated ground patch is large compared with the sizes of reflecting structures is readily seen: its size is approximately $\lambda R/a$, where R is the distance to the patch. In general, the transmitting antenna will be a conventional antenna for airborne radar, which means that a/λ will be on the order of 20 to 50. Even at a relatively short range such as 20 km, the cross-range dimension is several hundred meters. Thus, notwithstanding the excellent results described, it is possible that natural terrain and man-made target complexes will, from time to time, deny the system the ability to self-synchronize on this algorithm. The Doppler-aided technique described in Section 6.6 helps to solve that problem.

Figure 9.4 shows an aircraft flying at speed V illuminating a ground patch with a transmitting beam with beamwidth $\Delta\theta$ at an angle θ from the ground track. The width of the illuminated sector is $R\,\Delta\theta$. Each of the three large cross-section reflectors shown in the illuminated patch reradiates a spherical field that is essentially planar at the aircraft (Figure 9.5). The sum of the three fields has a highly nonplanar phasefront, which therefore renders it inadequate for the self-cohering algorithm described earlier. The radiation from the three targets, however, which originate from different bearing angles, have different Doppler shifts. Thus narrowband filtering, akin to synthetic aperture processing, can pass the echoes from, say, the central target while rejecting the

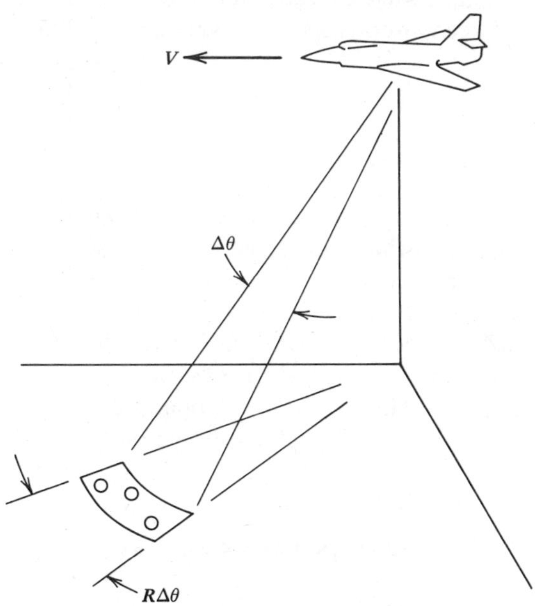

FIGURE 9.4. A narrowband filter in the receiver passes echoes from a narrow strip in the illuminated ground patch.

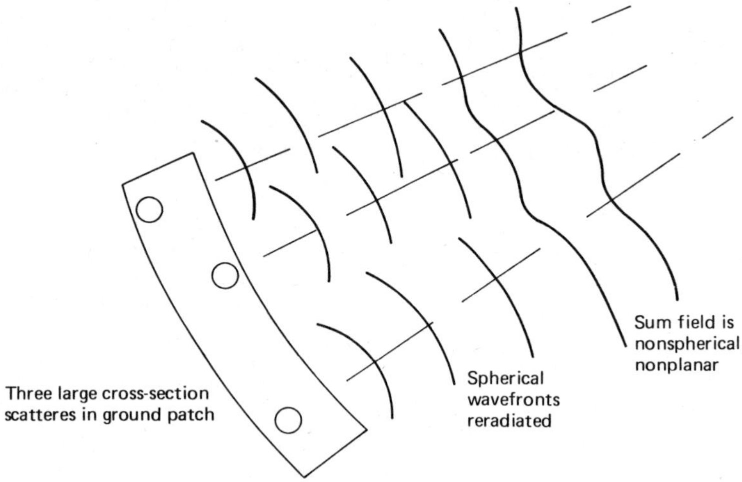

FIGURE 9.5. Ground patch detail.

radiation fields of the others. In this way a large, area-extensive clutter patch that contains many prominent reflectors can be reduced effectively to a small, confined patch reradiating a planar wavefront at the aircraft.

To ensure that the radar backscatter will be essentially planar across the array the effective cross-range dimension of the scattering region must be small enough so that the lobes of the reradiation pattern from the clutter, no matter how complex or irregular, will be large compared to the aircraft size. Given the Doppler relation between aircraft speed, wavelength, and target angle from the ground track

$$f_d = 2V\cos\frac{\theta}{\lambda} \tag{9.1}$$

the development in Scction 6.6 placed the following upper limit on the bandwidth W of the narrowband filter: $W = V/L$ where L is the aircraft size.

Narrowband filtering need be only in a single antenna element channel in the system. Figure 6.11 shows the minimal circuitry. The filtered output from one antenna element serves as a reference wave for measuring the phase distortion across the array. The self-correction procedure shown in that figure involves phaselock loops. The efficiency of operation is improved if narrowband filters are added to each channel, as shown dashed.

9.4. HYBRID SYSTEM

By the Doppler-aided technique echoes from every range bin will have a suitably low amplitude variance across the array to be useful for adaptive beamforming. The integration time $T \sim W^{-1}$ may, however, be excessive for certain applications, in which case a shorter integration time followed by the minimum echo amplitude variance test may be used. Let the maximum available integration time $T = kW^{-1}$, where k is, say, $\frac{1}{2}$, $\frac{1}{3}$, or $\frac{1}{4}$. The Doppler-reduced patch is then two, three, or four times too large to ensure a planar wavefront across the array, but it is still many times smaller than $R\,\Delta\theta$, the cross-range dimension of the patch illuminated by the transmitter. Although it is not small enough to ensure planar reradiation, the likelihood that the radiation is planar has materially increased. In other words, by integrating as long as is practical, the probability is maximized for achieving the excellent results demonstrated with the first algorithm.

9.5. MINIMUM CLUTTER BANDWIDTH CRITERION

The Doppler-aided method is appropriate when the terrain is "frozen." If the scatterers are wind-driven or the backscatter comes from the sea surface, the Doppler shift associated with each scattering center is related not only to its bearing angle from the radar but also to the radial component of its motion. This smearing of the Doppler signature by scatterer motion effectively spreads or enlarges the size of the clutter patch whose echoes pass through the Doppler filter. Calculations show that the Doppler spread associated with backscatter from the sea surface at wavelengths shorter than 0.3 m is too great to depend on the Doppler-aided method described in Section 9.3 [49].

A more complicated signal processor which accommodates the additional Doppler spread due to scatterer motion is described in this section. Conceptually, an adaptive array of the kind used to solve adaptive nulling problems is introduced. An LMS (least mean square) closed-loop circuit, which includes a band reject filter (BRF), is designed to set the complex weights of the antenna elements to minimize the spectral spread of the clutter echoes as they pass through the filter. Provided that the clutter is statistically homogeneous, which is typical more of the sea surface than ground clutter, this condition is equivalent to the formation of a beam with minimum beamwidth. The technique is described subsequently.

Figure 9.6 demonstrates the relationship between spatial properties of the beam and spectral properties of the clutter echoes. Because scatterers are tagged in angle by their Doppler shifts, the bandwidth of clutter echoes measured at the output of the array is monotonic with the beamwidth of the radiation pattern. Any circuit that minimizes the bandwidth of the clutter also minimizes the array beamwidth.

One circuit shown in Figure 9.7 is an adaptive subsystem for determining the weights. Following the adaptive weight-setting, the weights are frozen and the band reject filters are bypassed by the target echoes. This mode change is indicated by the SYNC/DETECTION switch (which delivers the array output to the weight-control circuit (SYNC Mode) or to the radar receiver) and the

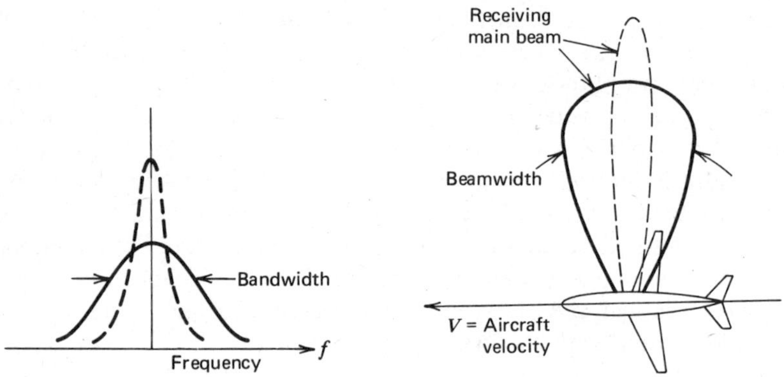

FIGURE 9.6. Array beam pattern and clutter spectrum; - - - - - after array is cohered; —— before array is cohered.

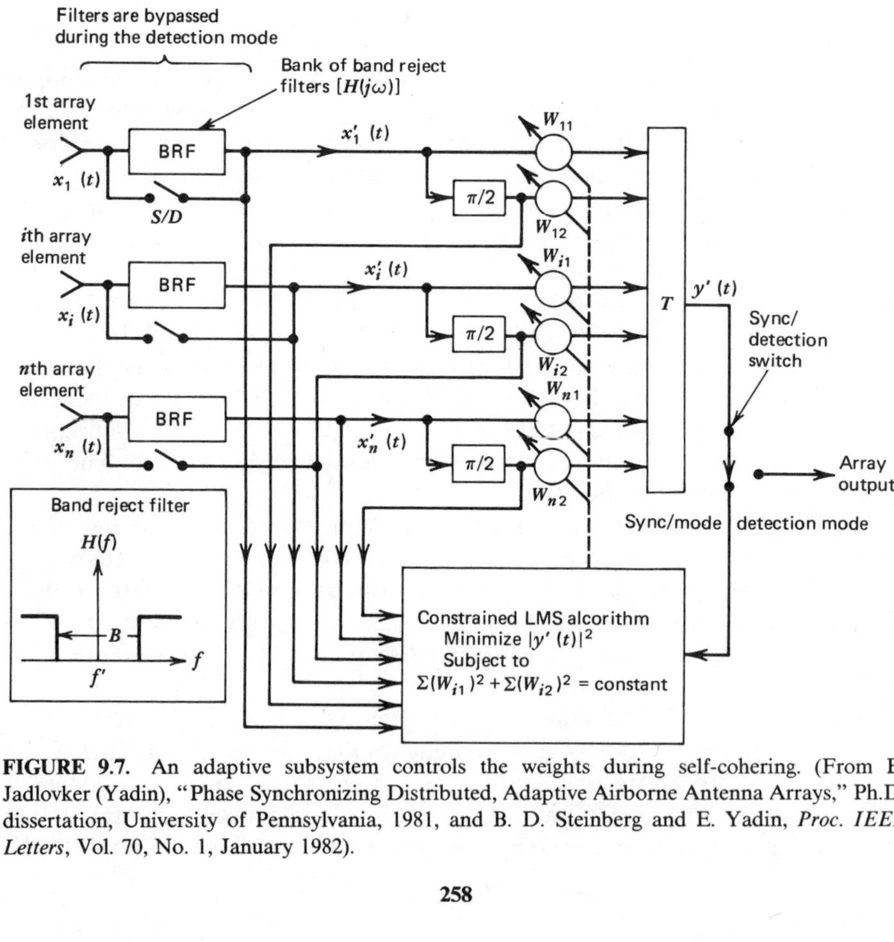

FIGURE 9.7. An adaptive subsystem controls the weights during self-cohering. (From E. Jadlovker (Yadin), "Phase Synchronizing Distributed, Adaptive Airborne Antenna Arrays," Ph.D. dissertation, University of Pennsylvania, 1981, and B. D. Steinberg and E. Yadin, *Proc. IEEE Letters*, Vol. 70, No. 1, January 1982).

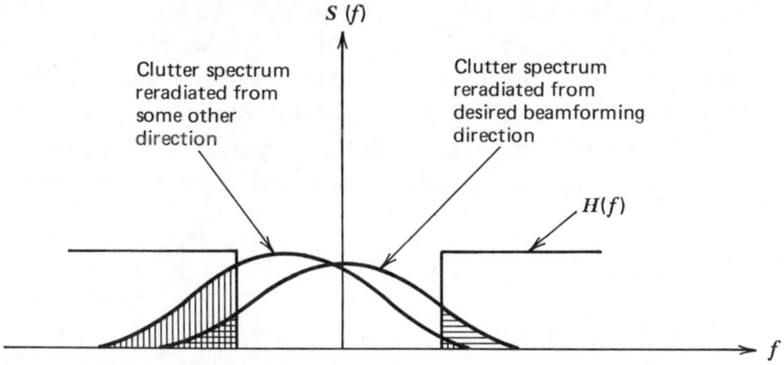

FIGURE 9.8. Ideal band reject filter and two echo spectra: ≡ clutter coming from desired direction; |||||| clutter coming from other direction.

bank of S/D switches at the inputs. The discussion that follows relates only to the self-cohering mode in which the inputs are sea-clutter echoes. The outcome of the process is a proper weighting vector. The input vector $\mathbf{X}^T = (X_1, X_2, \ldots, X_n)$ is applied to the bank of reject filters. The weighted sum $Y'(t)$ is applied to an LMS-constrained subsystem that constantly changes the weighting vector $\mathbf{W}$ to minimize the mean output power $E\{|Y'(t)|^2\}$ while maintaining the system gain $\Sigma_{i=1}^{N}|W_1|^2$ constant.† The gain-constancy constraint prevents the system from shutting itself off.

The operation of the system can be understood by considering the following discussion. The beamforming direction is determined by the notch in frequency of the band reject filters. The system is constrained not to shut itself off. Hence to minimize the output power $E\{|Y'|^2\}$ the system must form nulls toward the directions in which signals are coming. Signals arrive, however, from all directions within the azimuthal sector illuminated by the transmitter. The nulls are, therefore, deployed in those directions from which the clutter power is greatest. Figure 9.8 shows that the clutter power passing through a BRF (shaded areas) is smallest for echoes arriving from the desired-look direction, and the output power increases as the angle of arrival (relative to the desired look direction) increases. Because the clutter power passing through the BRF from the desired direction is smaller than from any other direction, no null is placed in that direction. Thus the process results in the formation of a main beam. A detailed analysis is given in [49].

This adaptive beamforming technique was computer-tested with simulated sea-clutter echoes and found to work successfully. The experiments involved a seven-element array 2.9 wavelengths long and a variety of system parameters. Uncertainties in element locations were accounted for by assuming that initially the array was weighted with random complex weights

† W_i is the complex weight $W_{i1} + jW_{i2}$. X_i' is phase-shifted by $\pi/2$ to produce quadratic components. Weighting the quadrature components by two real weights (W_{i1}, W_{i2}) is equivalent to weighting X_i' by the complex weight W_i.

$W_1(0), W_2(0), \ldots, W_n(0)$, which were independent of the element positions. The outcome of the self-cohering process was a set of weights $W_1, W_2, \ldots, W_n$. This set defined the radiation pattern of the array. During the synchronization process sea-clutter echoes are received by the different array elements. Clutter echoes were modeled as a sum of several independent Gaussian processes $g_1(t), g_2(t), \ldots, g_k(t)$, each of which represented echoes reradiated from a

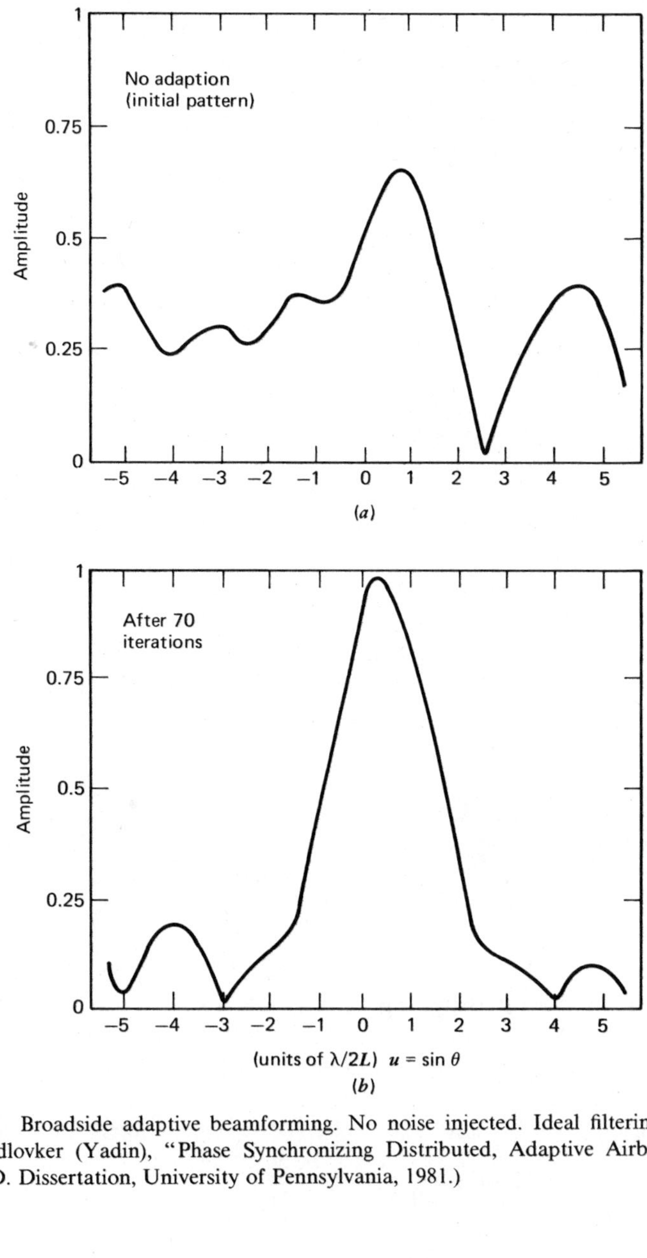

FIGURE 9.9. Broadside adaptive beamforming. No noise injected. Ideal filtering is assumed. (From E. Jadlovker (Yadin), "Phase Synchronizing Distributed, Adaptive Airborne Antenna Arrays," Ph.D. Dissertation, University of Pennsylvania, 1981.)

different direction. The processes have the same spectral shapes but each is shifted in frequency by an amount given by (9.1) because of its assumed direction of arrival. Coherent demodulation at the front end is also assumed. Thus clutter echoes arrive at the different elements with different phases that are due to the different geometrical locations of the elements and the directions of arrival of the clutter components. The simulated inputs were time samples

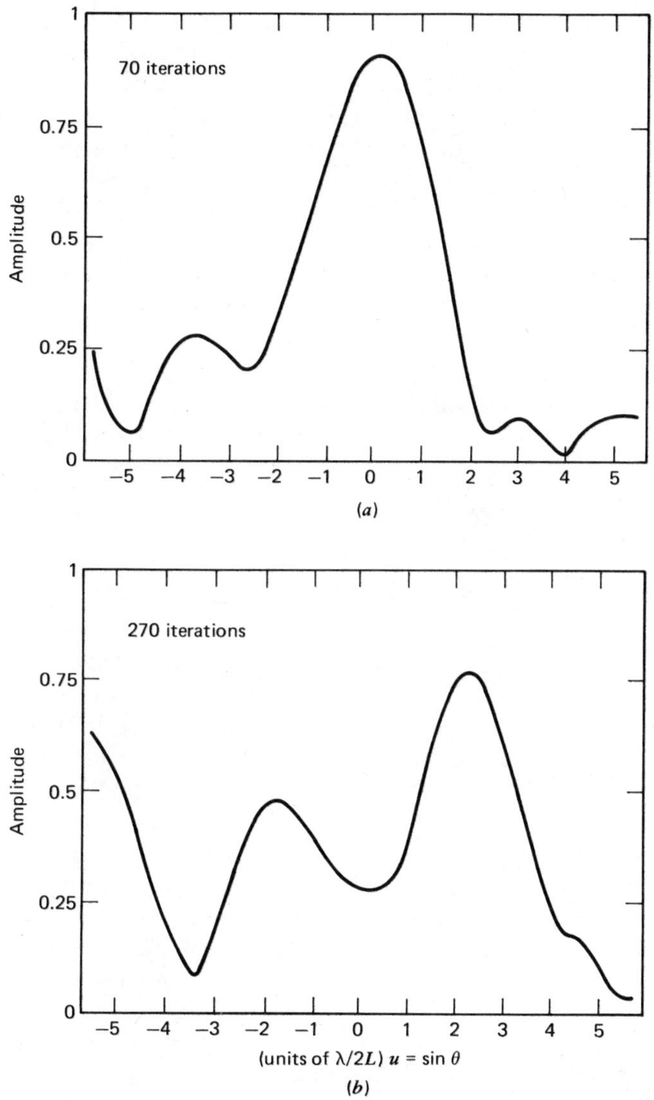

FIGURE 9.10. Broadside adaptive beamforming. Ideal filtering is assumed: (a) SNR = 10 dB; (b) SNR = 5 dB. (From E. Jadlovker (Yadin), "Phase Synchronizing Distributed, Adaptive Airborne Antenna Arrays," Ph.D. Dissertation, University of Pennsylvania, 1981.)

of the Gaussian processes. A gradient search technique based on the Widrow LMS algorithm [103], [29], [49] was used to find the optimum weighting vector **W**, which minimized the mean output power $E\{|Y'(t)|^2\}$, subject to the gain constraint $(\Sigma|W_i|^2 = 1)$. The process is iterative in **W** as given by

$$W(j+1) = W(j) - \mu Y'(j)^*[X'(j) - W(j)Y'(j)] \tag{9.2}$$

where j represents the iteration number and μ is a constant gain. It is shown in [49] that the iterative process converges to the desired solution in **W** for $0 < \mu < 2/p$, where p is the expected output power $E\{|Y'(t)|^2\}$.

Figure 9.9 shows the results of one of the simulated experiments in which an ideal BRF (Figure 9.8) was used. Figure 9.9*a* is the pattern associated with the initial random weights **W**(0). Figure 9.9*b* is the pattern achieved after 70 iterations of the adaptive process (9.2). The 3-dB beamwidth is $\sim 0.9\lambda/L$ and the main-beam gain toward the desired look direction is ~ -0.2 dB; 0 dB represents ideal coherent summation. The expected beamwidth based on the element distribution is $0.76\ \lambda/L$. Similar results were obtained with realistic, five-pole band reject filters.

Figure 9.10 describes the same problem with noise added. The upper figure shows successful convergence with SNR = 10 dB at the input to the BRFs. The lower figure shows significant gain-loss and rising sidelobes at SNR = 5 dB, notwithstanding a quadrupling of the allowed time for convergence. In many simulation experiments 10 dB SNR was a safe threshold for successful operation.

9.6. PERFORMANCE ADVANTAGE

Three questions are asked in Section 9.1. The first two deal with the problem of overcoming distortion in an airborne phased array the size of an aircraft due to nonrigidity of the skin and airframe. This problem is addressed in the earlier sections of this chapter. The third question deals with the utility of an airborne radio camera, which recognizes that its limited size precludes the extraordinary resolving power potentially available from a ground-based system. As the following analysis shows, a distributed array the size of an airplane can offer a significant increase in detection performance measured by SNR or detection range. The remainder of this chapter focuses on the comparative signal-to-noise ratio performances of three airborne array systems when compared with the conventional radar with a modest-sized, confined antenna used for transmitting and receiving. From the parametric relations that are developed, we can see the conditions under which advantages in detection performance can be obtained. Generally speaking, it is found that the larger the aircraft and the shorter the wavelength, the greater the potential benefit. In one X-band design a potential increase in detection range by a factor of 4 is possible. The analysis is taken from [97].

The reference system against which the distributed antenna systems are compared is the monostatic radar with rotating transmit-receive antenna, identified later as system I. Examples are the E-2C (Hawkeye) and E-3A (AWACS) airborne early warning (AEW) systems of the U.S. Navy and Air Force, respectively. The integrated output SNR can be easily shown to be [5]

$$\text{SNR} \triangleq r = \frac{PG_T A_R \sigma T_0}{(4\pi)^2 R^4 kTM} \tag{9.3}$$

where

P = average transmitted power (W)
G_T = transmitter antenna gain
A_R = receiving antenna area (m^2)
λ = wavelength (m)
T_0 = integration time (s)
σ = target cross section (m^2)
R = target to radar distance (m)
k = Boltzmann's constant (1.38×10^{-23} J/°K)
T = reference temperature (290°K)
M = total system and propagation loss factor (including receiver noise figure, integration loss, antenna efficiency loss, filter matching loss, etc.); subscripts are used in the text to distinguish between the systems

Other parameters used are

θ = azimuth beamwidth (rad)
θ_s = azimuthal surveillance sector (rad)
ϕ = elevation beamwidth (rad)
ϕ_s = elevation sector (rad)
D = width of high gain, rotating antenna (m)
L = length of aircraft (m)
W = width of the strip available on the fuselage for the distributed antenna array (m)
η = deployment efficiency (fraction of available fuselage area used as electromagnetic transducer)
T_s = scan time (s)
N = number of antenna elements
P_e = average power radiated by individual element (W)
A_e = element aperture (m^2)

The azimuth beamwidth $\theta = \lambda/D$, the antenna gain is given by $G_T = 4\pi/\theta\phi = 4\pi D/\lambda\phi$, and the receiving antenna area is given by $A = \lambda D/\phi$. It is assumed that the integration time T_0 is equal to the time on target (time during which the target is illuminated). That time is less than T_s by the ratio of azimuthal beamwidth to 2π, or $T_0 = \lambda T_s / 2\pi D$, where the scan time T_s is the time for a mechanically scanning antenna to rotate 2π rad. Making these substitutions yields

$$r = \frac{P\lambda\sigma D T_s}{8\pi^2 \phi_s^2 R^4 kTM_1} \tag{9.4}$$

where the subscript 1 has been given to M to designate system 1. The elevation beamwidth ϕ of the antenna is assumed to be equal to the desired elevation coverage ϕ_s.

9.7. DISTRIBUTED RECEIVING ARRAY AND HIGH-GAIN ROTATING TRANSMITTING ANTENNA (II)

Figure 9.11 shows how the reference system is modified. The high-gain rotating antenna is retained but only for transmission. The receiving system is a fuselage array consisting of N receiving elements distributed in a band along the length of the fuselage. Let the length of the fuselage be L and the width of the band, W.

The receiving array is distributed along the aircraft surface to form an aperture of length $L \gg D$; hence the receiving beamwidth is much smaller than the transmitting beamwidth. In order that the number of hits per target be undiminished, the signal processor simultaneously forms a group of adjacent receiving beams to fill the transmitting main beam. The number of receiving beams is $\theta_T/\theta_R = L/D$, where θ_T is the transmitting beamwidth and θ_R is the receiving beamwidth. The elevation beamwidth of the receiving array (λ/W) is usually much narrower than the required elevation coverage ϕ_s. Hence $W\phi_s/\lambda$ receiving beams in elevation must be formed to cover the desired sector ϕ_s. The total number of beams is $LW\phi_s/\lambda D$.

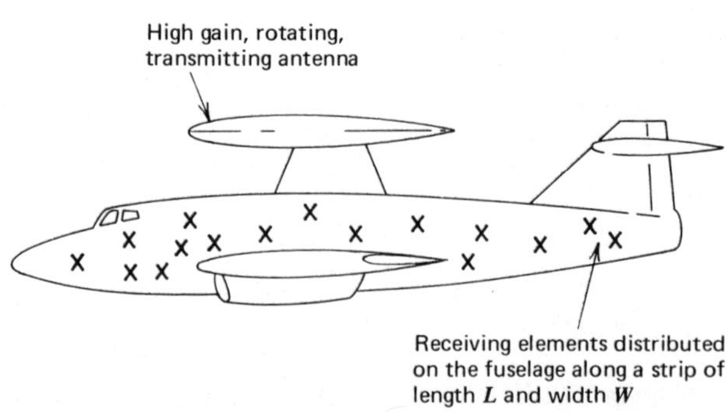

FIGURE 9.11. Distributed receiving array and high-gain rotating transmitting antenna.

Number of Receiving Elements

The element radiation pattern must cover the desired surveillance sector of the radar. Therefore it should have a beamwidth of θ_s in azimuth and ϕ_s in elevation. The width and length of each array element are therefore $D_1 = \lambda/\theta_s$ and $D_2 = \lambda/\phi_s$, respectively. The fuselage area available for deploying the elements is LW. If we assume that a fraction $\eta < 1$ of that area is used for the electromagnetic transducer, the number of antenna elements in the fuselage array becomes

$$N = \eta \frac{LW}{D_1 D_2} = \eta \frac{LW\theta_s\phi_s}{\lambda^2} \tag{9.5}$$

Equation 9.5 is evaluated for two aircraft in Table 9.1. One airplane is the Boeing 707, a large aircraft that could serve an AEW function; the other is a smaller, high-performance aircraft such as the General Dynamics F-16; L and W for these aircraft are approximately 45 and 2 m, and 10 and 1 m, respectively. In both cases $\eta = 0.5$ is assumed. The azimuthal surveillance sector $\theta_s = 2$ rad and the required elevation coverage $\phi_s = 1$ rad. The SNR is derived by substituting the proper area, which is given in (9.3) by NA_e. The effective element area A_e is given by $\lambda^2/\phi_s\theta_s$, G_T, by $4\pi D/\lambda\phi_s$, N, by (9.5), and T_0, by $\lambda T_s/2\pi D$. Substitution yields

$$r = \frac{\eta P\sigma LWT_s}{8\pi^2\phi_s R^4 kTM_2} \tag{9.6}$$

It is interesting to observe from (9.6) that the signal-to-noise power ratio is not affected by the transmitting azimuth beamwidth. This is true provided that the beamwidth does not get so large that it is no longer possible to integrate coherently across the beam. In that case integration must be partly coherent and partly noncoherent and the integration loss increases, a factor affecting the performance of system III. This is discussed numerically in a later section. Also discussed are the relative beam-shape losses. Another relevant factor in comparing the systems is that even though the SNR may not be influenced by the azimuth transmitting beamwidth the number of receiving beams is affected by

TABLE 9.1. Number of Elements Deployable Along One Side of the Fuselage $\eta = 0.5$

Wavelength (m)	Aircraft	N
0.3	Boeing 707	1000
0.1	Boeing 707	9000
0.03	General Dynamics F-16	11000

the beamwidth; hence the size of the transmitting antenna affects the complexity of the signal processor.

9.8. DISTRIBUTED RECEIVING ARRAY AND LOW-GAIN NONROTATING TRANSMITTING ANTENNA (III)

In this design the high-gain rotating transmitting antenna is eliminated and one of the array elements is substituted as a low-gain, wide-beam transmitter. The azimuthal width of the transmitting antenna pattern becomes equal to θ_s. Its elevation beamwidth ϕ_s is unchanged. Therefore $G_T = 4\pi/\phi_s\theta_s$. The receiving antenna is formed from N of these antenna elements. Therefore $A = NA_e = N\lambda^2/\theta_s\phi_s$. Substituting (9.5) for N yields $A = \eta LW$.

To achieve the maximum integration gain the signal processor simultaneously forms a group of adjacent receiving beams such that the azimuthal sector covered by the receiving beams is the same as that of the transmitting pattern. Hence the integration time T_0' will be longer than in designs I and II, which compensates for the low gain of the transmitting antenna. The relative efficiencies are discussed in Section 9.11. As a result of the increased integration time, the integration loss will also be increased. This increased loss is included in the system loss M_3. As in system II, $W\phi_s/\lambda$ receiving beams in elevation are formed simultaneously to cover the desired sector ϕ_s. For generality the transmitter power will be indicated by P'. These substitutions result in

$$r = \frac{\eta P'\sigma LWT_0'}{4\pi\theta_s\phi_s R^4 kTM_3} \tag{9.7}$$

9.9. DISTRIBUTED TRANSMITTING AND RECEIVING ARRAY (IV)

Here it is assumed that each element transmits and receives as in the systems described in Chapter 8. The average power transmitted per element is P_e and the element gain is G_e. The average power density at the target due to a single transmitter element is

$$W_e = \frac{P_e G_e}{4\pi R^2}$$

The electric field E_e is proportional to $W_e^{1/2}$. When the transmitting array is focused on a target the total electric field is $E = NE_e$ and the total average power density is $W = N^2 W_e$. Therefore the SNR at the receiver is

$$r = \frac{(N^2 P_e G_e)\sigma(NA_e)T_0''}{(4\pi)^2 R^4 kTM_4} \tag{9.8}$$

where NA_e is the total receiving aperture and T_0'' is the integration time. In this design electronic scan in azimuth and elevation for both transmit and receive beams is required. The ratio T_s/T_0'' is equal to the number of transmit-receive beams in the surveillance sector θ_s, ϕ_s. Therefore $T_0'' = \lambda^2 T_s/LW\theta_s\phi_s$. For this design integration losses can be ignored because $T_0'' \ll T_s$. (Care must be taken that T_0'' is not less than one interpulse period, a condition to which (9.8) may lead at the shorter wavelengths, in which case T_s must be lengthened or the other parameters decreased so that at least one hit per target is available to each beam.) The gain and aperture area of an element is the same as G_T, A_e in design III; N is given by (9.5). Hence $NA_e = \eta LW$ and $T_0'' = \eta T_s/N$. These substitutions yield

$$r = \frac{\eta^2 NP_e\sigma LWT_s}{4\pi\theta_s\phi_s R^4 kTM_4} \tag{9.9}$$

9.10. SNR IMPROVEMENT FACTOR

The ratios of (9.6), (9.7), and (9.9) to (9.4) are the relative SNRs of the three distributed array systems to the conventional monostatic radar. This ratio, designated the relative gain or improvement factor, is in the first column of Table 9.2.

The second column of Table 9.2 is based on four assumptions: (1) the angular sector θ_s is assumed to be about 2 rad; (2) the required elevation coverage is about 1 rad; (3) $T_0' = T_s$, which implies that the rate at which target data are delivered to the user is the same for all systems; (4) $P'/P = NP_e/P = \theta_s/2\pi$, which implies the use of the same power per angular sector for all systems. It is shown that for all the distributed designs the improvement increases linearly with $LW\phi_s/\lambda D$, LW is the fuselage area (one side) and

TABLE 9.2. SNR Gains of Systems II, III, and IV Relative to System I

System	SNR Improvement Factor	
	General case	Special case $\theta_s = 2$ rad, $\phi s = 1$ rad, $T_0' = T_s$, $P'/P = \dfrac{NP_e}{P} = \dfrac{\theta_s}{2\pi}$
II	$\dfrac{\eta M_1\phi_s LW}{M_2\lambda D}$	$\dfrac{\eta M_1 LW}{M_2\lambda D}$
III	$\dfrac{2\pi\eta M_1 P'T_0'\phi_s LW}{M_3 P\theta_s T_s\lambda D}$	$\dfrac{\eta M_1 LW}{M_3\lambda D}$
IV	$\dfrac{2\pi\eta^2 M_1 NP_e\phi_s LW}{M_4 P\theta_s\lambda D}$	$\dfrac{\eta^2 M_1 LW}{M_4\lambda D}$

$\lambda D/\phi_s$ is the aperture of the conventional antenna used in design I; $LW\phi_s/\lambda D$ is therefore the maximum potential increase in aperture. Because η is the fraction of the fuselage area used to deploy the elements, $\eta LW\phi_s/\lambda D$ is the factor by which the aperture increases. It is shown in Section 9.11 that system losses M_2, M_3, and M_4 are larger than the loss M_1 of the conventional design; that is, the fractions M_1/M_2, M_1/M_3, M_1/M_4 are smaller than unity. Hence the aperture gain $\eta LW\phi_s/\lambda D$ must be significantly higher than unity to make the distributed designs attractive.

9.11. RELATIVE SYSTEM LOSSES

The parameters M_2/M_1, M_3/M_1, and M_4/M_1 represent total system and propagation losses for designs II, III, and IV in relation to the conventional design I. The contributions to the system losses are compared subsequently. Propagation losses are unaffected by choice of design.

Beam-Shape Loss

As the beam scans over a target, the echo pulses modulated by the antenna pattern reach the maximum value only when the beam points directly at the target. Blake [7], [8] showed that antenna modulation represents a loss of SNR of 1.6 dB for an antenna scanning in one dimension and a 3.2-dB loss when a pencil beam is used to scan in two dimensions. When the number of hits per scan is low the average loss will increase because the target may be seen only at the edge of the beam. The pattern losses for these cases are given in [35].

For the purpose of this analysis it is assumed that the conventional design (system I) has a pattern loss of 1.6 dB (0.8 dB for transmission and 0.8 dB for reception).

Design II, using the mechanically scanned transmitting antenna and the electronically scanned multiple narrow beams for receiving, has the same modulation effect for transmitting and receiving. Therefore the pattern loss for this design is the same as that in system I.

System III also uses multiple receiving beams that are stationary, but, because they are narrow, aircraft and target motions cause target echoes to move through several receiving beams during the integration time. This is equivalent to scanning the receiving beams over the target. Therefore a pattern loss of 0.8 dB for reception is assumed.

The broad transmitting beam in system III is stationary also. Its power pattern $G_T(\phi)$ is angle-dependent. Its maximum value is at $\theta = 0$ and it drops by 3 dB for $\theta = \pm\theta s/2$. The beam-shape loss is defined as the additional SNR required to maintain the same average probability of detection P_d as in the loss-free case.

The signal-to-noise ratio depends linearly on $G_T(\theta)$. Because P_d is a function of SNR, it is also angle-dependent; that is, $P_d = g(\theta)$. If we assume that targets are uniformly distributed in angle throughout the surveillance sector θ_s, the average probability of detection is given by

$$\overline{P_d} = \frac{1}{\theta_s} \int_{\theta=-\theta_s/2}^{\theta=\theta_s/2} g(\theta)\, d\theta \tag{9.10}$$

It is assumed further that in the interval $-\theta_s/2 < \theta < \theta_s/2$ the one-way power pattern and therefore the signal-to-noise ratio can be approximated by a constant times $\cos^2(2\pi\theta/2\theta_s)$. The nonlinear dependence of P_d on SNR is calculated in many radar handbooks for different false alarm rates. Equation (9.10) was evaluated by using P_d (SNR) given in [78, Figure 4]. For a 2-rad surveillance sector and a constant probability of false alarm of 10^{-6} it was found that a signal-to-noise ratio of 14.3 dB is required to achieve an average probability of detection of 0.9. Comparing it with the required 13.2-dB SNR associated with the loss-free case and the same probability of detection, we conclude that the beam-shape loss for transmitting is 1.1 dB. Therefore a total of 1.9-dB beam-shape loss is assumed for design II.

Design IV involves two-dimensional scanning for which the loss is 3.2 dB, provided that there is more than one hit per scan. In addition, the transmitting gain for each element has the same angle-dependence as design III. Therefore an additional loss of 1.4 dB is added. A total beam-shape loss of 4.3 dB is assumed for this design. On the basis of this analysis, the estimated beam-shape losses (in relation to design I) are 0, 0.3, and 2.7 dB for designs II, III, and IV respectively.

Integration Loss

The integration losses are expected to be negligible for designs I, II, and IV because their integration times will be short enough to allow coherent integration with negligible losses. The integration time for each is limited by the beamwidth of the transmitting antenna or the transmitting array and by the required scanning time through the sector. In design III, however, the total sector is continuously illuminated by the broad beamwidth, nonrotating transmitting antenna. The integration time will be limited only by the required rate at which data are to be delivered to users. Because integration over a few thousand pulses (several seconds) is expected to take place in design III, some combination of coherent and noncoherent integration will be required. Hence integration loss in relation to ideal coherent integration cannot be ignored. As an example, an integration loss of ~ 2.5 dB is expected for a mixed integration process with a 2000-pulse integration time when coherent integration of only 100 pulses is possible [80].

Receiver Noise Figure

There is no reason why the receiver noise figures should differ in the four designs.

Adaptive Beamforming Loss

For the reasons given earlier it is expected that the three distributed array designs will require adaptive self-cohering techniques to compensate for fuselage vibrations. Experience with experimental adaptive beamforming and scanning systems at the Valley Forge Research Center indicates that the one-way loss can be held below 1 dB. This value is attributed to designs II and III and a 2-dB loss is attributed to design IV.

Combined Relative Losses

These estimates are used in the performance examples to follow. Expressed in decibels, the ratios of the loss factor M_1 to the losses in systems II, III, and IV are -1, -3.8, -4.7 dB, respectively.

9.12. INCREASES IN SNR AND DETECTION RANGE

In comparing the four systems it is tempting to assume that their average powers are equal or that their average powers per unit azimuthal angle are equal. These assumptions, however, are not necessarily realistic; P and NP_e are affected by and therefore limited by different physical and design phenomena; NP_e can be larger or smaller than P. Because of the large possible range in their ratio, it is fruitless to take it into account in comparing the relative merits of the systems. Thus only the effects of the enhanced aperture are compared.

Table 9.3 shows the SNR improvement factors due to the enlarged aperture of the distributed array and Table 9.4 shows the factors by which the detection ranges increase for $\theta_s = 2$ rad, $\phi_s = 1$ rad, $T_0' = T_s$, $P'/P = NP_e/P = \theta_s/2\pi$, $\eta = 0.5$ and for three sets of parameters: (1) $D = 20\lambda$, $L = 150\lambda$, $W = 7\lambda$; (2) $D = 50\lambda$, $L = 450\lambda$, $W = 20\lambda$; (3) $D = 20\lambda$, $L = 350\lambda$, $W = 35\lambda$. The first two sets are realistic for L- and S-band ($\lambda = 0.3$ and 0.1 m) AEW radar on a large aircraft like the Boeing 707; the third set is realistic for an X-band ($\lambda = 0.03$ m) radar on a small aircraft like the General Dynamics F-16. The value 0.5 chosen for η is based on examination of aircraft models and photographs and may prove somewhat optimistic. For $\eta = 0.25$ systems II and III lose 3 dB in SNR gains and system IV loses 6 dB.

TABLE 9.3. SNR Gains in Decibels of Systems II, III, and IV in Relation to System I Due to the Enlarged Aperture of Distributed Array: $\eta = 0.5$

System	*L*-Band AEW on Boeing 707	*S*-Band AEW on Boeing 707	*X*-Band on General Dynamics F-16
II	13.2	18.5	23.9
III	10.4	15.7	21.1
IV	6.5	11.8	17.2

TABLE 9.4. Ratios of Detection Ranges of Systems II, III, and IV to System I: $\eta = 0.5$

System	L-Band AEW on Boeing 707	S-Band AEW on Boeing 707	X-Band on General Dynamics F-16
II	2.14	2.90	3.96
III	1.82	2.47	3.37
IV	1.45	1.97	2.69

The SNR gains shown in Table 9.3 of systems II, III, and IV in relation to system I are also equal to the amounts by which the total transmitted powers of the distributed systems can be decreased while maintaining the same detection range as the conventional system.

9.13. AN AEW EXAMPLE

Consider the long range detection problem with the following radar and target parameters:

$$\sigma = 10\ \text{m}^2$$
$$f_r = 200\ \text{Hz}$$
$$\lambda = 0.1\ \text{m}$$
$$D = 50\lambda$$
$$L = 45\ \text{m}$$
$$kT = 4 \times 10^{-21}\ \text{W/Hz}$$
$$M_1 = 10\ \text{dB}$$
$$T_0' = T_s = 10\ \text{s}$$
$$\phi_s = 1\ \text{rad}$$
$$\theta_s = 2\ \text{rad}$$
$$W = 2\ \text{m}$$
$$\eta = 0.5$$
$$P'/P = NP_e/P = \theta_s/2\pi$$
$$M_1/M_2 = -1.0\ \text{dB}$$
$$M_1/M_3 = -3.8\ \text{dB}$$
$$M_1/M_4 = -4.7\ \text{dB}$$
$$\text{SNR} = 13\ \text{dB (corresponding to 0.9 probability of detection and } 10^{-6} \text{ probability of false alarm)}$$
$$R = 650\ \text{km (desired detection range)}$$

The average radiated transmitted power required to achieve a 650-km detection range for the four systems is shown in Table 9.5 for several different

TABLE 9.5. Average Radiated Transmitted Power Required to Achieve 650-km Detection Range

System	$\eta = 0.1$	$\eta = 0.2$	$\eta = 0.3$	$\eta = 0.4$	$\eta = 0.5$
I			83 kW (not a function of η)		
II	5.86 kW	2.93 kW	1.95 kW	1.46 kW	1.17 kW
III	11.15 kW	5.58 kW	3.72 kW	2.79 kW	2.23 kW
IV total	136.98 kW	34.25 kW	15.22 kW	8.56 kW	5.48 kW
per module	76 W	9.51 W	2.82 W	1.19 W	0.61 W

values of element deployment efficiency η. The differences are dramatic. Whereas an unrealistic 83 kW is required for the conventional, monostatic design, less than 5 kW suffices for a bistatic system with a large, distributed receiving array.

Implicit in this table is the assumption that the efficiencies of delivery of power from the transmitters to the antennas is the same for the four systems. This assumption penalizes system IV. Because a transmitter is located in each antenna module, the coupling loss is negligible compared with the several decibel loss encountered in the feed structure of a phased array driven by a single source of microwave power. This factor largely eliminates the differences in required generated power between the systems for the higher values of η.

9.14. SIDELOBE LEVEL OF A RANDOM ARRAY

One of the important parameters in system performance is the average sidelobe level (ASL) of the array in the receiving mode, which is $1/N$ (2.21). The pattern of the transmit-receive array is the square of the radiation pattern of the distributed receiving array; hence ASL for this case is $1/N^2$. The factor of N advantage is of considerable importance in detecting targets in clutter, which therefore makes system IV preferred over systems II and III. The improved two-way sidelobe pattern, however, is of no advantage with respect to jamming; all systems will perform according to the $1/N$ ASL of the receiving array. Because the one-way average sidelobe level is not likely to be less than -35 dB, other ECCM techniques must be designed into the system. Adaptive nulling is attractive in this regard [103], [29], [2].

9.15. OBSERVATIONS REGARDING SNR AND DETECTION RANGE

The attractive results shown in the tables indicate that the distributed airborne array will be useful when a large detection range with low transmitted power is required. In addition, the distributed airborne array is useful when adaptive nulling close to the beam axis and better angular resolution are desired.

In systems I, II, and III a single transmitter radiates the full power. System III, with a single low gain nonrotating antenna, is far superior to the other two mechanically, but it requires a more complicated signal processor to form simultaneously many receiving beams in azimuth and elevation. In addition, it must provide efficient integration over T_0. In SNR performance system III is slightly poorer (about 3 dB) than system II, but by not requiring the massive rotating antenna used in II the third design is more attractive. System IV radiates from the distributed array. It is a more complicated system than system III, but it offers a much lower two-way sidelobe level. It exhibits a poorer SNR and detection range than designs II and III for the same total radiated power but may be close to equivalence in terms of total generated power. SNR performance (in relation to designs II, III) decreases linearly with the deployment efficiency factor η. Differences of nearly 4 and 11 dB in SNR performances in relation to design III are estimated for $\eta = 0.5$ and $\eta = 0.1$, respectively. In general, design IV is preferred when the reduced two-way sidelobe level is needed.

CHAPTER TEN

Epilogue

The objective of designing and building a high-resolution microwave imaging system is to take pictures of targets or scenes much as we would with an optical camera. The pictures will be quite different from optical photographs because of the relatively long wavelength of the microwave system; the differences are discussed in Chapter 2. Although the detail found in an optical image will be lacking, target counting, target deployment, and the distinction between types of targets can be made. Furthermore, two-dimensional maps can be made at great distances in useful detail due to the high angular resolution available from the systems described in this book. An example is the Phoenixville street scene in Chapter 5.

Because of the unusual physical-size requirement placed on the aperture of a high-resolution microwave imaging system, this book is devoted primarily to means of achieving high resolution per se. Two other topics or activities are relevant to the making of quality pictures. The three stages of image processing fall into a natural sequence, shown schematically in Figure 10.1. The first stage (labeled I) converts the signals in the aperture to a primary or raw image by a transformation like the diffraction integral. High angular resolution demands a large aperture which in turn requires that some low-order intelligence be built into the first stage of the signal processor; for example, adaptive or self-cohering beamforming techniques. The second stage modifies or augments the signal processor based on the attributes of the primary image. It is a vernier or correcting process that operates within the realm of the coherent signal processor itself. It may use the complex primary image or merely its intensity by extracting information relating to the quality of the image and feeding back control signals to the signal processor to alter the image. The result is a secondary or improved quality image.

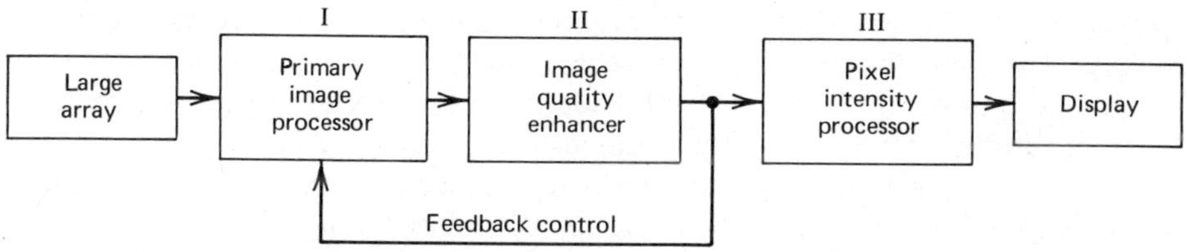

FIGURE 10.1. Image-processing stages.

The third stage is postprocessing image enhancement. Here pixel intensities are filtered (in the broadest sense) to produce a final, enhanced image.

Stage II, which was introduced in Section 4.5, and Stage III of image processing are not the subjects of this book. Stage II is not covered in detail because the field is in the early research stage. Stage III is ignored for quite the opposite reason; the field has matured during the last two decades and is the subject of many excellent texts [44], [46]. The postprocessing techniques reported in them and in myriad research papers are commended to the high-resolution microwave image system designer.

Several research topics that may be associated with the middle or interactive stage in image quality enhancement are worth noting. The first is diversity techniques. Any method that reduces the high sidelobes inherent to high-resolution system design of the nature described in this book will reduce the artifactual content of microwave images and the number of false targets that appear. Chapter 3 discusses several diversity techniques. The second is image feedback control. Although the Muller-Buffington technique [63] (see Section 4.5) is limited in its direct applicability, there is optimism that related research will prove useful for microwave imaging [3]. Third, there is some evidence that the effect of the gaps in the aperture of a thinned array can be mitigated. Instead of forming the image as the square magnitude of the Fourier transform of the samples of the electric field measured by the thinned array, all cross products $e_i e_j^*$ can be formed and a statistical interpolator such as maximum entropy may be applied to bridge the gaps between them. The result is a gapless estimate of the autocorrelation function of the radiation field across the aperture. The Fourier transform of the correlation estimate becomes the image [100], [11], [53]. A fourth method has proved successful in image-reconstruction techniques for speckle interferometry [26], [27]. As discussed in Section 4.5, heuristic information can be applied as constraints on both sides of the Fourier signal processor as the image and its correlation function are iterated around a loop. The schematic process was illustrated in Figure 4.12. In speckle interferometry the amplitude of the correlation function can be measured reasonably well but no information regarding its phase is available. A random phase function is constructed for it from the Fourier transform of a two-dimensional random noise image. The inverse Fourier transform, which is the first trial image, is constrained to nonnegative values. This modified image is retransformed and its phase is grafted onto the known or measured modulus of the correlation function. After a large number of iterations an acceptable image results. This work has already been reported in one radar context [61].

Last, the most significant potential source of knowledge for the optimum or near optimum processing of two-dimensional radiation fields has yet to be tapped. The most remarkable known image processor is the human (or other biological) visual system. It is a highly complicated, spatial, and temporal processor whose mechanisms are little understood and whose algorithms are totally unknown. In recent years, however, a phenomenological "understanding" of the process has been developing. Recent theoretical models by Buchsbaum appear to account completely for all the psychophysical experiments in color vision published in the last quarter-century [15], [16]. Therefore the time may not be too distant before some of the remarkable spatial processing properties of biological imaging systems will be ready for incorporation into high-quality microwave imaging systems.

References

1. S. P. Applebaum, "Adaptive Arrays," Syracuse University Research Corp., Report STLSPL TR66-1, August 1966.
2. ______, "Adaptive Arrays," *IEEE Trans. Antennas Propag.*, **AP-24**, 585–598 (September 1976).
3. E. H. Attia, "The Applicability of the Muller-Buffington Technique to Microwave Images," Valley Forge Research Center Quarterly Progress Report No. 41, The Moore School of Electrical Engineering, University of Pennsylvania, Philadelphia, November 15, 1982.
4. Elie J. Baghdady, Ed., *Lectures on Communication System Theory*, McGraw-Hill, New York, 1961.
5. R. S. Berkowitz, Ed., *Modern Radar Systems*, Wiley, New York, 1965.
6. ______ and E. N. Powers, "Tasar, a Thinned Adaptive Synthetic Aperture Radar," IEEE EASCON '78 Record, Arlington, Virginia, IEEE Publication 78 CH 1345-4 AES.
7. L. V. Blake, "The Effective Number of Pulses per Beamwidth for a Scanning Radar," *Proc. IRE*, **41**, 770–774 (June 1953).
8. ______, "Addendum to Pulses per Beamwidth for Radar," *Proc. IRE* (Correspondence), **41**, 1785 (December 1953).
9. I. F. Blake and W. C. Lindsey, "Level Crossing Problems for Random Processes," *IEEE Trans. Inf. Theory*, **IT-19**, 295–315 (May 1973).
10. M. Born and E. Wolf, *Principles of Optics*, 5th ed., Pergamon, New York, 1975.
11. S. B. Bowling and S. T. Lai, "The Use of Linear Prediction for the Interpolation and Extrapolation of Missing Data and Data Gaps Prior to Spectral Analysis," Technical Note 1979-46, Lincoln Laboratories, M.I.T., June 18, 1979.
12. D. G. Brennan, "Linear Diversity Techniques," *Proc. IRE*, **47**, 1075–1102 (June 1959).
13. L. E. Brennan and I. S. Reed, "Theory of Adaptive Radar," *IEEE Trans. Aerosp. Electron. Syst.*, **AES-9**(2), 237–252 (March 1973).
14. E. Brookner, *Radar Technology*, Artech House, Dedham, Massachusetts, 1977.
15. G. Buchsbaum, "The Retina as a Two-Dimensional Photodetector Array in the Context of Color Vision Theories and Signal Detection Theory," *Proc. IEEE*, **69**, 772–786 (1981).

16. _____ and A. Gottschalk, "Trichromacy, Opponent Color Coding and Efficient Information Transmission in the Human Retina," submitted for publication *Proc. R. Soc. London*.

17. D. L. Carlson, "The Phase Synchronization of Large Random Arrays Using Specular Aircraft Radar Return Signals," M.S. thesis, University of Pennsylvania, Philadelphia, 1975.

18. R. Chernoff, "Large Active Retrodirective Arrays for Space Applications," *IEEE Trans. Antennas Propag.*, **AP-27**(4), 489–496 (July 1979).

19. L. J. Cutrona, "Synthetic Aperture Radar," Chapter 23, *Radar Handbook*, M. I. Skolnik, Ed., McGraw-Hill, New York, 1970. For application to space-based radar see *Science*, **218** (4576) (December 3, 1982); C. Elachi et al., "Shuttle Imaging Radar Experiment," pp. 996–1003, and J. F. McCauley et al., "Subsurface Valleys and Geoarcheology of the Eastern Sahara Revealed by Shuttle Radar," pp. 1004–1020.

20. M. B. DonVito and S. A. Kassam, "Characterization of the Random Array Peak Sidelobe," *IEEE Trans. Antennas Propag.*, **AP-27**, 379–385 (May 1979).

21. C. Nelson Dorny, *A Vector Space Approach to Models and Optimization*, Wiley, New York, 1975.

22. _____, "Self-Cohering by Self-Surveying," Valley Forge Research Center Quarterly Progress Report No. 22, The Moore School of Electrical Engineering, University of Pennsylvania, Philadelphia, August 1977.

23. _____, "A Self-Survey Technique for Self-Cohering of Antenna Systems," *IEEE Trans. Antennas Propag.*, **AP-26**, 877–881 (November 1978).

24. _____, "Forward Looking Synthetic Aperture Radar," Valley Forge Research Center Quarterly Progress Report No. 30, The Moore School of Electrical Engineering, University of Pennsylvania, Philadelphia, August 1979.

25. _____ and B. S. Meagher, Jr., "Experimental Beamforming and Scanning of an 80-ft Random Antenna Array Operating at 1 GHz," Valley Forge Research Center Report No. 107, The Moore School of Electrical Engineering, University of Pennsylvania, Philadelphia, March 1977.

26. J. R. Feinup, "Space Object Imaging Through the Turbulent Atmosphere," *Opt. Eng.*, **18**(5) 529–534 (September–October 1979).

27. _____, "Iterative Method Applied to Image Reconstruction and to Computer-Generated Holograms," *Opt. Eng.*, **19**(3), 297–305 (May–June 1980).

28. P. B. Fellgett, and E. H. Linfoot, *Philos. Trans. R. Soc. London*, **247**, 269 (1955).

29. W. Gabriel, "Adaptive Array—An Introduction," *Proc. IEEE*, **64**, 239–272 (February 1976).

30. _____, "Spectral Analysis and Adaptive Array Superresolution Techniques," *Proc. IEEE*, **68**, 654–666 (June 1980).

31. F. M. Gardner, *Phaselock Techniques*, 2nd ed., Wiley, New York, 1979.

32. R. W. Gerchberg and W. O. Saxton, "A Practical Algorithm for the Determination of Phase From Image and Diffraction Plane Pictures," *Optik*, **35**, 237–246 (1972).

33. P. E. Glazer, "Power from the Sun: Its Future," *Science*, **162**, 857–861 (November 22, 1968).

34. T. F. Godlove, and V. L. Granatstein, "Relativistic Electron Beam Sources," *Soc. Photo-Optical Instrumentation Eng.*, **105**, Far Infrared/Submillimeter Wave Technology/Applications, 1977.

35. W. M. Hall and D. K. Barton, "Antenna Pattern Loss Factor for Scanning Radars," *Proc. IEEE* (Correspondence), **53**, 1257–1258 (September 1965).

36. H. J. Hall and H. K. Howell, Eds., *Photographic Considerations for Aerospace*, ITEK Corp., Boston, 1965.

37. J. P. Hamaker, J. D. O'Sullivan, and J. E. Noordam, "Image Sharpness, Fourier Optics, and Redundant-Spacing Interferometry," *J. Opt. Soc. Am.*, **67**(8), 1122–1123 (August 1977).

38. R. C. Hansen, "Communications Satellites Using Arrays," *Proc. IRE*, **49**, 1066–1074 (June 1961).

39. R. C. Hansen, Ed., *Microwave Scanning Antennas*, Vol. 3, Academic, New York, 1964. See Chapter 5 by D. L. Margerum.

40. R. O. Harger, *Synthetic Aperture Radar Systems*, Academic, New York, 1970.

41. R. M. Hjellming, *An Introduction to the NRAO Very Large Array*, 1st ed., NRAO, Socorro, New Mexico, November 1978.

42. _____ and R. C. Bignell, "Radio Astronomy with the Very Large Array," *Science*, **216**, (4552), 1279–1285 (June 18, 1982).

43. P. A. Howels, "Intermediate Frequency Sidelobe Canceler," U.S. Patent No. 3-202-990 (August 1965).

44. T. S. Huang, Ed., "Picture Processing and Digital Filtering," Vol. 6, *Topics in Applied Physics*, Springer-Verlag, New York, 1975.

45. F. O. Huck, and S. K. Park, *Appl. Opt.*, **14**, 2508 (1975).

46. B. R. Hunt, "Digital Image Processing," Chapter 4 of *Applications of Digital Signal Processing*, A. V. Oppenheimer, Ed., Prentice-Hall, Englewood Cliffs, New Jersey, 1978.

47. *IEEE Transactions on Antennas and Propagation*, **AP-12**(2), March 1964. This special issue on active and adaptive antennas contains 13 papers on retrodirectivity.

48. E. Jadlovker (Yadin), "A Study of the Effect of Multipath and Scattering upon Scanning with a Radio Camera," M.S. thesis, School of Engineering, Department of Electronics, Tel Aviv University, 1978.

49. _____, "Phase Synchronizing Distributed, Adaptive Airborne Antenna Arrays," Ph.D. Dissertation, Department of Systems Engineering, School of Engineering and Applied Science, University of Pennsylvania, Philadelphia, 1981.

50. R. C. Jones, *J. Opt. Soc. Am.*, **50**, 1166 (1960).

51. _____, *J. Opt. Soc. Am.*, **51**, 1159 (1961).

52. J. Klapper and J. T. Frankle, *Phase-Locked and Frequency Feedback Systems: Principles and Techniques*, Academic, New York, 1972.

53. J. S. Lim and N. A. Malik, "Maximim Entropy Power Spectrum Estimation for Signals with Missing Correlation Points," *IEEE Trans. Acoust., Speech Signal Process.*, **ASSP-29**(6), 401–403 (December 1981).

54. W. C. Lindsey, *Synchronization Systems in Communications and Control*, Prentice-Hall, Englewood Cliffs, New Jersey, 1972.

55. _____ and A. V. Kantak, "Automatic Phase Control in Solar Power Satellite Systems," prepared for NASA/JSC, TR-7802-0977, February 1978, LinCom Corp., Los Angeles, California; also W. C. Lindsey and A. V. Kantak, "Time and Frequency Transfer by the Master-Slave Retunable Timing System," *Proceedings of the International Telemetering Conference*, Volume XV, 1979, San Diego, California, pp. 187–201.

56. W. C. Lindsey and R. C. Tausworthe, "Bibliography of the Theory and Application of the Phase-Lock Principle," Tech. Rep. 32-1581, Jet Propulsion Laboratory, California Institute of Technology, Pasadena, April 1973.

57. E. H. Linfoot, *J. Opt. Soc. Am.*, **45**, 808 (1955).

58. _____, *J. Photogr. Sci.*, **9**, 188 (1961).

59. Y. T. Lo, "A Mathematical Theory of Antenna Arrays with Randomly Spaced Elements," *IRE Trans. Antennas Propag.*, **AP-12**, 257–268 (May 1964).

60. A. K. Luthra and B. D. Steinberg, "Wideband Array Processing," submitted to *IEEE Trans. Antennas Propag.*

61. D. L. Mensa, *High Resolution Radar Imaging*, Artech House, Dedham, Massachusetts 1981.

62. R. H. Mitchell and S. Marder, "Synthetic Aperture Radar (SAR) Image Quality Considerations," *Proc. SPIE, Image Quality*, **310**, 58–68 (1981).

63. R. A. Muller and A. Buffington, "Real-Time Correction of Atmospherically Degraded Telescope Images Through Image Sharpening," *J. Opt. Soc. Am.*, **64**(9), 1200–1210 (September 1974).

64. F. Nathanson, *Radar Design Principles*, McGraw-Hill, New York, 1969.

65. P. Panter, *Communication System Design: Line-of-Sight and Troposcatter Systems*, McGraw-Hill, New York, 1972.

66. A. Papoulis, *Probability, Random Variables, and Stochastic Processes*, McGraw-Hill, New York, 1965.

67. ______, "A New Algorithm in Spectral Analysis and Bandlimited Signal Extrapolation," *IEEE Trans. Circuits Syst.*, **CAS-22**, 735–742 (September 1975).

68. E. N. Powers, "Adaptive Arrays for Microwave Imaging," Ph.D. Dissertation, University of Pennsylvania, Philadelphia, 1975.

69. A. C. S. Readhead, "Radio Astronomy by Very Long Baseline Interferometry," *Sci. Am.*, 53–61 (June 1982).

70. S. O. Rice, "Mathematical Analysis of Random Noise," *Bell Syst. Tech. J.*, **23**, 282–332 (July 1944); **24**, 46–156 (January 1945).

71. S. D. Robertson, "Targets for Microwave Radar Navigation," *Bell Syst. Tech. J.*, **26**, 852–869 (October 1947).

72. George T. Ruck, Ed., *Radar Cross Section Handbook*, Plenum, New York, 1970.

73. "Satellite Power System-Concept Development and Evaluation Program," USDOE and NASA, Report No. DOE/ER-0023, available from National Technical Information Service, U.S. Department of Communications, January 1979.

74. M. Schwartz, W. R. Bennett, and S. Stein, *Communication Systems and Techniques*, McGraw-Hill, New York, 1966.

75. C. Shannon, *Bell Syst. Tech. J.*, **27**, 379 (1948), or C. Shannon and W. Weaver, *The Mathematical Theory of Communications*, University of Illinois Press, Urbana, 1964.

76. E. D. Sharp and M. A. Diab, "Van Atta Reflector Array," *IRE Trans. Antennas Propag.*, **AP-8**, 436–438 (July 1960).

77. R. Shaw, *Photogr. Sci. Eng.*, **6**, 281 (1962).

78. M. I. Skolnik, Ed., *Radar Handbook*, McGraw-Hill, New York, 1970.

79. L. G. Sodin, "Statistical Analysis of Nonequidistant Linear Antenna Arrays," *Radio Eng. Electro. Phys.*, **11**, 1715–1720 (November 1966).

80. B. D. Steinberg, "Radar Feedback Filters," Chapters 3 and 4, Part 4 of *Modern Radar*, R. S. Berkowitz, Ed., Wiley, New York, 1965, pp. 506–511.

81. ______, "The Peak Sidelobe of the Phased Array Having Randomly Located Elements," *IEEE Trans. Antennas Propag.*, **AP-20**, 129–136 (March 1972).

82. ______, "Comparison between the Peak Sidelobe of the Random Array and Algorithmically Designed Random Arrays," *IEEE Trans. Antennas Propag.*, **AP-21**, 366–370 (May 1973).

83. ______, "Design Approach for a High Resolution Microwave Imaging Radio Camera," *J. Franklin Inst.* (December 1973).

84. ______, "The Effects of Relative Source Strength and Signal-to-Noise Ratio on Angular Resolution of Antennas," *Proc. IEEE*, **62**(6), 758–762 (June 1974).

85. ______, *Principles of Aperture and Array System Design*, Wiley, New York, 1976.

86. B. D. Steinberg, "High Angular Microwave Resolution from Distorted Arrays," *Proc. 1980 Int. Computing Conf.*, **231**, 150–156 (April 1980).

87. ______, "Radar Imaging from a Distorted Array: The Radio Camera Algorithm and Experiments," *IEEE Trans. Antennas Propag.*, **AP-29**(5), 740–748 (September 1981).

88. ______, "Adaptive Microwave Holography," *Opt. Acta*, **29**(4), 363–369 (1982).

89. ______, "Properties of Phase Synchronizing Sources for a Radio Camera," *IEEE Trans. Antennas Propag.*, **AP-30**(6), 1086–1092 (November 1982).

90. ______, "Phase Synchronizing a Nonrigid, Distributed, Transmit-Receive Radar Antenna Array," *IEEE Trans. Aerospace Electron. Syst.*, **18**(6), 609–620 (September 1982).

91. B. D. Steinberg and E. H. Attia, "Sidelobe Reduction of Random Arrays by Element Position and Frequency Diversity," to be published in *IEEE Trans. Antennas Propag.*

92. B. D. Steinberg and A. K. Luthra, "Asymmetries in the Diffraction Pattern of an Asymmetrical Aperture," *IEEE Trans. Antennas Propag.*, **AP-29**(4), (July 1981).

93. ______, "A Simple Theory of the Effects of Medium Turbulence upon Scanning with an Adaptive Phased Array," *J. Acoust. Soc. Am.*, **71**(3), 630–634 (March 1982).

94. B. D. Steinberg and E. N. Powers, "High Angular Resolution Radar," *IEEE Int. Radar Conf. Proc.*, Arlington, Virginia, April 1975.

95. B. D. Steinberg, E. N. Powers, D. Carlson, B. S. Meagher, Jr., R. S. Berkowitz, C. N. Dorny, and S. H. Seeleman, "First Experimental Results from the Valley Forge Radio Camera Program," *Proc. IEEE Lett.*, **67**(9), 1370–1371 (September 1979).

96. B. D. Steinberg and E. Yadin, "Radio Camera Experiment with Airborne Radar Data," *Proc. IEEE Lett.*, **70**(1), 96–98 (January 1982).

97. ______, "Distributed Airborne Array Concepts," *IEEE Trans. Aerospace Electron. Syst.*, **AES-18**(2), 219–227 (March 1982).

98. B. D. Steinberg and E. Yadin-Jadlovker, "Self-Cohering an Airborne Radio Camera," *IEEE Trans. Aerospace Electron. Syst.*, **19**(4), (July 1983).

99. B. D. Steinberg and E. Yadin, "Effect of Multipath and Scattering on Array Gain of a Radio Camera," submitted to *IEEE Trans. Antennas Propag.*

100. Harish M. Subbaram, "Maximum Entropy Interpolation for Highly Thinned Arrays," Valley Forge Research Center Quarterly Progress Report No. 41, The Moore School of Electrical Engineering, University of Pennsylvania, Philadelphia, September 1982.

101. S. Hassan Taheri and B. D. Steinberg, "Tolerances in Self-Cohering Antenna Arrays of Arbitrary Geometry," *IEEE Trans. Antennas Propag.*, **AP-24**(5), 733–739 (September 1976).

102. A. J. Viterbi, *Principles of Coherent Communication*, McGraw-Hill, New York, 1966.

103. B. Widrow, P. E. Mantey, L. J. Griffiths, and B. B. Goode, "Adaptive Antenna Systems," *Proc. IEEE*, **55**(12), 2143–2159 (December 1967).

104. R. E. Ziemer and W. H. Tranter, *Principles of Communications*, Houghton-Mifflin, Boston, 1976.

INDEX